Advanced Mathematics

Personal Study Notes

Part II

By

Mohamed F. El-Hewie

BRIEF TABLE OF CONTENTS

TABLE OF CONTENTS

PREFACE

This is Part II of My Personal Study Notes in advanced mathematics which encompasses theoretical analysis, solved examples, and exercises on the following topics:

Each topic is explained in concise theoretical details and supported by solved examples. Plenty of exercises and their answers follow each topic that cover all details of the presented matter.

CHAPTER 9: THEORY OF FUNCTIONS OF A COMPLEX VARIABLE

9.1. Introduction to complex numbers:

Not every algebraic equation can be solved in terms of real numbers. It is therefore necessary to extend the domain of real numbers so that basic algebraic operations can always be employed. Complex numbers are just such an **extension of the domain of real numbers**.

The introduction of complex numbers and functions of a complex variable is convenient when:

1. integrating elementary functions,
2. solving differential equations,
3. electrical engineering,
4. radio engineering,
5. electrodynamics, hydrodynamics and aerodynamics,
6. theory of elasticity,
7. and other natural sciences.

9.2. Theory of functions of a complex variable

9.2.1. The real and imaginary parts of a complex number
A complex number z is characterized by a pair of real numbers (a, b) having an established sequential order of the numbers a and b. This is stated in the notation z (a, b). The first number a of the pair (a, b) is called the real part of the complex number z and is denoted by the symbol a = Re z; the second number b of the pair (a, b) is called the imaginary part of the complex number z and is symbolized by b= Im z.

16 \ 10 \ 1971. . Theory of Function of a complex Variable

1. Complex numbers (Revision).

   Any number in the form $x + iy$. where x and y are both real and $i = \sqrt{-1}$ is called a complex number for example.

   e.g. $6 - 4i$

9.2.2. Complex number as vectors

Any complex number $x + iy$ can be expressed in the alternative form $r(\cos\theta + i\sin\theta)$

$$x + iy = r(\cos\theta + i\sin\theta)$$

$$\therefore \left.\begin{array}{l} x = r\cos\theta \\ y = r\sin\theta \end{array}\right\} \qquad x^2 + y^2 = r^2$$

$$\boxed{\left.\begin{array}{l} r = +\sqrt{x^2 + y^2} \\[4pt] \cos\theta = \dfrac{x}{\sqrt{x^2 + y^2}} \\[8pt] \sin\theta = \dfrac{y}{\sqrt{x^2 + y^2}} \end{array}\right\}}$$

9.2.3. Modulus and Argument of a complex number

$r = +\sqrt{x^2 + y^2}$ is called the "modulus" of the complex number $x + iy$, and if we put $z = x + iy$ then modulus z is denoted by $|z|$ the angle θ which satisfies the above two equation is called the "argument" or "amplitude" of the complex number. If we keep the angle θ to satisfy in addition the inequality

$$-\pi < \theta \leq \pi$$

the θ is called the "principle value" of the argument or amplitude.

9.2.4. Cartesian and Polar representation of complex numbers

2. Representation of complex number on the argand plane

The complex number $z = x + iy = r(\cos\theta + i\sin\theta)$ is represented on the argand plane by the point

$$\underset{\text{cartesian}}{(x, y)} \qquad or \qquad \underset{\text{Polar}}{(r, \theta)}$$

9.2.5. Pure Real and Imaginary numbers

Real number is located on the x axis where as purely imaginary number are represented by points on the y axis there is why the x axis is usually refered to as the real axis and the y axis as the imaginary axis.

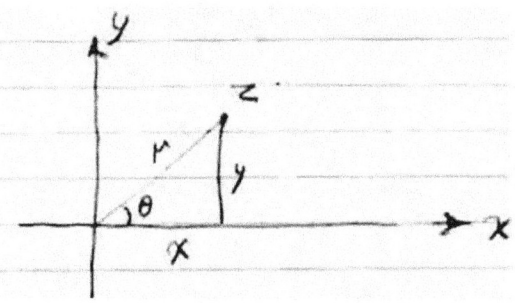

9.2.6. Representation of complex numbers on a **Riemann Sphere**

3. Representation of complex number on a Riemann Sphere. (Germany)

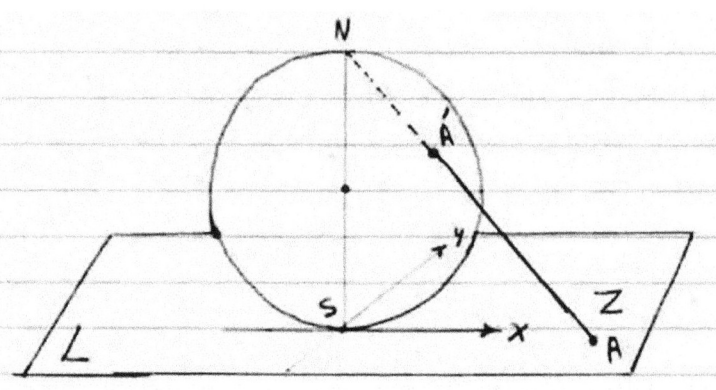

Let L the complex plane of z and all is the argend. A is any point in the z plane representing a complex number z.

now draw a sphere touching the plane L at the point o and let SN' the diameter of this sphere through o. Join A to N and let it intersect the sphere in A' hence corresponding to every complex number in the z plane. there is on point on the sphere if A is at infinity in the z plane. then N'A is parallel the plane and N corsponds to the point at infinity to the z plane.

All the points in the plane L together with the points at infinity are said to inform the entire the complexplane.

Two complex numbers $z1(a1, b1)$ and $z2 (a2, b2)$ are equal only when both the real and imaginary parts are equal, that is, $z1 = z2$, only when $a1 = a2$, and $b1 = b2$.

9.3. Algebraic operations on complex numbers.
9.3.1. Addition of complex numbers

④ Fundamental operation in the complex numbers:-

I. Addition of complex numbers:-

Let $z_1 = x_1 + iy_1$
$z_2 = x_2 + iy_2$
$z_1 + z_2 = x_1 + x_2 + i(y_1 + y_2).$

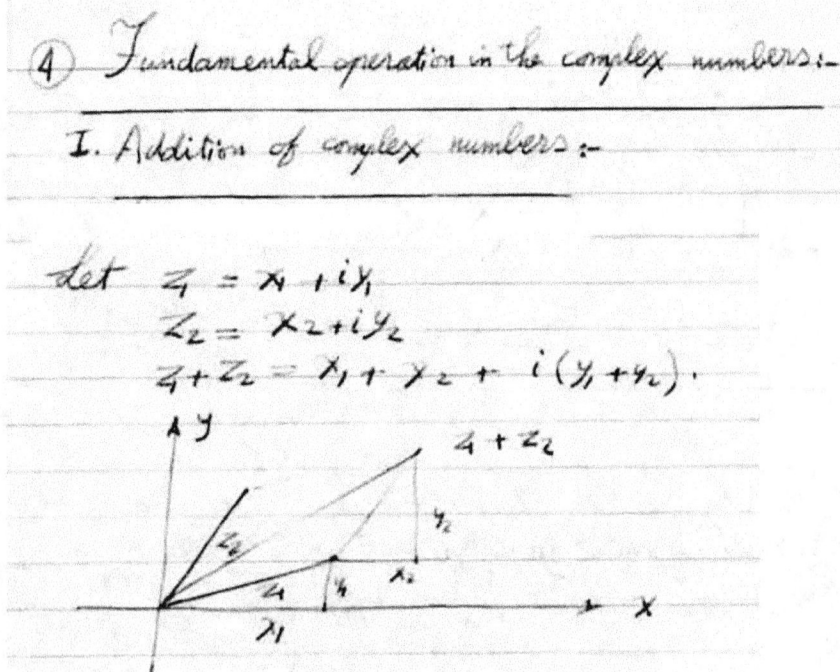

The addition of complex number followed the law of vectors addition.

now since any side of a triangle cannot exceed the sum of the two other sides. then

$$|z_1 + z_2| \leq |z_1| + |z_2|$$

now $z_1 = (z_1 + z_2) + (-z_2)$

$$\therefore |z_1| \leq |z_1 + z_2| + |z_2|$$

$$\therefore |z_1 + z_2| \geq |z_1| - |z_2|$$

similarly $|z_1 + z_2| \geq |z_2| - |z_1|$

$$\therefore |z_1 + z_2| \geq ||z_1| - |z_2||$$

9.3.2. Reciprocal of a complex number

II. Reciprocal of a complex number.

Let $z = r(\cos\theta + i\sin\theta)$

$$\frac{1}{z} = \frac{1}{r(\cos\theta + i\sin\theta)} = \frac{1}{r} \cdot \frac{\cos\theta - i\sin\theta}{\cos^2\theta + \sin^2\theta}$$

$$= \frac{1}{r}(\cos\theta - i\sin\theta) = \frac{1}{r}[\cos-\theta + i\sin-\theta]$$

$$\left|\frac{1}{z}\right| = \frac{1}{r} = \frac{1}{|z|}$$

arg. $\frac{1}{z} = -\theta = -arg.z$

9.3.3. The product of the complex numbers

II Multiplication of Complex numbers :—

Let $z_1 = x_1 + iy_1 = r_1(\cos\theta_1 + i\sin\theta_1)$

$z_2 = x_2 + iy_2 = r_2(\cos\theta_2 + i\sin\theta_2)$

$z_1 z_2 = x_1 x_2 - y_1 y_2 + i(x_1 y_2 + x_2 y_1)$

$$Z_1 Z_2 = r_1 r_2 (\cos\theta_1 + i\sin\theta_1)(\cos\theta_2 + i\sin\theta_2)$$
$$= r_1 r_2 \left[\cos\theta_1 \cos\theta_2 - \sin\theta_1 \sin\theta_2 + i(\sin\theta_1 \cos\theta_2 + \cos\theta_1 \sin\theta_2) \right]$$

$$= r_1 r_2 \left[\cos(\theta_1 + \theta_2) + i\sin(\theta_1 + \theta_2) \right]$$

$$\therefore \quad |Z_1 Z_2| = r_1 r_2 = |Z_1||Z_2|$$
$$\text{arg. } Z_1 Z_2 = \theta_1 + \theta_2 = \text{arg. } Z_1 + \text{arg. } \theta_2$$

9.3.4. Division of complex numbers

IV. The quotient of two complex numbers

$$\frac{Z_1}{Z_2} = \frac{r_1}{r_2} \cdot \frac{(\cos\theta_1 + i\sin\theta_1)}{(\cos\theta_2 + i\sin\theta_2)} = \frac{r_1}{r_2}(\cos\theta_1 + i\sin\theta_1)(\cos-\theta_2 + i\sin-\theta_2)$$
$$= \frac{r_1}{r_2}\left[\cos(\theta_1 - \theta_2) + i\sin(\theta_1 - \theta_2) \right]$$

$$\left|\frac{Z_1}{Z_2}\right| = \frac{r_1}{r_2} = \frac{|Z_1|}{|Z_2|}$$

$$\text{arg. } \frac{Z_1}{Z_2} = \theta_1 - \theta_2 = \text{arg. } Z_1 - \text{arg. } Z_2$$

9.3.5. The Nth roots of complex numbers

V. The nth roots of a complex numbers:-

$$Z^{\frac{1}{n}} = r^{\frac{1}{n}}(\cos\theta + i\sin\theta)^{\frac{1}{n}}$$
$$= r^{\frac{1}{n}}\left[\cos(\theta + 2S\pi) + i\sin(\theta + 2S\pi) \right]^{\frac{1}{n}}$$

$$= r^{\frac{1}{n}}\left[\cos\frac{\theta + 2S\pi}{n} + i\sin\frac{\theta + 2S\pi}{n} \right]$$

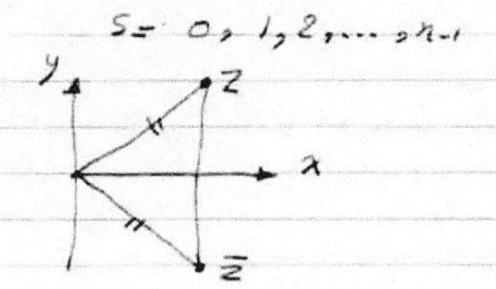

$$S = 0, 1, 2, \ldots, 2n-1$$

9.3.6. Conjugate complex numbers

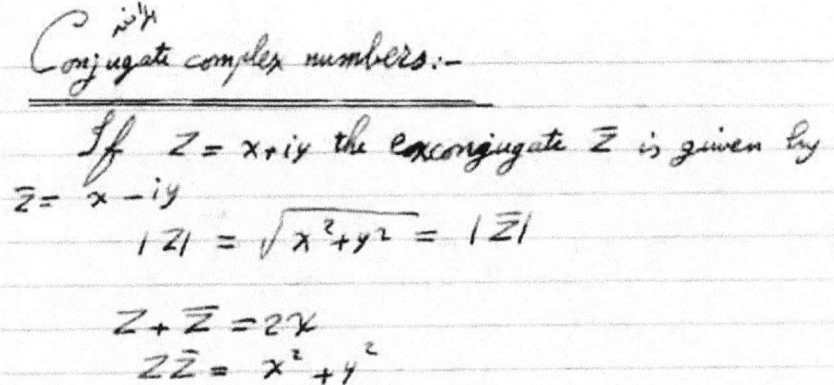

Conjugate complex numbers:-

If $Z = x + iy$ the conjugate $\bar{Z}$ is given by
$$\bar{Z} = x - iy$$
$$|Z| = \sqrt{x^2 + y^2} = |\bar{Z}|$$

$$Z + \bar{Z} = 2x$$
$$Z\bar{Z} = x^2 + y^2$$

This means that the sum and product of two conjugate complex numbers are real.

The elementary functions (f^{ns}) of a complex variable:-

9.3.7. Polynomials of complex numbers

I. Polynomials or rational integral f^{ns} of Z:-

These are given by $a_0 Z^n + a_1 Z^{n-1} + \cdots + a_{n-1} Z + a_n$ where n is a +ve integer and $a_0, a_1, \ldots, a_n$ are in general complex.

9.3.8. Rational Functions of complex numbers

13

II. Rational f^{ns} (in z)

These are $\dfrac{P(z)}{Q(z)}$ where $P(z)$ and $Q(z)$ are polynomials in z.

9.3.9. Exponential Functions of complex numbers

III. The exponential f^n of a complex variables

Def. $e^z = 1 + z + \dfrac{z^2}{2!} + \dfrac{z^3}{3!} + \cdots$

$$e^{z_1} \cdot e^{z_2} = \left(1 + z_1 + \dfrac{z_1^2}{2!} + \cdots\right)\left(1 + z_2 + \dfrac{z_2^2}{2!} \cdots\right)$$

$$= 1 + z_1 + z_2 + \dfrac{1}{2!}\left(z_1^2 + 2 z_1 z_2 + z_2^2\right) \cdots$$

$$= 1 + z_1 + z_2 + \dfrac{1}{2!}\left(z_1 + z_2\right)^2$$

$$= e^{z_1 + z_2}$$

$\therefore e^z = e^{x+iy} = e^x \cdot e^{iy} = e^x(\cos y + i \sin y)$

(Euler's formula $e^{i\theta} = \cos\theta + i\sin\theta$)

$\therefore |e^z| = e^x$, arg. $e^z = y$

$$e^{z+2\pi i} = e^z \cdot e^{2\pi i} = e^z(\cos 2\pi + i\sin 2\pi)$$
$$= e^z$$

Hence the exponential function e^z is periodic and of period $2\pi i$.

If a is a real positive constant then $a^z = e^{z \log a}$.

9.3.10. Hyperbolic Functions of complex numbers

IV. The hyperbolic f^{ns} of a complex variables :—

$$\cosh z = \frac{e^z + e^{-z}}{2}, \quad \sinh = \frac{e^z - e^{-z}}{2}$$

$$\tanh z = \frac{\sinh z}{\cosh z} = \frac{e^z - e^{-z}}{e^z + e^{-z}}$$

Ex: $\cosh(\alpha + i\beta) = \dfrac{e^{\alpha + i\beta} + e^{-\alpha - i\beta}}{2}$

$$= \frac{1}{2}\left[e^{\alpha} \cdot e^{i\beta} + e^{-\alpha} \cdot e^{-i\beta} \right]$$

$$= \frac{1}{2}\left[e^{\alpha}(\cos\beta + i\sin\beta) + e^{-\alpha}(\cos\beta - i\sin\beta)\right]$$

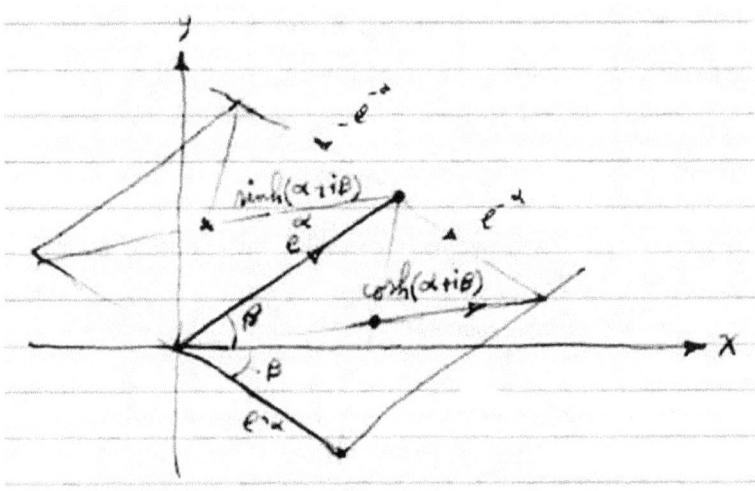

9.3.11. Trigonometric Functions of complex numbers

23-10-71

V. Trigonometric f^{ns} of complex variables:—

$$i \sin z = \frac{e^{iz} - e^{-iz}}{2}, \quad \cos z \; \frac{e^{iz} + e^{-iz}}{2}, \dots$$

$$\cosh iz = \cos z \qquad \qquad \sin iz = i\sinh z$$
$$\cos iz = \cosh z \qquad \qquad \sinh iz = i\sin z$$

15

ex: 1   prove that all zeroes of $\sin z$ are real and find them. by a zero of $\sin z$ we mean a value of z which makes $\sin z = 0$

$$\therefore \frac{e^{iz} - e^{-iz}}{2i} = 0$$

$$e^{iz} = e^{-iz} \qquad \therefore \quad \frac{e^{2iz}}{1} = e^{2ik\pi}$$

$$\therefore \quad 2iz = 2ik\pi$$

$$\therefore \quad z = k\pi = 0, \pm \pi, \pm 2\pi \ldots \ldots$$

ex. 2   prove that $|\sin z|^2 = \sin^2 x + \sinh^2 y$

$$\sin z = \sin(x + iy) = \sin x \cos iy + \cos x \sin iy$$

$$= \sin x \cosh y + i \cos x \sinh y$$

$$|\sin z|^2 = \sin^2 x \cosh^2 y + \cos^2 x \sinh^2 y$$

$$= \sin^2 x (1 + \sinh^2 y) + (1 - \sin^2 x) \sinh^2 y$$

magnitude $\qquad = \sin^2 x + \sinh^2 y$

The above example shows the use of both trigonometric functions and hyperbolic functions through the theory of complex numbers.

9.3.12. Logarithmic Functions of complex numbers

من اهم دوال المتغير المركب هي الدوال اللوغارتمية وتعريفها.

VI. Logarithmic f^n of a complex variables:-

let $\quad w = u + iv = \log z$

$$\therefore z = e^{u+iv} = e^u \cdot e^{iv} = e^u (\cos v + i \sin v)$$

let $z = r(\cos\theta + i \sin\theta)$

$$\therefore r(\cos\theta + i \sin\theta) = e^u (\cos v + i \sin v)$$

$$e^u = r$$

$$\therefore u = \log_e r \qquad\qquad v = \theta + 2k\pi$$

16

$$\log_e z = \log_e r + i(\theta + 2k\pi).$$

This shows that $\log z$ is a many-valued f^n if we take the angle θ as the principall value of the argument then $\log_e z = \log_e r + i\theta$ is called the principale value of $\log z$.

ex: Find $\log(1+i)$

$$1+i = \sqrt{2}\left(\frac{1}{\sqrt{2}} + \frac{i}{\sqrt{2}}\right) = \sqrt{2}\left(\cos\frac{\pi}{4} + i\sin\frac{\pi}{4}\right)$$

$$\log(1+i) = \frac{1}{2}\log 2 + i\frac{\pi}{4}$$

9.3.13. Inverse Hyperbolic Functions of complex numbers

VII. Inverse hyperbolic f^{ns}:-

$$\sinh^{-1} z \ , \ \cosh^{-1} z, \ldots$$

Simmilar to the real variables we have

$$\sinh^{-1} z = \log\left(z + \sqrt{z^2+1}\right)$$
$$\cosh^{-1} z = \log\left(z + \sqrt{z^2-1}\right)$$
$$\tanh^{-1} z = \frac{1}{2}\log\frac{1+z}{1-z}$$

9.3.14. Inverse Trigonometric Functions of complex numbers

VIII. Inverse trigonometric f^{ns}:-

$$\sin^{-1} z \ , \ \cos^{-1} z, \ldots$$

let $w = \sin^{-1} z$

$$\therefore z = \sin w = \frac{e^{iw} - e^{-iw}}{2i}$$

$$e^{iw} - e^{-iw} = 2iz$$

$$e^{2iw} - 2iz\,e^{iw} - 1 = 0$$

17

$$e^{iw} = \frac{2iz + \sqrt{4i^2z^2 + 4}}{2}$$

$$= iz + \sqrt{1 - z^2}$$

$$\therefore \quad iw = \log\left(iz + \sqrt{1 - z^2}\right)$$

$$w = \sin^{-1} z = \frac{\log\left(iz + \sqrt{1 - z^2}\right)}{i}$$

9.3.15. Complex Exponent Functions of complex numbers

IX. Z^x where x is complex:–

$$Z^x = e^{x \log z}$$

9.3.16. Algebraic and Transcendental Functions of complex numbers

X. Algebraic & transcendental funs:–

Any solution w of the polynomial eqⁿ

$$P_0(z)w^n + P_1(z)w^{n-1} + \cdots + P_{n-1}(z)w + P_n(z) = 0$$

where n is a positive integer.

and $P_0(z)$, $P_1(z)$, ..., $P_n(z)$ are polynomials in z and is an algebraic fⁿ of z.

e.g. $\quad w = \sqrt{z}$

$$w^2 - z = 0$$

Other funs are non-algebraic or transcendental.

e.g. $\log z$, $\sin z$, ...

funs derived from fⁿˢ I – IX (1–9) by performing a finite number of operations of addition substraction, multiplication, division and extraction of roots are called elementary fⁿˢ.

as $\sin z + 2\cosh^{-1} z$ \qquad elementary fⁿˢ.

18

9.3.17. Definition of Limit of complex functions

Definition of a limit:-

Meaning of $\lim_{z \to z_0} f(z) = \ell$

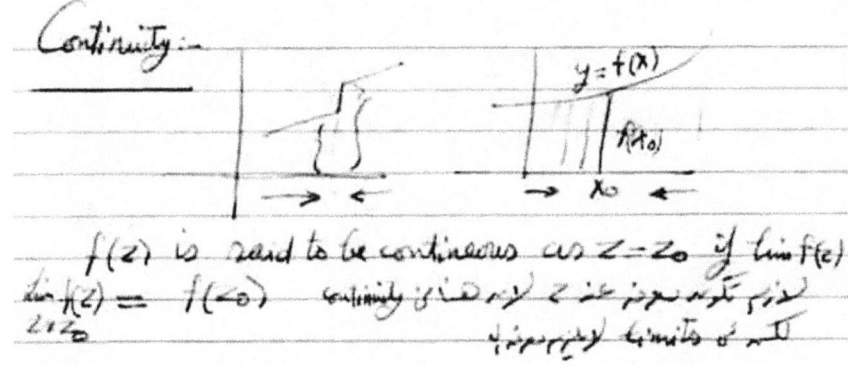

eg. $\lim_{x \to x_0} f(x)$

(Arabic text)

Let $f(z)$ be a single-valued f^n of z defined from all points in the immediate neighbourhood of z_0 itself. we say that $f(z) \to \ell$ as $z \to z_0$ If given any +ve number ϵ however small there exists $\delta = \delta(\epsilon)$ that is dependent on ϵ. such that $|f(z) - \ell| < \epsilon$

for all z satisfying $|z - z_0| < \delta$ irrespective of the way in which z approaches its limit z_0

eg. $\lim_{x \to c} \dfrac{\sin x}{x} = 1$ [*]

(Arabic text)

9.3.18. Continuity of complex functions

Continuity:-

$y = f(x)$

$f(x_0)$

x_0

$f(z)$ is said to be continuous as $z = z_0$ if $\lim f(z)$
$\lim_{z \to z_0} f(z) = f(z_0)$ (Arabic text continuity)

(Arabic text) limits of ...

19

This implies that $f(z)$ should be defined at $z = z_0$. A f^n $f(z)$ is said to be continuous in a given region of it is continuous at all points of that ratio.

The p^t. at ∞

This is the p^t in the z plane which transforms into the origin when we make the transformation $\omega = \frac{1}{z}$

9.3.19. Definition of a domain of complex numbers

Sat.   30 \10\ 1971

Definition of a domain :- ‪تعريف المجال‬

A domain is a surface of points D satisfying the following two properties :-

① if P and Q are any two points of a domain then we can join P and Q by a curve. all of those points belong to the domain.

② If A is a point of the domain then with A as centre we can draw some circles such that all p^{ts} in the interior of the circle belong to the domain.

points of surfaces are not in the domain   not a domain   domain

9.3.20. Definition of derivatives of complex functions

20

Definition of derivatives of a f^n of z:-

$$f'(z) = \lim_{\Delta z \to 0} \frac{f(z+\Delta z) - f(z)}{\Delta z}$$

Irrespective of the way in which $\Delta z = 0$.
⊞ suppose that Q approach P
parallel to the x-axis
then $\Delta z = \Delta x$
and suppose that $f(z)$ ....
$$= f(z) = u(x,y) + iv(x,y) \quad — ①$$

u & v are conjugate f^{ns}

$$f'(z) = \lim_{\Delta x \to 0} \frac{u(x+\Delta x, y) + iv(x+\Delta x, y) - u(x,y) - iv(x,y)}{\Delta x}$$

$$\boxed{f'(z) = \frac{\partial u}{\partial x} + i \frac{\partial v}{\partial x}} — Ⓘ$$

because u f^n of x and y. and y still unchanged.

⊞ Next suppose that we approach D in a direction parallel
to the y axis
$$\because z = x + iy$$
$$\Delta z = \Delta x + i\Delta y$$
as $\Delta x = 0$
$$\therefore \Delta z = i\Delta y \quad — ①$$

$$\therefore f'(z) = \lim_{\Delta y \to 0} \frac{u(x, y+\Delta y) + iv(x, y+\Delta y) - u(x,y)}{i\Delta y}$$
$$iv(y,x)$$

$$f'(z) = \frac{1}{i}\left(\frac{\partial u}{\partial y} + i\frac{\partial v}{\partial y}\right)$$

$$\boxed{f'(z) = \frac{\partial v}{\partial y} - i\frac{\partial u}{\partial y}} \rightarrow ⒾⒾ$$

Equating 1 and 2

$$\frac{\partial u}{\partial x} + i \frac{\partial v}{\partial x} \longrightarrow f'(z) \Bigg\}$$

$$\frac{\partial v}{\partial y} - i \frac{\partial u}{\partial y} = f'(z) \Bigg\}$$

real = real

imaginary = imaginary

9.3.20.1. Cauchy-Riemann Equations

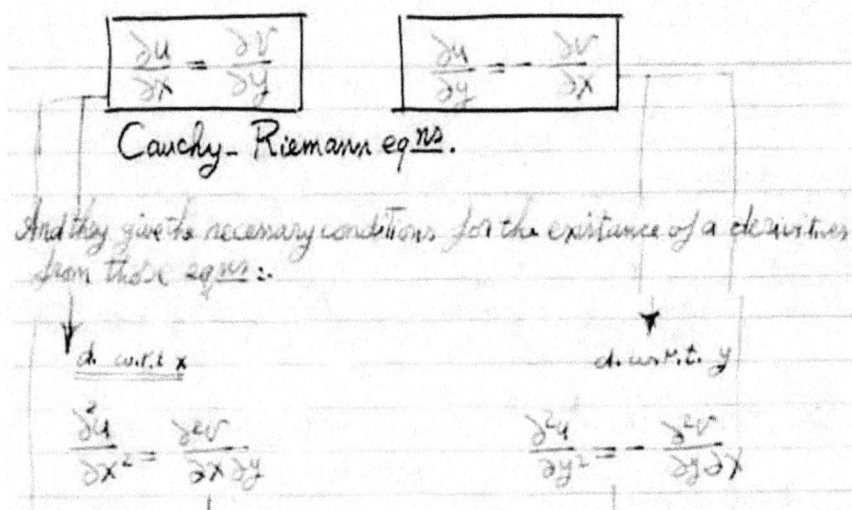

$$\boxed{\frac{\partial u}{\partial x} = \frac{\partial v}{\partial y}} \qquad \boxed{\frac{\partial u}{\partial y} = - \frac{\partial v}{\partial x}}$$

Cauchy - Riemann eq^{ns}.

And they give the necessary conditions for the existance of a derivatives from those eq^{ns}:-

d w.r.t x

$$\frac{\partial u}{\partial x^2} = \frac{\partial v}{\partial x \partial y}$$

d w.r.t. y

$$\frac{\partial^2 u}{\partial y^2} = - \frac{\partial^2 v}{\partial y \partial x}$$

9.3.20.2. Laplace Equation

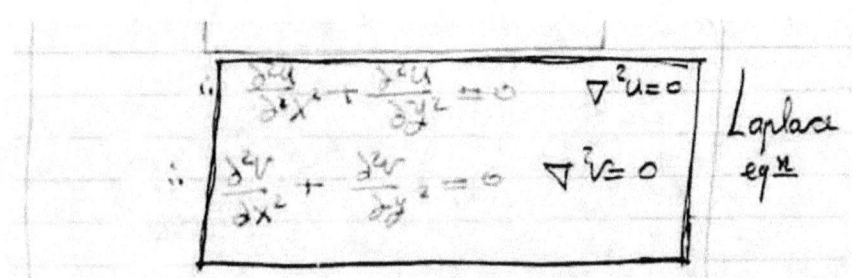

$$\boxed{\begin{array}{ll} \text{i.} \quad \dfrac{\partial^2 u}{\partial x^2} + \dfrac{\partial^2 u}{\partial y^2} = 0 & \nabla^2 u = 0 \\[2mm] \text{ii.} \quad \dfrac{\partial^2 v}{\partial x^2} + \dfrac{\partial^2 v}{\partial y^2} = 0 & \nabla^2 v = 0 \end{array}} \quad \begin{array}{l} \text{Laplace} \\ \text{eq}^n \end{array}$$

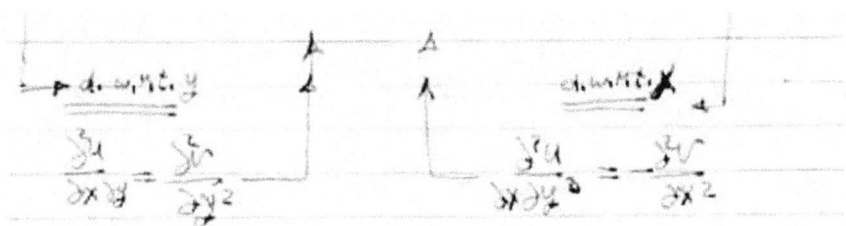

$$\frac{\partial u}{\partial x^2} = \frac{\partial v}{\partial y^2} \qquad\qquad \frac{\partial u}{\partial x} = \frac{\partial v}{\partial x^2}$$

Note in fluid Mechanics $\nabla^2 q = 0$ but here
& $q = +U + iV$ $\nabla^2 U = 0$
$\nabla^2 V = 0$

That is each of the conjugate f^{ns} u, v, satisfy Laplace
eq^n . f^{ns} which ~~tan~~ satisfy Laplace's eq^{ns}
are called <u>Harmonic f^{ns}</u>.
we shall now show that curves of <u>constant</u> u cut
curves of <u>const.</u> v at right angles

$$U(X, y) = C \longrightarrow \textcircled{1}$$

• that means the component of the velocity parallel to y axis
= 0 or $du = 0$. or change of $u = 0$
• $u(x, y)$ or u depend to x and to y as the following

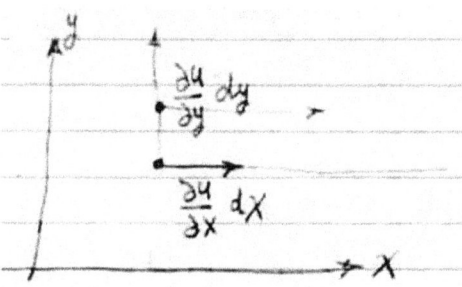

$$\therefore dU = \frac{\partial u}{\partial x} dX + \frac{\partial u}{\partial y} dy = 0$$

$$\therefore \frac{dy}{dx} = - \frac{\partial u / \partial x}{\partial u / \partial y} \quad \text{from Laplace} \longrightarrow = \frac{\partial v / \partial y}{\partial v / \partial x}$$

$y \quad v(x, y) = c \longrightarrow \boxed{2}$

$$dv = \frac{\partial v}{\partial x} dx + \frac{\partial v}{\partial y} dy = 0$$

$$\frac{dy}{dx} = -\frac{\frac{\partial v}{\partial x}}{\frac{\partial v}{\partial y}} \quad \text{from Laplace} = \frac{\frac{\partial u}{\partial y}}{\frac{\partial u}{\partial x}}$$

The product of the two slopes $= -1$. Hence the family of curves $v = c'$ cut the family of curves $u = c$ at right angles wherever they meet.

9.3.21. Regular and Analytic functions

Regular and analytic fns:-

A f^n $f(z)$ which is single-valued and which has a derivative at all pts of a given domain is called Regular or analytic fn in that domain.

ex. 1

$$w = z^2 \text{ where } w = u + iv$$
$$z = x + iy$$

$$\therefore u + iv = (x + iy)^2$$
$$= x^2 - y^2 + 2ixy$$

24

$$U = x^2 - y^2$$
$$V = 2xy$$

$$\frac{\partial^2 u}{\partial x^2} = 2$$
$$\frac{\partial^2 u}{\partial y^2} = -2$$

$$\boxed{\frac{\partial^2 u}{\partial x^2} + \frac{\partial^2 u}{\partial y^2} = 0}$$

$$\frac{\partial u}{\partial x} = 2x$$

$$\frac{\partial u}{\partial y} = -2y$$

$$\frac{\partial V}{\partial x} = 2y$$

$$\frac{\partial V}{\partial y} = 2x$$

$$\frac{\partial^2 V}{\partial x^2} = 0$$
$$\frac{\partial^2 V}{\partial y^2} = 0$$

$$\boxed{\frac{\partial^2 V}{\partial x^2} + \frac{\partial^2 V}{\partial y^2} = 0}$$

<u>answer 1</u>   Suring if it follows Laplace.

<u>U = const</u>

$$x^2 - y^2 = const \longrightarrow \textcircled{1}$$
$$2x - 2y \frac{dy}{dx} = 0$$

$$\boxed{\frac{dy}{dx} = \frac{x}{y}}$$

<u>V = const</u>

$$2xy = const \longrightarrow \textcircled{2}$$
$$2y + 2x \frac{dy}{dx} = 0$$

$$\boxed{\frac{dy}{dx} = \frac{-y}{x}}$$

25

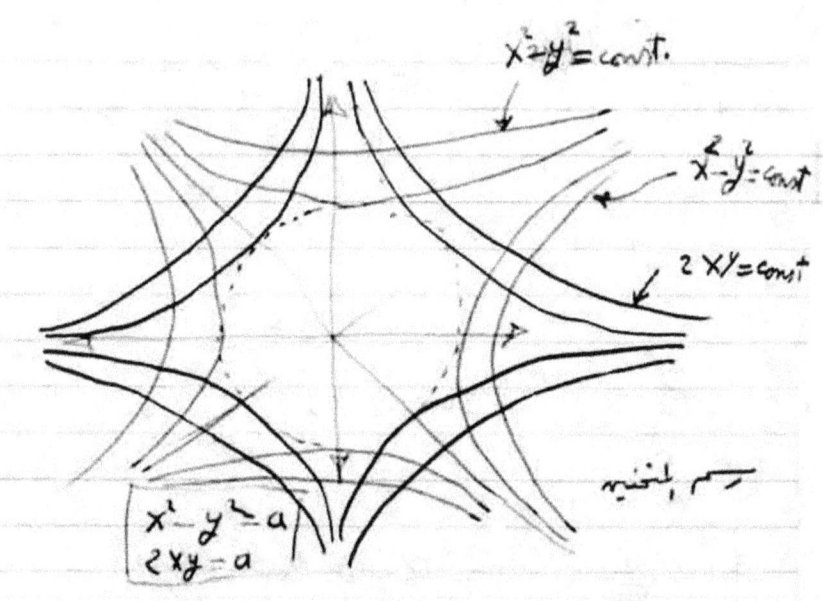

$x^2+y^2=$ const.

$x^2-y^2=$ const

$2xy=$ const.

$x^2-y^2=a$
$2xy=a$

tangents in infinity of $x^2-y^2=a$ $\qquad$ ∴ $2x-2y\frac{dy}{dx}=0$

$\frac{dy}{dx}=\frac{x}{y}$ $\quad$ eq^n of tg^t $\qquad$ $\frac{y-y_1}{x-x_1}=\frac{dy}{dx}=\frac{x}{y}$

$(y-y_1)y=(x-x_1)x$

$\quad y^2-y_1y=x^2-x_1x$ $\quad$ ∴ $x^2-y^2=\boxed{x_1x-y_1y=a}$

$x_1x-y_1y=a \quad \longleftarrow \quad ①$

as $x_1=y_1=0 \qquad a=0$

∴ $x_1x-y_1y=0$

$\cdot$ Slope $=\frac{x_1}{y_1}=45$

$2xy=a$

	x'	y'
x	$\cos\alpha$	$-\sin\alpha$
y	$\sin\alpha$	$\cos\alpha$

$$2[x'(\cos 45) + y' \sin 45)][x' \sin 45 + y' \cos 45] = 0$$

$$2\left[\frac{x'}{\sqrt{2}} - \frac{y'}{\sqrt{2}}\right]\left[\frac{x'}{\sqrt{2}} + \frac{y'}{\sqrt{2}}\right] = a$$

$$2\left[\frac{x'^2}{2} - \frac{y'^2}{2}\right] = a$$

$$\underline{\underline{x'^2 - y'^2 = a}} \longrightarrow \boxed{2}$$

Ex. 2.　$\omega = \dfrac{A}{z}$ 　　　$\therefore \dfrac{A}{x+iy} = A\,\dfrac{x-iy}{x^2+y^2}$

$\therefore \omega = \dfrac{A(x-iy)}{x^2+y^2} = u+iv$

$u = \dfrac{Ax}{x^2+y^2}$ 　　　$\therefore \left| \begin{array}{l} u(x^2+y^2) - Ax = 0 \quad \}\!-\!\text{①} \\[2mm] v(x^2+y^2) + Ay = 0 \quad)\!-\!\text{②} \end{array} \right.$

$v = -\dfrac{Ay}{x^2+y^2}$

- Eq^n 1 represent a family of circles all of which touch the y axis at the origin since if you put $x=0$ we get $y=0$
- Eq^n 2 represents a family of circles touch the x-axis at origin

$u(x^2+y^2) - Ax = 0$ 　　centre $\left(+\dfrac{A}{2}, 0\right) \}\!-\!\text{①}$
　　　　　　　　　　　as $x=0$　$y=0$

$v(x^2+y^2) + Ay = 0$ 　　centre $\left[+0, \pm\dfrac{A}{2}y\right) \}\!-\!\text{②}$
　　　　　　　　　　　$x=0$　$y=0$

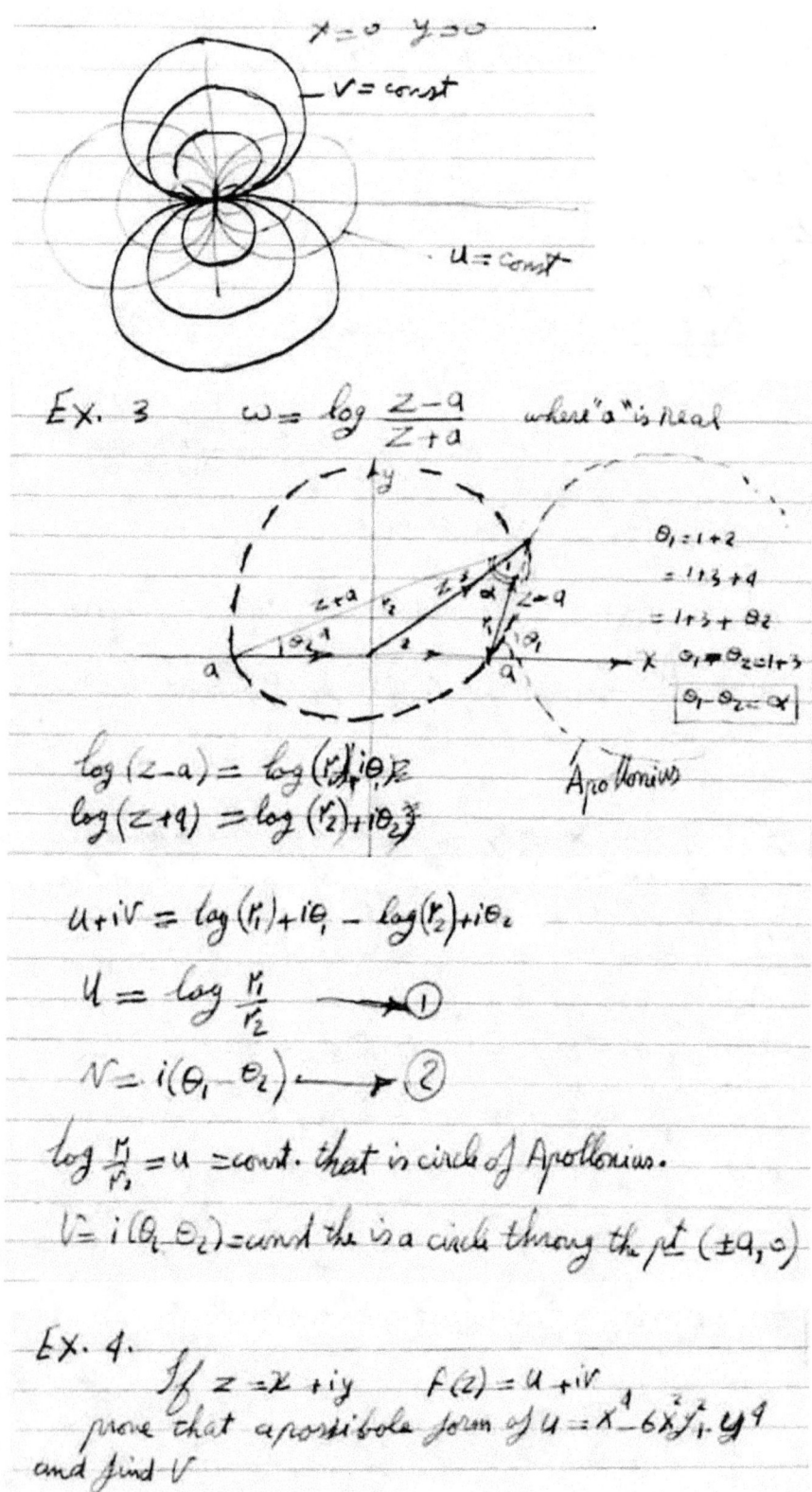

$x=0$ $y=0$

— $V = $ const

· $u = $ const

EX. 3 $\omega = \log \dfrac{z-a}{z+a}$ where "a" is real

$\theta_1 = 1+2$
$= 1+3+4$
$= 1+3+\theta_2$
$\theta_1 - \theta_2 = 1+3$

$\boxed{\theta_1 - \theta_2 = \alpha}$

Apollonius

$\log(z-a) = \log(r_1) + i\theta_1$
$\log(z+a) = \log(r_2) + i\theta_2$

$u + iv = \log(r_1) + i\theta_1 - \log(r_2) + i\theta_2$

$u = \log \dfrac{r_1}{r_2}$ ———→ ①

$v = i(\theta_1 - \theta_2)$ ———→ ②

$\log \dfrac{r_1}{r_2} = u = $ const. that is circle of Apollonius.

$v = i(\theta_1 - \theta_2) = $ const the is a circle through the pt ($\pm a$, 0)

EX. 4.
   $\mathscr{I}\!f$ $z = x + iy$ $f(z) = u + iv$
   prove that a possible form of $u = x^4 - 6x^2y^2 + y^4$
and find V

28

9.3.22. Example on Laplace Equations

■ Laplace eqn

$\cdot \dfrac{\partial u}{\partial x} = 4X^3 - 12XY^2$ $\cdot \dfrac{\partial u}{\partial y} = -12X^2Y + 4Y^3$

$\dfrac{\partial^2 u}{\partial x^2} = 12X^2 - 12Y^2$ $\dfrac{\partial^2 u}{\partial y^2} = -12X^2 + 12Y^2$

$$\dfrac{\partial^2 u}{\partial x^2} + \dfrac{\partial^2 u}{\partial y^2} = 0$$

Hence u can be the real part of $f(z)$

■ Finding of V

$$\dfrac{\partial u}{\partial x} = \dfrac{\partial v}{\partial y} = 4X^3 - 12XY^2 \quad\text{——①}$$

$$\dfrac{\partial u}{\partial y} = -\dfrac{\partial v}{\partial x} = -12X^2Y + 4Y^3 \quad\text{——②}$$

By integration the 2nd eqn w.r.t. x

$$V = 4X^3Y - 4Y^3X + \phi(y)$$

$$\dfrac{\partial v}{\partial y} = \text{①} \quad eqn = 4X^3 - 12Y^2X + \phi'(y)$$

$$\therefore \phi'(y) = 0 \quad \text{from ①}$$

This should be $4X^3 - 12Y^2X$

$$\therefore \boxed{V = 4X^3Y - 4Y^3X}$$

$$f(z) = u + iv = X^4 - 6X^2Y^2 + Y^4 + i(4X^3Y - 4Y^3X)$$

$$\therefore (x+iy)^4 = X^4 + 4X^3(iy) + \dfrac{4(4-1)}{2!}X^2(iy)^2 + \dfrac{4(4-1)(4-2)}{3!}X(iy)^3$$

$$+ \frac{4(4-1)(4-2)(4-3)\,(iy)^4}{4!}$$

$$= x^4 + i4x^3y - 6x^2y^2 - ix y^3 \frac{4\times3\times2}{3\times2} + \frac{4\times3\times2\times1 y}{4\times3\times2\times1}$$

$$(x+iy)^4 = x^4 - 6x^2y^2 + y^4 + i(4x^3y - 4xy^3).$$

9.4. Conformal Mapping

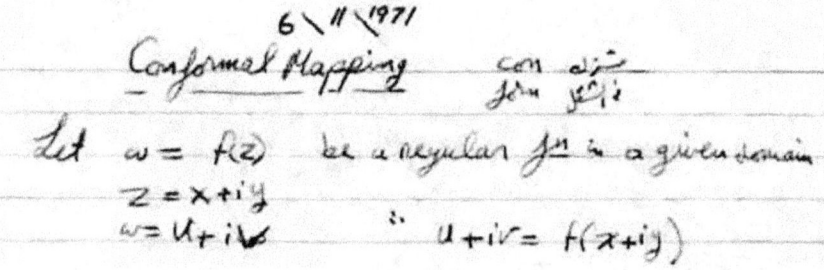

6 \ 11 \ 1971

Conformal Mapping   con المخطط
                         في الكتاب

Let $w = f(z)$ be a regular f^n in a given domain

$z = x + iy$

$w = u + iv$ ∴ $u + iv = f(x+iy)$

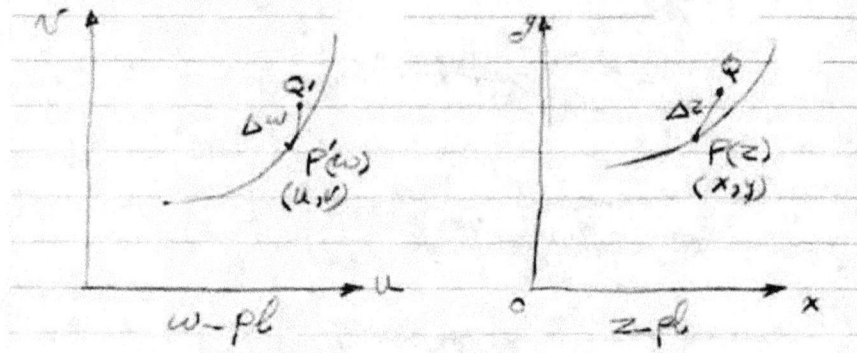

Every pt $P(z)$ in the z-pl determined by (x,y) as a corresponding عنصر on $P'(w)$ in the w-pl Used ordinates (u, v). When P traces a curve, P' will traces a corresponding curve.

let Q a point and PQ be Δz. in the w-pl let Q' correspond to Q and QP' be Δw.

now $\Delta w = \frac{dw}{dz} \Delta z$ approx. as $w = f(z)$

∴ $|\Delta w| = \left|\frac{dw}{dz}\right| |\Delta z|$

$\arg \Delta w = \arg \frac{dw}{dz} + \arg \Delta z$

30

Thus Δw is obtained from Δz by multiplying the length PQ $\left|\frac{dw}{dz}\right|$ and adding a arg. of PQ to arg. $\frac{dw}{dz}$

This is true for infinitesimal vectors issuing from P. ie they are all magnified or deminished in the sam ratio $\left|\frac{dw}{dz}\right| : 1$ & they are rotating through the same angle arg $\left|\frac{dw}{dz}\right|$. from this it follow that.

1. Angle of intersection of curves in the z-pl. is equal to angle of intersection of a corresponding curves in the w-plane

ii) infinitesmal fig2 in z-plane transformes into similar infinitesimal fig2, it due to above properties that the Transformation is set to be conformal.

Note: $\dfrac{\text{Area of element in w-pl}}{\text{Area of element in z-pl}} = \left(\dfrac{dw}{dz}\right)^{3}$

9.4.1. Jacobian or rates of change of x, y, on u, v

<u>Rate of areas of corresponding elements in the z- and w planes</u>

Let $z = \phi(w^-)$

$z = \phi(w)$ $\Delta z = \dfrac{dz}{dw}\,\Delta w$

i.e. $x + iy = \phi(u + iv)$

$\dfrac{\partial x}{\partial u} + i\dfrac{\partial y}{\partial u} = \phi'(w)\dfrac{\partial w}{\partial u} = \phi'(w)$

$\dfrac{\partial x}{\partial v} + i\dfrac{\partial y}{\partial v} = \phi'(w)\dfrac{\partial w}{\partial v} = i\,\phi'(w)$

$\therefore \quad \dfrac{\partial x}{\partial u} = \dfrac{\partial y}{\partial v}$ $\left.\begin{array}{c} \\ \\ \end{array}\right\}$ <u>Area of element in z-pl</u>

$\dfrac{\partial x}{\partial v} = -\dfrac{\partial y}{\partial u}$ $= \left|\dfrac{dz}{dw}\right|^2 = |\phi'(w)|^2$

$\left|\dfrac{dz}{dw}\right|^2 = \left[\dfrac{\partial x}{\partial u}\right]^2 + \left(\dfrac{\partial y}{\partial u}\right)^2 = \dfrac{\partial x}{\partial u}\cdot\dfrac{\partial x}{\partial u} + \dfrac{\partial y}{\partial u}\cdot\dfrac{\partial y}{\partial u}$

$= \dfrac{\partial x}{\partial u}\dfrac{\partial y}{\partial v} - \dfrac{\partial y}{\partial u}\dfrac{\partial x}{\partial v} = \begin{vmatrix} \dfrac{\partial x}{\partial u} & \dfrac{\partial y}{\partial u} \\ \dfrac{\partial x}{\partial v} & \dfrac{\partial y}{\partial v} \end{vmatrix} = \dfrac{\partial(x,y)}{\partial(u,v)}$

The determinants known as

Jacobian of x & y w.r.t u & v.

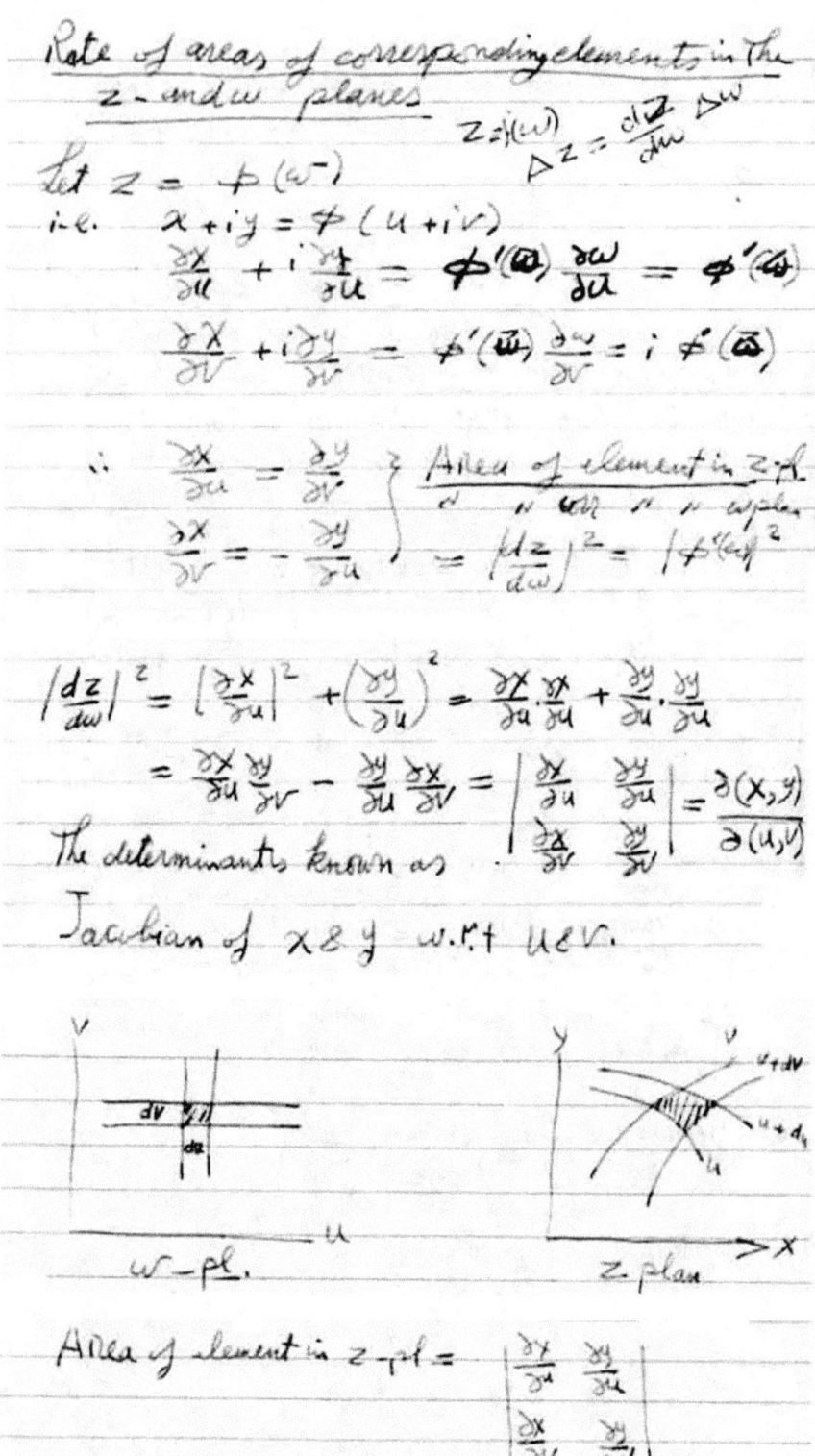

$w - pl.$ z plan

Area of element in z-pl $= \begin{vmatrix} \dfrac{\partial y}{\partial u} & \dfrac{\partial y}{\partial u} \\ \dfrac{\partial x}{\partial v} & \dfrac{\partial y}{\partial v} \end{vmatrix}$

9.4.2. Example of mapping complex functions z(x , y) on a quadratic function w (u, v)

Ex. If $z = w^2$ prove that curves of const u & curves of const v are conformal parabola & calculate the area bet. the two parab, $u = 3$, $v = 2$.

Soln

maybe $x + iy = (u + iv)^2 = u^2 - v^2 + 2ivu$

$$x = u^2 - v^2$$
$$y = 2uv \quad \Big\} \quad \text{—} \quad ①$$

$\underline{u = \text{const}}$ if x & y are f^{ns} of a single parameter v → eliminat v.

$$\therefore v = \frac{y}{2u} \quad \& \quad x = u^2 - \frac{y^2}{4u^2}$$

$$\text{i.e} \quad \boxed{y^2 = -4u^2(x - u^2)} \quad y^2 = 4ax$$

which is a parabola with $(u^2, 0)$ latus rectum

$4u^2$ & whose focus is the origin

$\underline{v = \text{const}}$: $u = \frac{y}{2v}$

$$\therefore x = \frac{y^2}{4v^2} - v^2$$

$$\boxed{y^2 = 4v^2(x + v^2)}$$

which is a parabola with $(-v^2, 0)$ vertex. L.R $4v^2$

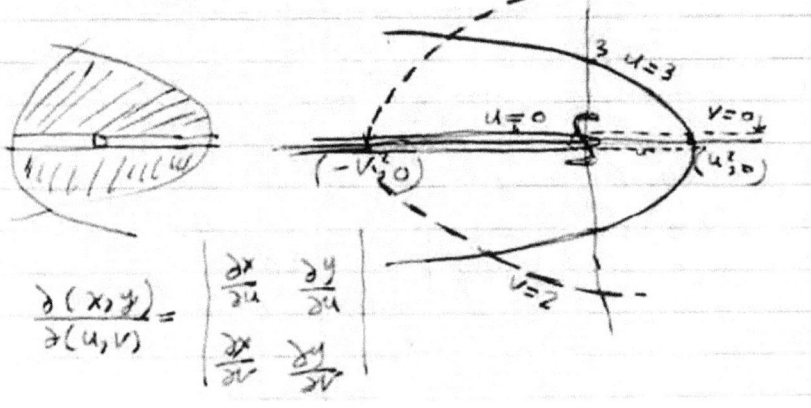

$$\frac{\partial(x, y)}{\partial(u, v)} = \begin{vmatrix} \dfrac{\partial x}{\partial u} & \dfrac{\partial y}{\partial u} \\[2mm] \dfrac{\partial x}{\partial v} & \dfrac{\partial y}{\partial v} \end{vmatrix}$$

from eq^n ①

$$= \begin{vmatrix} 2u & 2v \\ -2v & 2u \end{vmatrix} = 4u^2 + 4v^2$$

Area bet the 2 parabolas $= 2\int_{u=0}^{3}\int_{v=0}^{2} 4(u^2+v^2)\,du\,dv$

$$= 8\int_0^3 \left(u^2 v + \frac{v^3}{3}\right)_0^2 du$$

$$= 8\int_0^3 \left(2u^2 + \frac{8}{3}\right) du = 8\left[\frac{2u^3}{3} + \frac{8}{3}u\right]_0^3$$

$$= 8[18 + 8] = \underline{\underline{208}}$$

9.4.3. The Transformation of w (u, v) = exp (z(x, y))

I. The transformation $w = e^z$ on the infinite strip in the z-pl lines (bounded) $y=0$ $y=2\pi$

$$\overline{w = e^z} = e^{x+iy} = e^x e^{iy} = e^x(\cos y + i \sin y)$$

$$|w| = e^x \qquad \text{arg } w = y.$$

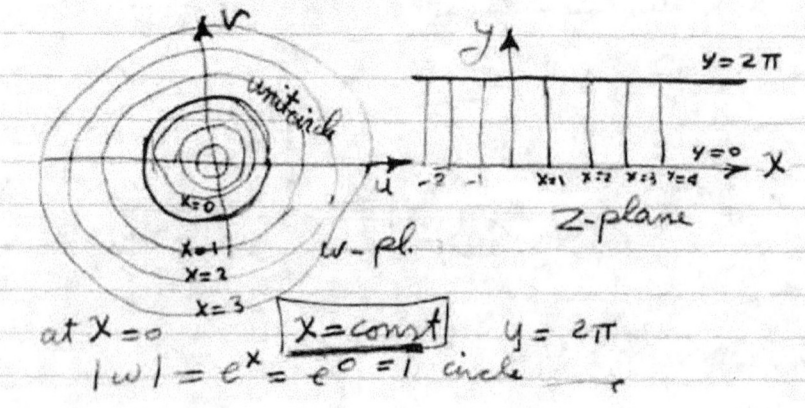

at $X = 0$ $\boxed{X = \text{const}}$ $y = 2\pi$

$|w| = e^x = e^0 = 1$ circle

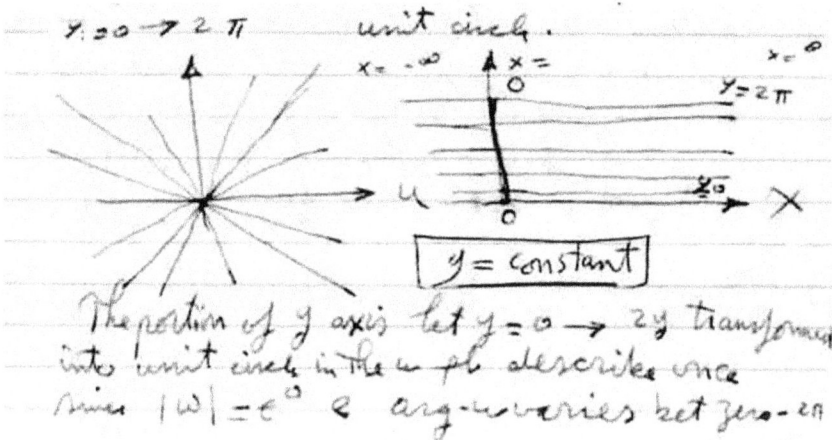

y: 0 → 2π   unit circle.

y = constant

The portion of y axis let y = 0 → 2y transformed into unit circle in the w-pl described once since $|w| = e^0$ & arg-w varies bet zero - 2π

Lines parallel to y-axis and to the right of which are in the strip are transformed into circles outside the unit circle in the w-pl. in other hand lines parallel to y-axis & to left are transformed into circles inside the unit circle. The lines y were const are transform into rays emanating from the origin of the w-plane hence the infinite strip in the z-pl. let. y = 0 → 2π is transformed into the whole of w-pl. mapped once if we take a second strip let 2π < w < 2 w = 4π Then the w-pl is mapped once again and so on

$$dx\,dy = \frac{\partial(x,y)}{\partial(u,v)}\,du\,dv$$

Jacobian

$$dx\,dy = \begin{vmatrix} \dfrac{\partial x}{\partial u} & \dfrac{\partial y}{\partial u} \\[2mm] \dfrac{\partial x}{\partial v} & \dfrac{\partial y}{\partial v} \end{vmatrix} du\,dv$$

9.4.4. The Transformation of w (u, v) = cosh (z(x, y))

13\12\71

The transformation $w = \cosh z$

$$u + iv = \cosh(x + iy) = \cosh x \cosh iy + \sinh x \sinh iy$$
$$= \cosh x \cos y + i \sinh x \sin y$$

$\therefore \; u = \cosh x \cos y$
$v = \sinh x \sin y$ $\qquad\Big\} \longrightarrow$ ①

x = const

$$\frac{u^2}{\cosh^2 x} + \frac{v^2}{\sinh^2 x} = 1 \longrightarrow ②$$

which presented for various values of x conical
ellipses having $(\pm 1, 0)$ since the distance
bet. the centers and a focal is $\sqrt{\cosh^2 - \sinh^2 x = 1}$

y = const

$$\frac{u^2}{\cos^2 y} - \frac{v^2}{\sin^2 y} = 1 \longrightarrow ③$$

which are conical hyperbolas having for the
common the same pts $(\pm 1; 0)$

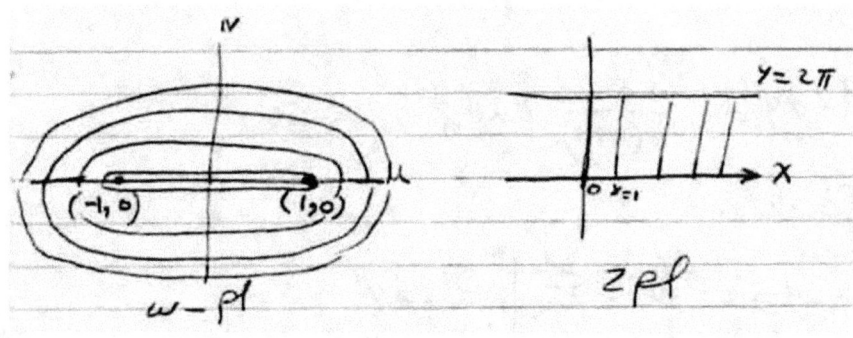

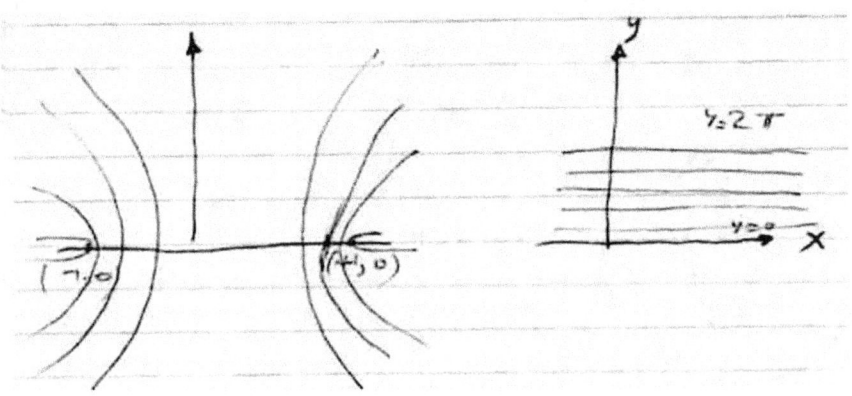

9.5. Inverse Transformation

The fundamental properties of

Inversion التحويل

Here we have a circle of inversion whose centre O is the centre of origin of inversion & whose radius k is the radius of inversion. we defined the inverse of the pt (P') to be the pt (P) such that

i) O, P, P' lay on a striaght line.

ii) $OP \cdot OP' = k^2$

When P traces a curve, P' traces the inverse curve. The nature of the inverse curve depends on the origin of inverse (the origin k) affects only the position and sides of the inverse curve.

37

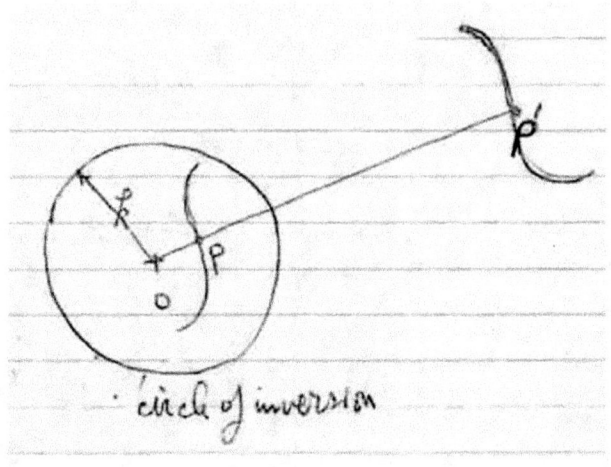

circle of inversion

9.5.1. Inverse transformation of a line on a point on same line

inv = inverse

Theorem I

The inverse of a straight line w.r.t a pt on it is the same straight line. o, P, P' lay on the same line $OP \cdot OP' = k^2$

9.5.2.. Inverse transformation of a line on a point not on same line

Theorem II

Then inv of the st. line w.r.t a pt o not on the line, is a circle

Drop the perpendicular ⊥ OA on the given line, let A' be the inv of A. P is any pt on the line and P' its inv. By Def. $OA \cdot OA' = OP \cdot OP' = k^2$

i.e A'APP' is a cyclic quadrilateral,

∴ $O\hat{P'}A = O\hat{A}P = 90°$

Hence the locus of P' is a circle.

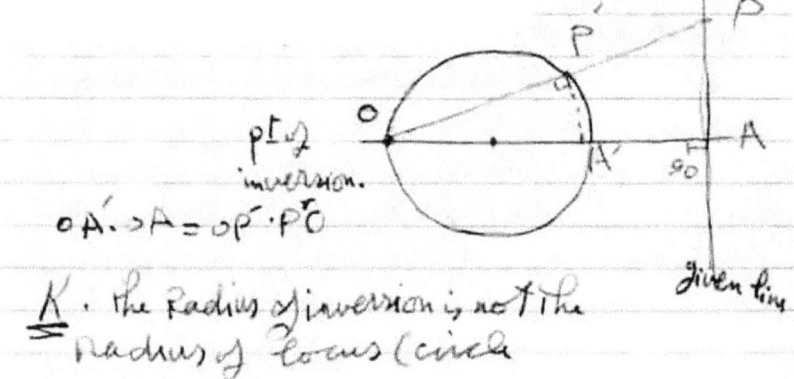

pt of inversion.

$OA' \cdot OA = OP' \cdot PO$

N: the Radius of inversion is not the Radius of locus (circle

9.5.3. Inverse transformation of a circle on a point on same circle

Theorem III The inverse of a circle w.r.t a pt on its circumference is a straight line

proof is similliar to II but steps are reversed

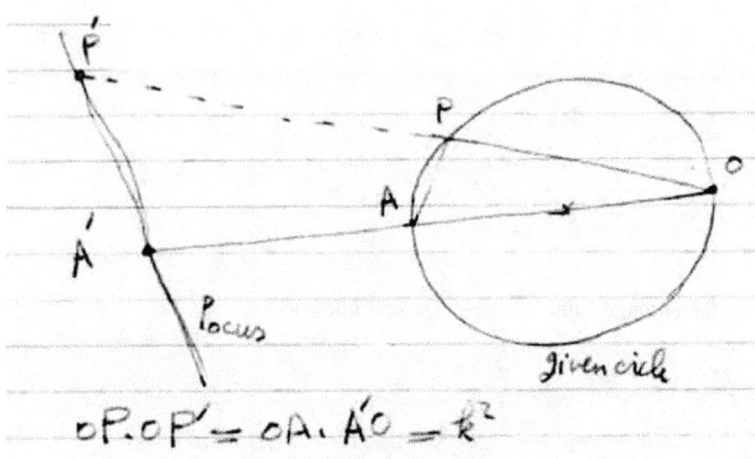

$$oP \cdot oP' = oA \cdot A'o = k^2$$

9.5.4. Inverse transformation of a circle on a point not on same circle

Theorem IV The inverse of a circle w.r.t a pt which does not lay on its circonference is a circle.

Take the origin of inversion as the origin of coordinates let eqn of the given circle ① be

$$x^2 + y^2 + 2gx + 2fy + c = 0$$

or in polar form.

$$r^2 + 2r(g\cos\theta + f\sin\theta) + c = 0 \longrightarrow ①$$

Change r into $\dfrac{k^2}{r}$

$$\therefore \frac{k^4}{r^2} + 2\frac{k^2}{r}(g\cos\theta + f\sin\theta) + c = 0$$

i.e $cr^2 + 2k^2r(g\cos\theta + f\sin\theta) + k^4 = 0$

i.e $c(x^2 + y^2) + 2k^2(gx + fy) + k^4 = 0 \longrightarrow ②$

which is a circle.

conel if ① path through that $c = 0$ and ② is straight line. O is the pt of intersection of two common tgt two circles that is a centre of similitude

9.5.5. Inverse transformation of angle between two curves

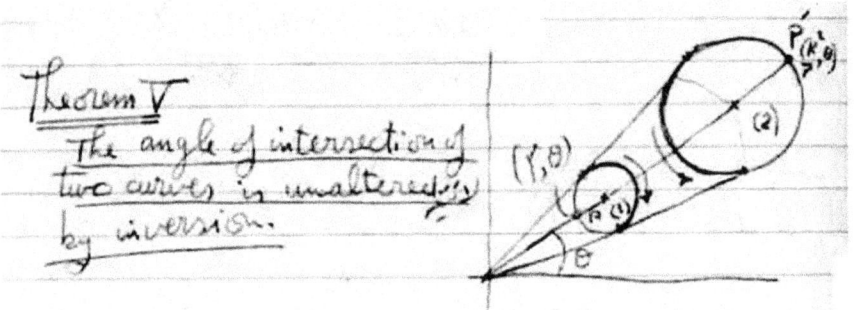

Theorem V

The angle of intersection of two curves is unaltered by inversion.

$$OP \cdot OP' = OQ \cdot OQ'$$
$$\therefore \widehat{OPQ} = \widehat{OQ'P} \longrightarrow ①$$
$$OP \cdot OP' = OR \cdot OR'$$
$$\therefore \widehat{OPR} = \widehat{OR'P} \longrightarrow ②$$
$$\text{Subt.} \quad \widehat{QPR} = \widehat{Q'P'R'} \longrightarrow ③$$

Rotate OQQ' to coincide ultimately with OPP' then in the limit the angle bet. the two tangents at = angle bet two tangents at P'

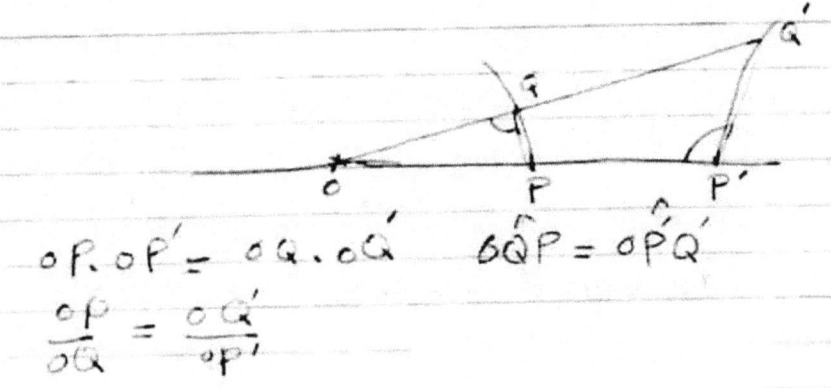

$$OP \cdot OP' = OQ \cdot OQ' \qquad \widehat{OQP} = \widehat{OP'Q'}$$
$$\frac{OP}{OQ} = \frac{OQ'}{OP'}$$

$$OP \cdot OP' = OR \cdot OR'$$
$$\frac{OP}{OR} = \frac{OR'}{OQ} \qquad \therefore \widehat{OPR} = \widehat{OR'P'}$$

41

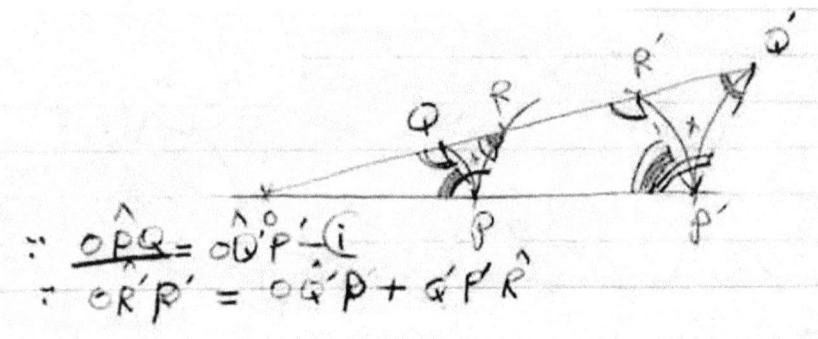

$$\therefore \quad O\hat{P}Q = O\hat{Q'}P' - (i$$
$$\therefore \quad O\hat{R'}P' = O\hat{Q'}P + Q'\hat{P'}R'$$

$$= O\hat{P}R = Q\hat{P}R + O\hat{P}Q \quad - ii)$$
from (i) & ii)

$$\therefore \quad O\hat{Q'}P' + Q'\hat{P'}R' = Q\hat{P}R + O\hat{Q'}P$$

$$\therefore \quad \boxed{Q'\hat{P'}R' = Q\hat{P}R}$$

write this in the form $\quad \omega = \dfrac{\dfrac{bc-ad}{c^2}}{z + \dfrac{d}{c}} + \dfrac{a}{c}$

because that $= \dfrac{\dfrac{bc-ad}{c^2} + \dfrac{az}{c} + \dfrac{ad}{c^2}}{z + \dfrac{d}{c}}$

$$= \dfrac{\cancel{\pm} c(b+az)}{(zc+d)\,\underset{c}{c^2}} = \dfrac{za+b}{zc+d}$$

• We 1st perform the transformation $z_1 = z + \dfrac{d}{c}$ as in $\underline{\text{I}}$

• Then the transformation

$$z_2 = \dfrac{k}{z_1} \quad as \quad k = \dfrac{bc-ad}{c^2}$$

as in ∇ $\quad z_1 = \cancel{\pm} re^{i\theta}, \quad \dfrac{1}{z_1} = \dfrac{1}{r}e^{-i\theta}$

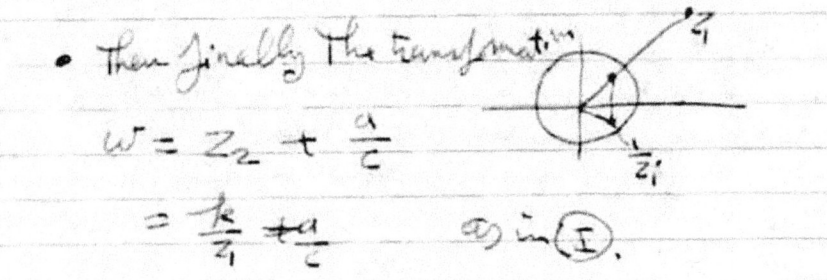

- Then finally The transformation

$$w = z_2 + \frac{a}{c}$$

$$= \frac{k}{z_1} + \frac{a}{c} \qquad \text{as in } \boxed{I}.$$

9.5.6. Operations involved in the inverse transformation of curves

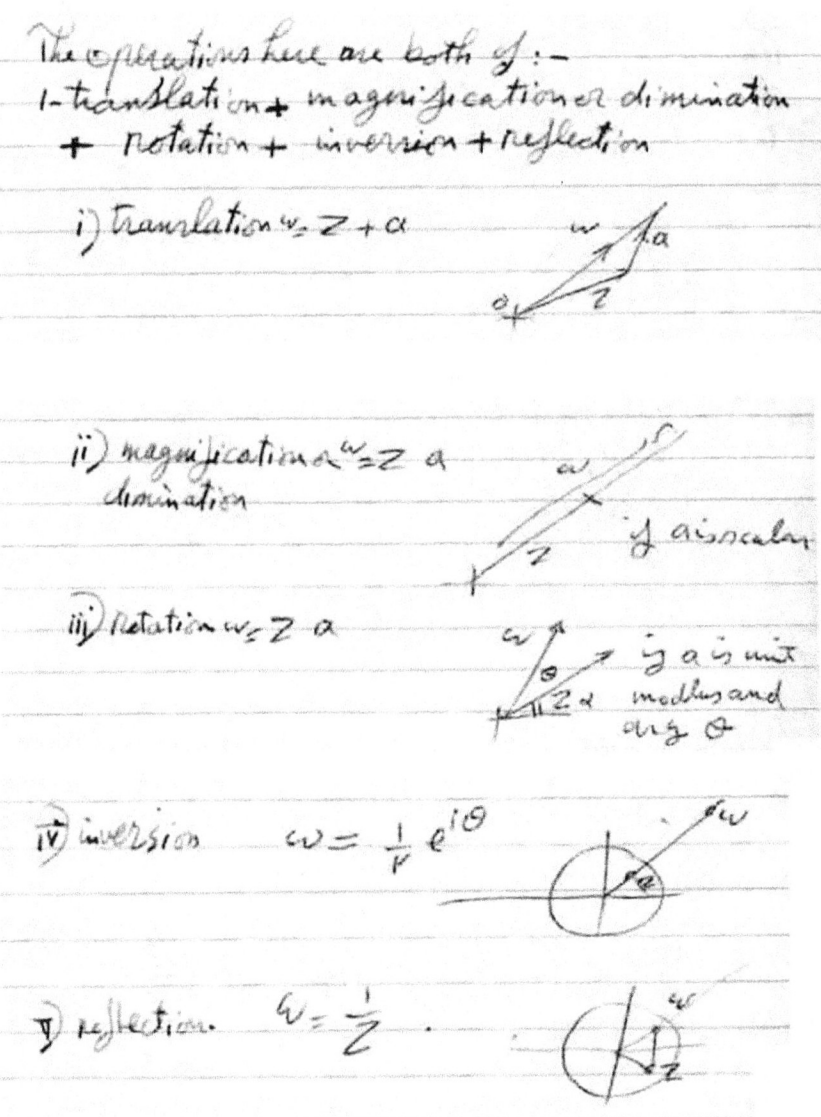

The operations here are both of :-
1- translation + magnification or diminution
    + rotation + inversion + reflection

i) Translation $w = z + a$

ii) magnification or
    diminution $w = z\,a$

    if a is scalar

iii) Rotation $w = z\,a$

    if a is unit modulus and arg θ

iv) inversion $\quad w = \frac{1}{r} e^{i\theta}$

v) reflection. $\quad w = \frac{1}{z}$.

9.5.7. Summary of bilinear transformation of lines and circles

43

The only operation which may effect the nature of the curve that of inversion. now we have shown that circles and straight lines inverted to either straight lines or circles. hence as bilinear transformation into circles st lines either into st circles.

In a bilinear trans. there are corresponding pts that is the trans. of each pt is the same pt the two pts are obtained by putting

$$z = \frac{az + b}{cz + d} \qquad cz^2 + dz = az + b$$

i.e $cz^2 + (d-a)z - b = 0$ which is quadradent in z The Bilinear transf can transformed 3 distinct pts into another three distinct pts.

$$\frac{(w-w_1)(w_2-w_3)}{(w-w_3)(w_2-w_1)} = \frac{(z-z_1)(z_2-z_3)}{(z-z_3)(z_2-z_1)}$$

9.6. Series of complex numbers

27\VII\1971

Series of complex terms .

This are given by $a_1 + a_2 + \cdots + a_n$. i.e Σa_n where (a) is complex that is;

$$a_n = u_n + i v_n.$$

9.6.1. Convergence of a series of complex numbers

The defination of convergence for a series of complex terms is the same as the of the series of real terms i.e we find A_n of first n terms

$$A_n = a_1 + a_2 + \cdots + a_n$$

we then let $n \to \infty$ if A_n terms of a finite limit we say that the series $\sum a_n$ is convergent (cgt) otherwise it is divergent.

$$\sum a_n = \sum U_n + i \sum V_n$$

The convergence of $\sum a_n$ implies the convergent of $\sum U_n$ & $\sum V_n$

Absolute & conditional convergence

1. A cgt series $\sum a_n$ is said to be absolutely cgt if $\sum |a_n|$ is cgt.
2. A cgt series $\sum a_n$ is said to be conditionally cgt if $\sum |a_n|$ is divergent (dgt)

We shall now show that if $\sum |a_n|$ is cgt. Then $\sum a_n$ is also cgt

Let $\quad a_n = U_n + i V_n$

$$|a_n| = \sqrt{U_n^2 + V_n^2}$$

Now $\quad 0 \leq |U_n| \leq |a_n|$

$$0 \leq |V_n| \leq |a_n|$$

45

Since $\sum |a_n|$ is cgt. The $\sum |u_n| \leq \sqrt{v_n}$ are also convergent by the comparision test.

Now from series of real part:

$$\sum u_n + i \sum V_n \text{ are convergent and hence}$$
$$\sum (u_n + i V_n) \text{ is cgt } \therefore$$

$$\therefore \sum a_n \text{ is cgt}$$

9.6.2. Power Series of complex variables

Power series of a complex variable

This is given by $c_0 + a_1 z + a_2 z^2 + \cdots + a_n z^n + \cdots$

i.e

$$\sum a_n z^n$$

where $z = x + i y$ and a_n in general complex

we shall now show that if $\sum a_n z^n$ is cgt for $z = z_0$. Then $\sum a_n z^n$ is cgt for absolutely cgt for all z satisfying $|z| < |z_0|$

The geometrical meaning of this is that if $\sum a_n z^n$ is cgt for some pts z_0 then its absolutely convergent for all pts inside

The circle centre origin passing through pt z_0

Since $\sum a_n z^n$ is cgt at $z = z_0$ That is $\sum a_n z_0^n$ is cgt. Then $a_n z_0^n \to 0$ as $n \to \infty$

i.e. $|a_n z_0^n| \to 0$ as $n \to \infty$

hence

$$|a_n z_0^n| < 1 \text{ for all } n > n_0 \qquad n < n_0 < \infty$$

$$|a_n z^n| = |a_n z_0^n| \left| \left(\frac{z}{z_0}\right)^n \right| < \left|\frac{z}{z_0}\right|^n$$

for all $n > n_0$

for all pts inside the circle $|z_0| > |z|$

46

i.e $\left|\dfrac{z_i}{z_0}\right| < 1$

hence $\sum \left|\dfrac{z}{z_0}\right|^n$ is cgt being a geometric series
with common ratio is less than unity,

∴ $\sum \left|a_n z^n\right|$ is cgt

i.e. $\sum a_n z^n$ is absolutely cgt.

9.6.3. Circle of convergence

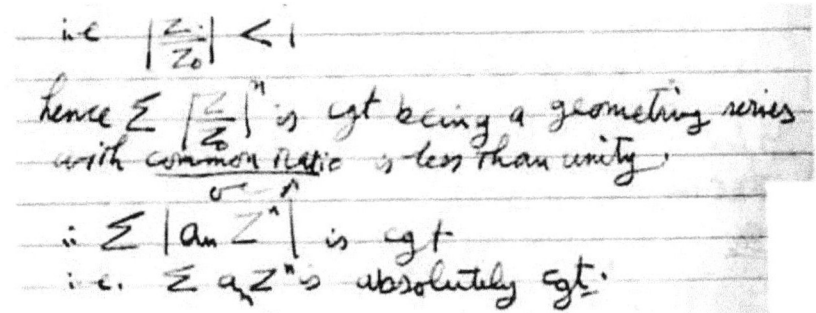

Circle of cgce

This is a circle, centre origin, having the following
properties:-

1. The series $\sum a_n z^n$ is absolutely cgt for all pts
   inside the circle.

2. The series is not cgt for any pt outside the ⊙

3. The series maybe cgt or not cgt for pts on
   the ⊙ itself. The radius of this ⊙ is called the
   Radius of cgce.

Radius of convergence is determined in a manner
similarly to that of the ratio test. The following rule
can be adopted:-

can be adopted:-
we find $\displaystyle\lim_{n\to\infty}\left|\dfrac{a_{n+1}}{a_n}\right|$

$$\boxed{\begin{array}{c} a_n \text{ coeff}^t \text{ of } z^n \\ a_n z^n \end{array}}$$

Denote this limit by "ℓ"

i.e $\displaystyle\lim_{n\to\infty}\left|\dfrac{a_{n+1}}{a_n}\right| = \ell$

that the Radius of cgce $= \dfrac{1}{\ell}$
or.

ex. Consider the exponential series
$$e^z = 1 + z + \frac{z^2}{2!} + \cdots + \frac{z^n}{n!} + \cdots$$

soln.
$$a_n = \frac{1}{n!}$$

$$a_{n+1} = \frac{1}{(n+1)!}$$

$$\lim_{n \to \infty} \frac{a_{n+1}}{a_n} = \lim \frac{n!}{(n+1)!}$$

$$= \lim_{n \to \infty} \frac{n!}{(n+1) \, n!} = \lim_{n \to \infty} \frac{1}{n+1} = 0$$

as $n = \infty$ $l = 0$ radius $= \infty$.

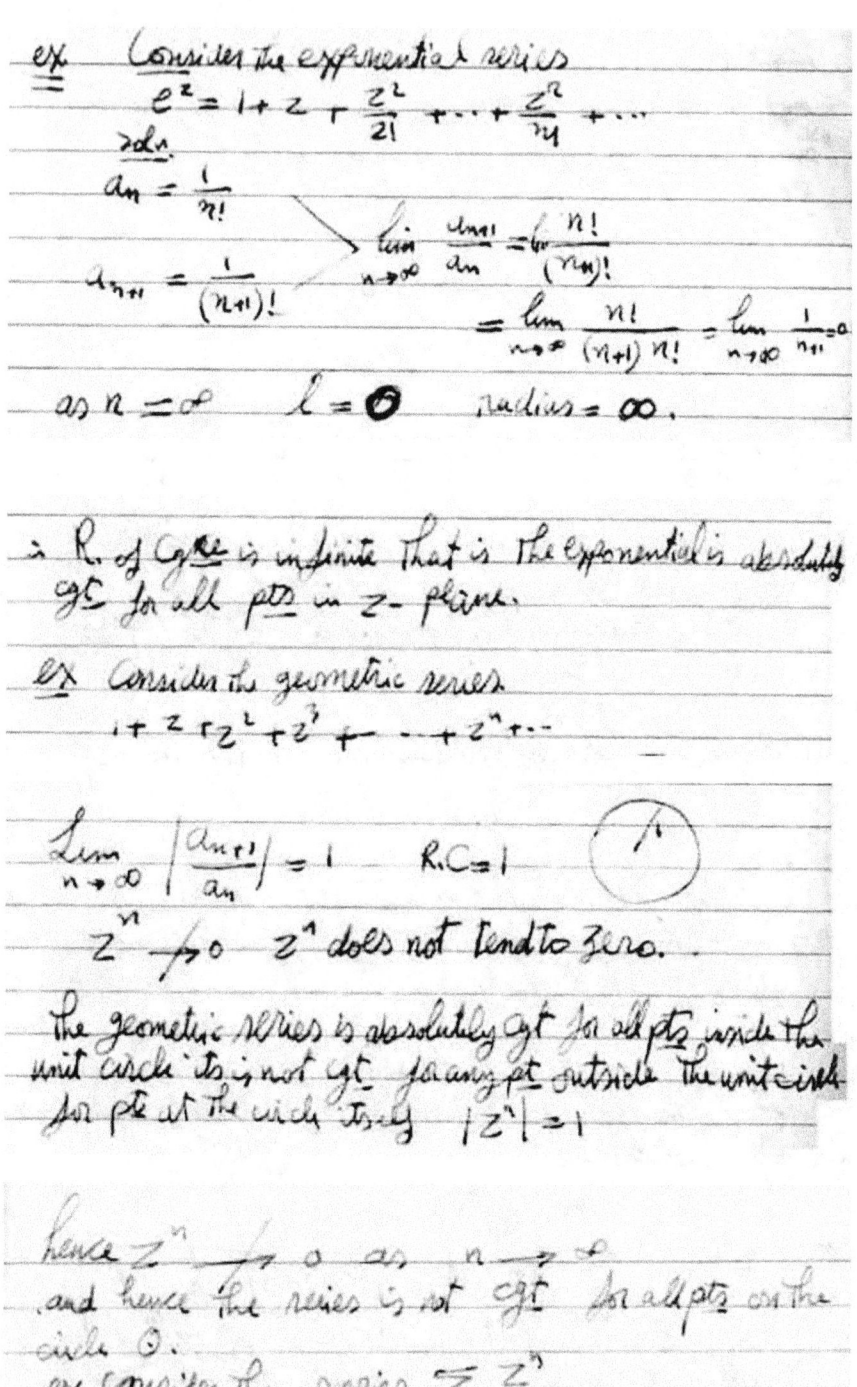

∴ R. of cgce is infinite That is the exponential is absolutely cgt for all pts in z-plane.

ex Consider the geometric series
$$1 + z + z^2 + z^3 + \cdots + z^n + \cdots$$

$$\lim_{n \to \infty} \left| \frac{a_{n+1}}{a_n} \right| = 1 \qquad R.C = 1$$

$$z^n \not\to 0 \quad z^n \text{ does not tend to zero.}$$

The geometric series is absolutely cgt for all pts inside the unit circle its is not cgt for any pt outside the unit circle for pts at the circle itself $|z^n| = 1$

hence $z^n \not\to 0$ as $n \to \infty$
and hence the series is not cgt for all pts on the circle O.

ex. Consider the series $\sum \frac{z^n}{n^2}$.

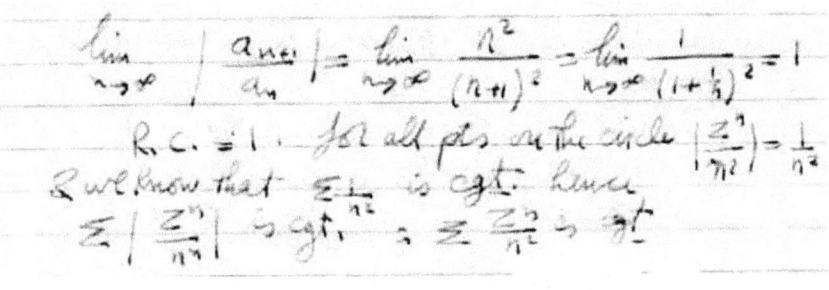

$$\lim_{n \to \infty} \left| \frac{a_{n+1}}{a_n} \right| = \lim_{n \to \infty} \frac{n^2}{(n+1)^2} = \lim_{n \to \infty} \frac{1}{(1+\frac{1}{n})^2} = 1$$

R. C. = 1. for all pts on the circle $\left| \frac{z^n}{n^2} \right| = \frac{1}{n^2}$

& we know that $\sum \frac{1}{n^2}$ is cgt. hence

$\sum \left| \frac{z^n}{n^2} \right|$ is cgt, $\therefore \sum \frac{z^n}{n^2}$ is cgt.

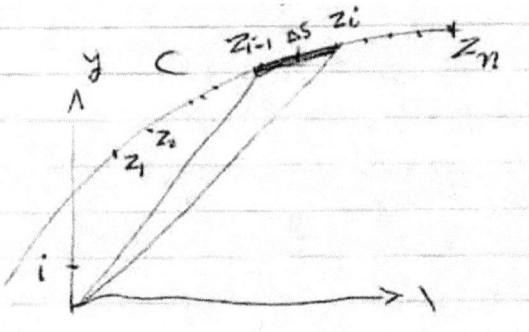

$$(z_{i-1} - z_i) = \Delta S$$

$$\int_C f(z)\,dz = \lim_{n \to \infty} \sum_{1}^{n} f(\xi_i)(z_{i-1} - z_i)$$

9.6.4. Line Integral of a function of complex variables

Line integral of a fn of complex variables :—

Let C be a curve represented parametrically by the eqn

$x = \phi(t)$

$y = \psi(t)$

where ϕ & ψ are differentiable fns of the real variable (t).

Now let $f(z)$ be a single-valued continuous fn of z for all pts on the curve C.

Subdivide the arc C into a large numbers of small subdivisions $z_0, z_1, \ldots z_n$ In each subdivision we take any pt on the arc Thus for the subdivision joining z_i & z_{i-1} take any pt. ξ_i η_i and form the sum

$$\left| \sum_{i=1}^{n} f(\xi_i)(z_i - z_{i-1}) \right|$$

Then we take the limit of this sum when n tends to the infinity in such a way that the areas are indifinitely Thus limit denoted by

$$\int_C f(z)\,dz$$

And hence $\int f(z)\,dz = \lim_{n \to \infty} \sum_{i=1}^{n} f(\xi_i)(z_i - z_{i-1})$

Now suppose that

$$f(z) = u(x,y) + i\,v(x,y)$$

conjugate f z

$z = x + iy$
$dz = dx + i\,dy$

$$\int_C f(z)\,dz = \int_C (u+iv)(dx+idy)$$

$$= \int_C u\,dx - v\,dy + i \int_C v\,dx + u\,dy$$

Thus we see that, the line integral of f of a complex variable dependent on the curvelinean integrals of real f^{ns}.

Note

     conjugate u Satisfies.
     (1) Laplace's eq^n
     (2) Cauchy - Rieman eq^n.

9.7. Cauchy's Theorem

Cauchy's theorem

let c be a simple closed contour. by the word of simple we mean that the curve never cross-it-self.

Let $f(z)$ be a regular or analytic f^{ns} in a domain containing c to gather with its interior then

$$\int_c f(z) \, dz = 0 \qquad \text{prove as z is exact.}$$

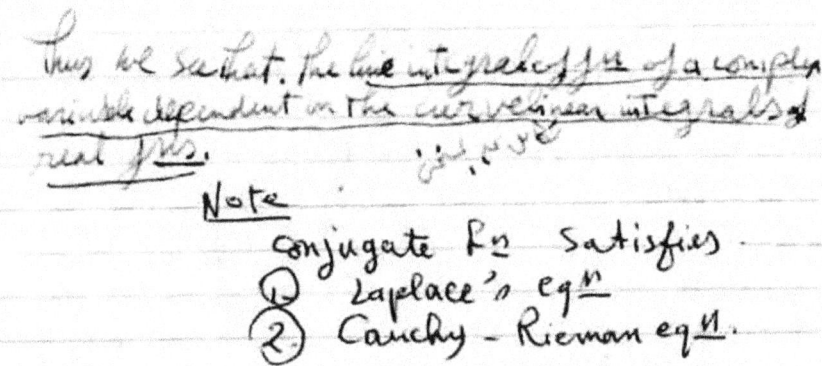
contour

$$\therefore \int f(z) \, dz = \int (u+iv)(dx+idy)$$

$$= \int_c u \, dx - v \, dy + i \left(\int v \, dx + u \, dy \right)$$

Now we will ready know from Green's theorem.

$$\iint_S \left(\frac{\partial Q}{\partial x} - \frac{\partial P}{\partial y} \right) dx \, dy = \int_c P \, dx + Q \, dy$$

hence $\int_C u\,dx - v\,dy = \int\!\!\int_A \left(-\frac{\partial v}{\partial x} - \frac{\partial u}{\partial y}\right) dy\,dx.$

$\int_C v\,dx + u\,dy = \int\!\!\int_B \left(\frac{\partial u}{\partial x} - \frac{\partial v}{\partial y}\right) dy\,dy$

Now from Th Cauchy and Green eqns   Laplace

Riemann

$$\frac{\partial u}{\partial x} = \frac{\partial v}{\partial y}$$

$$\frac{\partial u}{\partial y} = -\frac{\partial v}{\partial x}$$

Hence double integral vanishes and consequently each curvilinear integral vanishes and hence.

$$\boxed{\int_C f(z)\,dz = 0}$$

Corollary of Cauchy's Theorem :-

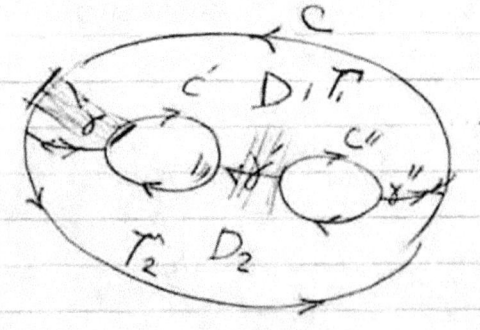

Let C (the) be the closed contour inclosing a finite No of closed contours $C', C'', \ldots$ & let $f(z)$ be a $\int^n$ regular in a domain containing C, C', C'' & the areas bet. them

$$\int_C f(z)dz = \int_{C'} f(z)dz + \int_{C''} f(z)dz - \ldots = Z$$

Defined the area into two domains P_1, P_2 by a set of curves $\gamma, \gamma', \gamma''$ and let the ~~domain~~ bounding contours of the domains of D_1 & D_2 are T_1, T_2.

Since $f(z)$ is regular in each of the domains D_1, D_2 then $\int_{T_1} f(z)dz = 0$ & $\int_{T_2} f(z)dz = 0$

According to Cauchy's theory. hence.

$$\int_{T_1} f(z)dz + \int_{T_2} f(z)dz = 0$$

An integrals on the signals γ, γ' & γ'' cancell one other. since they entered twice the opposite directions. That is with opposite signs.

hence. $\int_C f(z)dz = \int_{C'} f(z)dz = \int_{C''} f(z)dz$

$\therefore \int_C f(z)dz = \int_{C'} f(z)dz + \int_{C''} f(z)dz \ldots = 0$

The integral now all positive.

$9 \setminus 12 \setminus 71$

Ex.1 Evaluate. $\int_c \frac{dz}{z}$ where c is a closed contour

origin being its interior.

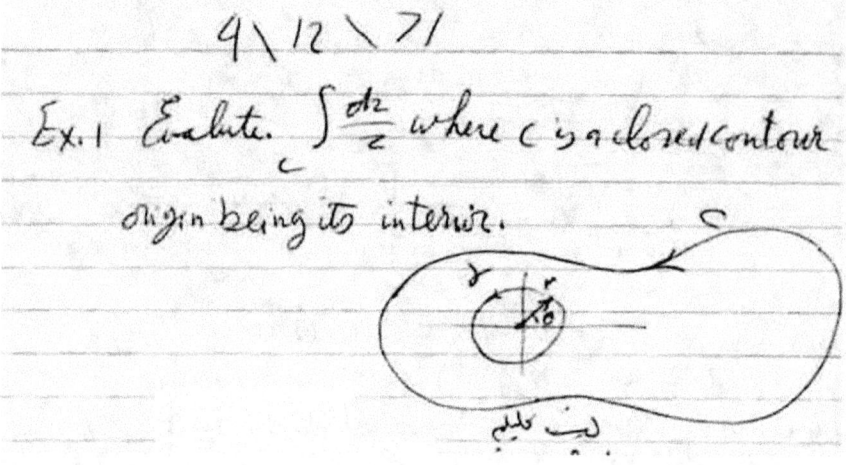

Here the $f^n \frac{1}{z}$ has a singularity at $z=0$ hence we cannot apply Cauchy's theory we Therefore isolate the origin by drawing a circle γ centre 0, Radius r & which lies inside c apply Cauchy's theorem c γ

$$\int_c \frac{dz}{z} = \int_\gamma \frac{dz}{z} \quad —①$$

on which. the circle γ we have $z = re^{i\theta} \quad —②$

$dz = ire^{i\theta} d\theta.$

$$\therefore \int_\gamma \frac{dz}{z} = \int_0^{2\pi} \frac{ire^{i\theta} d\theta}{re^{i\theta}} \quad ③ = \int_0^{2\pi} i\, d\theta$$

$$\int_\gamma \frac{dz}{z} = \underline{\underline{2\pi i}}$$

ex.2   Evaluate $\int_c \left(\frac{z^2+1}{z}\right)dz$ where c is a

closed contour having the origin in its
interior.

Sol$\underline{^n}$

$$\int_c \left(\frac{z^2+1}{z}\right)dz = \int_c z\,dz + \int_c \frac{1}{z}dz$$

$$\underset{\underset{\text{chauchy.}}{\to 0}}{}$$

The 1$\underline{^{st}}$ integral vanishes by chauchy's
theorem. z be a regular f^n?
The 2$\underline{^{nd}}$ integral can be evaluated
as the previous example by drawing
a circle γ with centre origin

$$\int_c \frac{z^2+1}{z}\,dz = 2\pi i$$

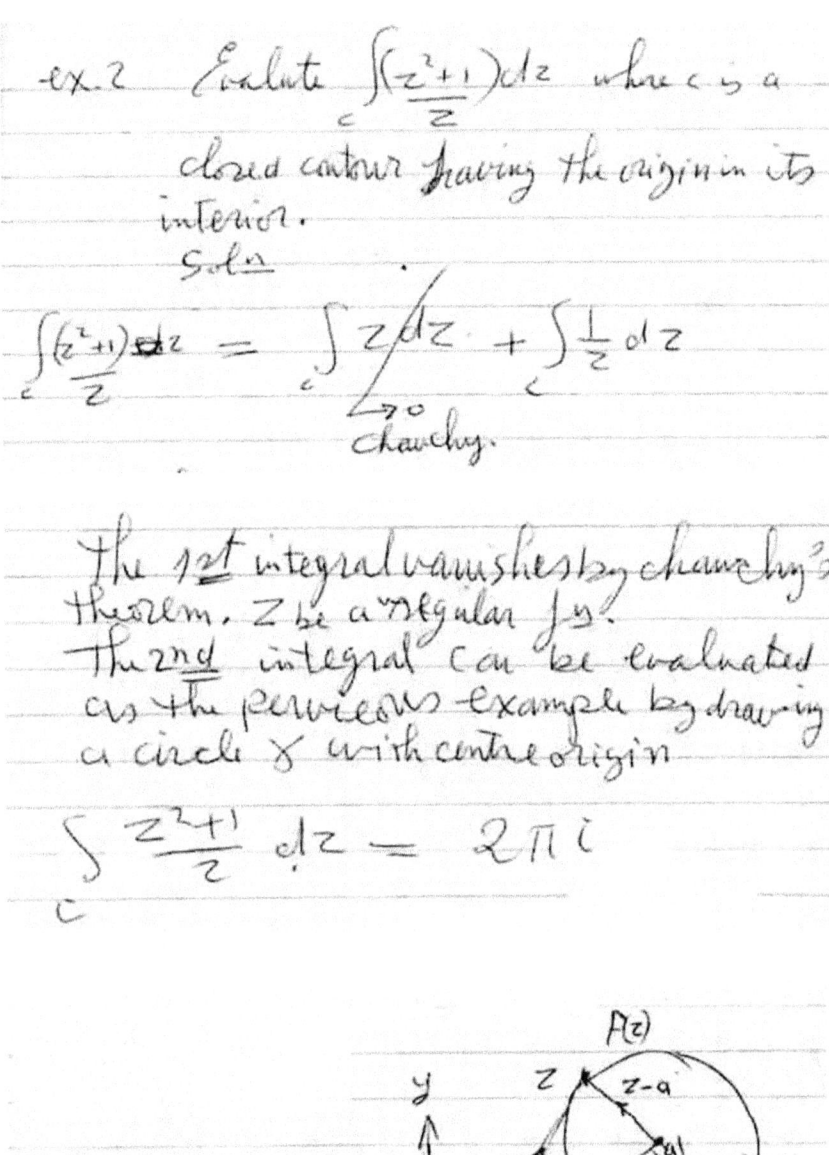

physical meaning

9.7.2. Poles of singular functions

55

<u>Poles</u>

Suppose that we have a f_2 z regular in circular domain except at the centre (a). $(a$ is not radius$)$ Suppose that $f(z)$ can be expended in the neighbourhood of $z = 0$ in the form.

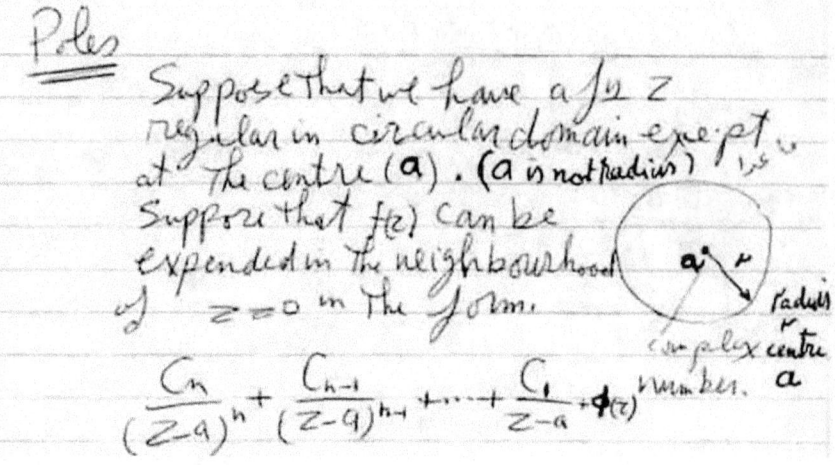

radius

complex centre number. a

$$\frac{c_n}{(z-a)^n} + \frac{c_{n-1}}{(z-a)^{n-1}} + \cdots + \frac{c_1}{z-a} + \phi(z)$$

9.7.2.1. Poles of simple order

where $\phi(z)$ is regular γ at $z = a$
Here we have a pole at $z = a$ of order n
& <u>residue</u> is
$n \geq 1$ $c_1 = Residue$
$c_n \neq 0$ at $z = a$
            $\phi(z) = 0$

<u>Case. I.</u>   Suppose we have a single pole that is the pole of the 1st order at $z_0 = a$
then if $f(z) = \frac{c_1}{z-a} + \phi(z)$

$\therefore (z-a) f(z) = c_1 + (z-a) \phi(z)$
let $z \to a$

$\therefore c_1 = \lim_{z \to a} (z-a) f(z)$

$(z-a) \left[f(z) - \phi(z) \right]$

9.7.2.2. Poles of n- order

Case II Suppose that we have a polar order $\frac{n}{2}$ at $z = a$

write the $\int z \, f(z)$ in the form

$$\frac{\phi(z)}{(z-a)^n}$$

where $\phi(z)$ consists of the Numerator and & the all factors of the denominator $f(z)$, except the factor $(z-a)^n$.

Now let us expand $\phi(z)$ by Taylor's theory in the neighbourhood of $z = a$

$$\phi(z) = \phi(a + \overline{z-a})$$
$$= \phi(a) + \phi'(a)(z-a) + \frac{(z-a)^2}{2!}\phi''(a) \dots$$
$$\dots + \frac{(z-a)^{n-1}}{(n-1)!}\phi^{n-1}(a) + \frac{(z-a)^n}{n!}\phi^n(a)$$

Hence dividing by $(z-a)^n$ we get

9.7.2.3. Residues at the nth pole

where $H(z)$ is some $\int us$ regular at $z = a$
Hence the residue at $z = \frac{1}{(n-1)!} \phi^{n-1}(a)$

i.e $\dfrac{1}{(n-1)!} \left[\dfrac{d^{n-1}}{dz^{n-1}} \phi(z) \right]_{z=a}$

$$\boxed{\text{Residue} = \frac{1}{(n-1)!}\left[\frac{D^{n-1}}{Dz^{n-1}}\ \phi(z)\right]_{z=a}}$$

Very important.

ex. 1 $\dfrac{z+1}{z} = 1 + \dfrac{1}{z}$

here we have a simple of order ① we residue(1)

ex. 2. $\dfrac{e^{z}}{z} = \dfrac{1}{z}\left(1 + z + \dfrac{z^{2}}{2!} + ..\right)$

$\qquad = \dfrac{1}{z} + 1 + \dfrac{z}{2!} + ...$

$\qquad = \dfrac{1}{z} + \phi(z)$ where $\phi(z)$ is regular

here we have origin $z=0$, a simple pole of residue
①

ex. 3. $\dfrac{1}{z(z-1)^{2}}$.   here we have two poles, a pole of
order 1 at $z=0$ & a pole of
order $\frac{3}{2}$ at $z=1$

$\underline{\underline{z=0}}$ Residue $=$
$\qquad \displaystyle\lim_{z \to 0} z\,\frac{1}{z(z-1)^{2}} = 1$

$\underline{\underline{z=1}}$ Residue $= \dfrac{1}{(2-1)!}\left[\dfrac{d}{dz}\ \dfrac{1}{z}\right)_{1} = \left[-\dfrac{1}{z^{2}}\right]_{z=1} = -1$

The residue at pole $z=1$ which is of order(2) can be evaluated also as follows :—

we expand the given f^n in the neighbourhood of $z=1$ put $z-1=\xi$; $\therefore z=1+\xi$

$$\therefore \frac{1}{z(z-1)^2} = \frac{1}{\xi^2(1+\xi)}$$

$$= \frac{1}{\xi^2}\left(1 - \xi + \xi^2 - \xi^3 \ldots\right)$$

$$= \frac{1}{\xi^2} - \frac{1}{\xi} + \phi.(\xi).$$

$$= \frac{1}{(z-1)^2} - \frac{1}{z-1} + \psi(z)$$

where $\psi(z)$ is regular at $z=1$, the Residue is -1

9.7.3. Cauchy's Theorem of Residue

Cauchy's Theorem of Residue :

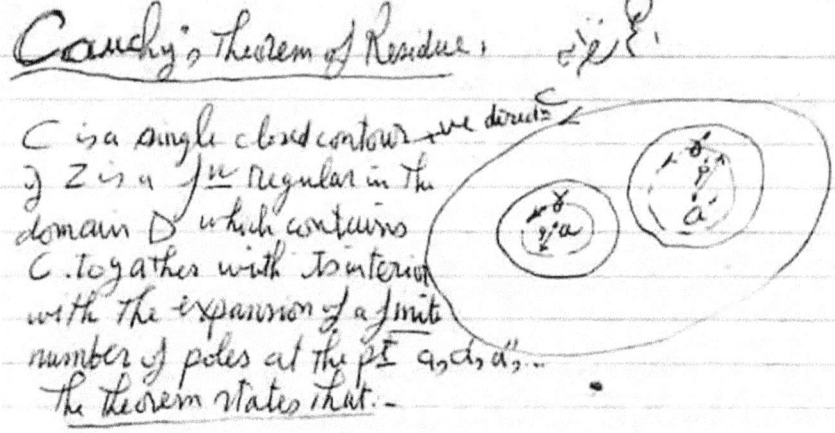

C is a single closed contour we denote $\leftarrow$ if Z is a f^n regular in the domain D which contains C. together with its interior with the expansion of a finite number of poles at the $p^{\underline{t}}$ $a_1, a_2, a_3 \ldots$
    The theorem states that :—

$$\int_c f(z)\,dz = 2\pi i \left(\left(\text{Sum of residues of } f(z) \text{ at its poles inside } c.\right)\right)$$

Consider the poles of $f(z)$ at $z=a$ & suppose that $f(z)$ can be expanded round a in the form :-

$$f(z) = \frac{C_n}{(z-a)^n} + \frac{C_{n-1}}{(z-a)^{n-1}} + \cdots + \frac{C_1}{z-a} + \phi(z)$$

where $\phi(z)$ is regular in the same circular domain centre "a" & **radius** r similarly for a', a'', ...etc. Now by taking r, r', r'' small enough we can ensure that they do not overlap & they all lies inside C with a centre, draw a circle γ of radius ρ as $\rho < r \lessgtr r_0$ for the other circles.

According to Cauchy's theorem cor

$$\int_c f(z)\,dz = \int_\gamma \phi(z)\,dz + \int_\gamma f(z)\,dz + \cdots$$

where all integrals are in the positive signs that is i.e. contour ^{unti} clock wise

Now consider $\int_\gamma f(z)\,dz$

$$\int_\gamma f(z)\,dz = \sum_{m=1}^{n} C_m \int_\gamma \frac{dz}{(z-a)^m} + \int_\gamma \phi(z)\,dz$$

Now the last integral $\int \phi(z)\,dz = 0$ since $\phi(z)$ is regular. for pts on the circle γ we can put

$$z = \rho e^{i\theta} + a$$
$$dz = i\rho\, e^{i\theta} d\theta$$

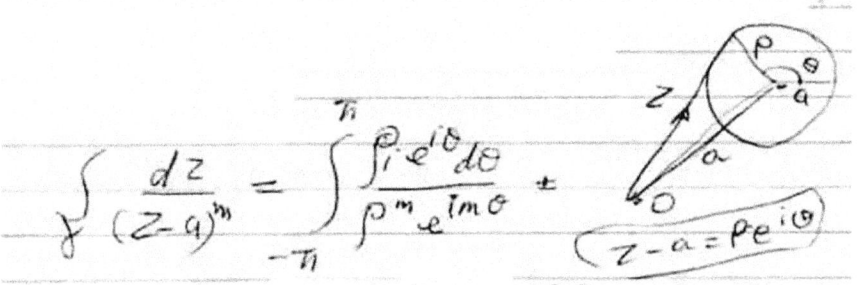

$$\oint \frac{dz}{(z-a)^m} = \int_{-\pi}^{\pi} \frac{\rho_i e^{i\theta} d\theta}{\rho^m e^{im\theta}} \pm$$

$$\left(z - a = \rho e^{i\theta}\right)$$

$$= \frac{i}{\rho^{m-1}} \int_{-\pi}^{\pi} e^{i(1-m)\theta} d\theta$$

$$= \frac{i}{\rho^{m-1}} \left[\frac{e^{i(1-m)\theta}}{i(1-m)} \right]_{-\pi}^{\pi}$$

$$= \frac{1}{\rho^{(m-1)}[1-m]} \left[e^{i(1-m)\pi} - e^{-i(1-m)\pi} \right] = 0$$

$$m \neq 1$$

but $e^{i\theta} - e^{-i\theta} = 2i \sin\theta$

when $m = 1$ (one pole of order (1))

$$\iint f(z) dz = c_1 \oint \frac{dz}{z-a} = c_1 \cdot 2\pi i \longrightarrow ②$$

because $\int f(z) dz = \int \frac{i \rho e^{i\theta}}{\rho e^{i\theta}} d\theta = i \int_{-\pi}^{\pi} d\theta = 2\pi i \, c_1$

Similarly for integrals around $\gamma', \gamma'' \dots$

$$\int_c f(z) dz = 2\pi i (c_1 + c_2 + c_3 +)$$

$$= 2\pi i \left(\text{Sum of Residues}\right)$$

9.7.3.1. Examples on Cauchy's Theorem of Residues

ex.1 Evaluate $\int_c \frac{5z-2}{z(z-1)} dz$ where is $|z| = 2$

Describe contour c.w (clockwise)

soln:

Here we have two simple poles at $z=0$, $z=1$ and they both lies inside C because the radius is 2.

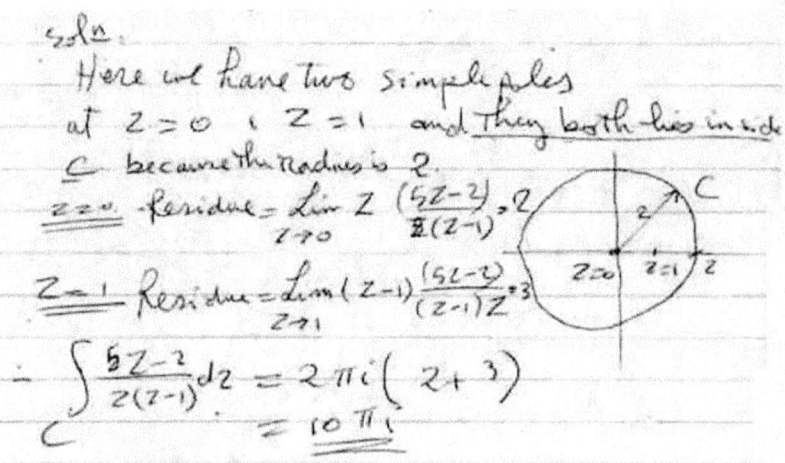

$z=0$. Residue $= \lim\limits_{z \to 0} z \dfrac{(5z-2)}{z(z-1)} = 2$

$z=1$ Residue $= \lim\limits_{z \to 1} (z-1)\dfrac{(5z-2)}{(z-1)z} = 3$

$\therefore \displaystyle\int_C \frac{5z-2}{z(z-1)}dz = 2\pi i \left(2+3 \right)$

$\qquad\qquad = \underline{10\pi i}$

Here we can also resolve the given fraction into its partiale fractions.

$$\int_C \frac{5z-2}{z(z-1)} dz = \int_C \frac{2}{z} dz + \int_C \frac{3}{z-1} \cdot dz$$

$$\qquad = 2 \times 2\pi i + 3 \times 2\pi i = \underline{10\pi i}$$

ex. 2. Find the value of $\displaystyle\int_C \frac{dz}{z^3(z+4)}$ where c is the circle

i) $|z| = 2$

ii) $|z+2| = 3$

iii) $|z-4) = 1$

Here we have a simple pole at $z=-4$ and pole of order ③ at $z=0$

$\underline{z=-4}$ Res. $= \lim\limits_{z \to -4} (z+4) \dfrac{1}{z^3(z+4)} = -\dfrac{1}{64}$

$$\underline{Z=0} \qquad \text{Res.} = \frac{1}{2!}\left[\frac{d^2}{dz^2}\left(\frac{1}{z+4}\right)\right]_{z=0}$$

$$= \frac{1}{2!}\left\{\frac{[-1 \times -2]}{(z+4)^3}\right\}_{z=0} = \frac{1}{64}$$

$$\frac{1}{4z^3} - \frac{1}{4z^2} \quad \text{and } z=4 \text{ gives}$$
$$\text{no pole.} \quad \text{but}$$
$$\oint_C \frac{1}{4\cdot 4z} dz$$

<u>another method</u>

$$\frac{1}{z^3(z+4)} = \frac{1}{4z^3(1+\frac{z}{4})} = \frac{1}{4z^3}\left[1 - \frac{z}{4} + \frac{z^2}{16} - \cdots\right]$$

$$\text{Res.} = \frac{1}{64}$$

-4	$\frac{1}{64}$
0	$\frac{1}{64}$

1) $|z| = 2$

$$\int_C \frac{dz}{z^3(z+4)} = 2\pi i\left(\frac{1}{64}\right) = \boxed{\frac{\pi i}{32}}$$

ii) Here the two poles lies inside the circle

$$\therefore \int_C \frac{dz}{z^3(z+4)} = 2\pi i\left(-\frac{1}{64} + \frac{1}{64}\right) = 0$$

iii) The two poles lies outside the circle
$\therefore$ the integral $= 0$

$$\int_C \frac{dz}{z^3(z+4)} = 0$$

ex. Evaluate. $\displaystyle\int_C \frac{dz}{z^2 e^z}$ where C is the circle, centre origin & radius unity. Here we have a pole of 2nd order at $z=0$

at $z=0$

$$\frac{1}{z^2 e^z} = \frac{1}{z^2} e^{-z}$$

$$= \frac{1}{z^2}\left(1 - z + \frac{z^2}{2!} - \frac{z^3}{3!}\right) \quad z, pl.$$

$$= \frac{1}{z^2} - \frac{1}{z} + \phi(z)$$

$\underbrace{\qquad}_{because\ z=0}$

where $\phi(z)$ is Regular at $z=0$

Residue $= -1$

$$\int_C \frac{dz}{z^2 e^z} = 2\pi i\,(-1) = -2\pi i$$

check

$$Res. = \frac{1}{(2-1)!}\frac{d}{dz}\left(\frac{1}{e^z}\right)_{z=0} = -1 \checkmark$$

or

$$Res = \frac{1}{?} \quad or\quad Res = \lim_{z\to 0} e^z \frac{1}{z^2 e^z} = --$$

ex. Evaluate $\displaystyle\int_0^\infty \frac{dx}{(x^2+1)^2}$

$$= \frac{1}{2}\int_{-\infty}^{\infty} \frac{dx}{(x^2+1)^2}$$

this follows from the fact that the integral z-pl. is an even f^n of x

Now consider $\int_C \frac{dz}{(z^2+1)^2}$

where c is a contour consisting of the Real part of the x-axis extending from $-R \to +R$ and a semi circle C_R in the upper half of the z-pl centre origin & Radius R when $R > 1$.

hence $\int_C \frac{dz}{(z^2+1)^2} = \int_{-R}^{R} \frac{dx}{(x^2+1)^2} + \int_{C_R} \frac{dz}{(z^2+1)^2}$

$$= 2\pi i \times \text{Residues at } z = i$$

we now calculate the residue at $z = R$

$$\frac{1}{(z^2+1)^2} = \frac{1}{(z-i)^2(z+i)^2}$$

Res. at $z = i$ is
Res at $z = -i$ is out the contour
there is no Residue.

$$\frac{1}{1!}\left[\frac{d}{dz}\frac{1}{(z+i)^2}\right]_{z=i} = \frac{-2}{(i+i)^3} = \frac{2}{8i^3} = \frac{1}{4i}$$

$$\therefore \int_C \frac{dz}{(z^2+1)^2} = 2\pi i \cdot \frac{1}{4i} = \frac{\pi}{2}.$$

Now consider $\int_{C_R} \frac{dz}{(z^2+1)^2}$

$$\left|\int_{C_R} \frac{dz}{(z^2+1)^2}\right| \leq \int_{C_R}\left|\frac{dz}{(z^2+1)^2}\right| < \int_{C_R} \frac{|dz|}{(R^2-1)^2} = \frac{\pi R}{(R^2-1)^2}$$

as $|z_1 + z_2| \leq |z_1| + |z_2|$ $\int dz = \pi R$

Modulus of summation   Summation of moduli

8   $|z_1 + z_2| \geqslant |z_1| - |z_2|$ $R \to \infty$

$\therefore \displaystyle\int_{-\infty}^{\infty} \frac{dx}{(x^2+1)^2} = \frac{\pi}{2}$ Let $x = \tan \theta$
                                        $dx = \sec^2\theta \, d\theta$
$\therefore \displaystyle\int_{0}^{\infty} \frac{dx}{(x^2+1)^2} = \frac{\pi}{4}$ $\displaystyle\int \frac{\sec^2\theta \, d\theta}{\sec^4\theta} = \int \cos^2\theta \, d\theta$

Let $x = \tan\theta$
$dx = \sec^2\theta \, d\theta$
$\displaystyle\int \frac{\sec^2\theta \, d\theta}{\sec^4\theta} = \int \cos^2\theta \, d\theta$

$= \displaystyle\int \frac{1}{2}(1 + \cos 2\theta) \, d\theta$ $\theta = \tan^{-1}x$
$= \frac{1}{2}\theta + \sin 2\theta$
$= \frac{1}{2}\theta + 2 \sin\theta\cos\theta = \frac{\pi}{2} + \frac{\pi}{2}$
$= \left[\frac{1}{2}\tan^{-1}x + 2 \cdot \frac{x}{x^2-1} \right]_{-\infty}^{\infty}$
$= \frac{\pi}{4} + \frac{\pi}{4} = \frac{\pi}{2}$

ex. 2 $\displaystyle\int_{0}^{\infty} \frac{\cos x \, dx}{x^2+1}$

$\displaystyle\int_{0}^{\infty} \frac{\cos x \, dx}{x^2+1} = \frac{1}{2}\int_{-\infty}^{\infty} \frac{\cos x}{x^2+1} \, dx$

impossible $\cos z$
but e^{iz}

$\displaystyle\int_{c} \frac{e^{iz}}{z^2+1} \, dz$ where c is the same contour as in
                                       the previous.

$\displaystyle\int_{-R}^{R} \frac{e^{ix}}{x^2+1} \, dx + \int_{C_R} \frac{e^{iz}}{z^2+1} \, dz$ ———①

$= 2\pi i \times \text{Residue at } z = i$

66

$$\therefore \frac{e^{iz}}{z^2+1} = \frac{e^{iz}}{(z-i)(z+i)} \quad \text{Residue } z=i$$

$$\therefore \lim_{z \to i} (z-i) \frac{e^{iz}}{(z-i)(z+i)} = \frac{e^{-1}}{2i} = \frac{1}{2ie} \quad - \textcircled{2}$$

$$\therefore \int_C \frac{e^{iz}}{z^2+1} = 2\pi i \frac{1}{2ie} = \frac{\pi}{e} \quad - \textcircled{3}$$

$$\left| \int_C \frac{e^{iz}}{z^2+1} dz \right| \le \int \left| \frac{|e^{iz}|}{|z^2+1|} \right| |dz|$$

$$|e^{iz}| = |e^{i(x+iy)}| = |e^{ix}| e^{-y} = |\cos x + i \sin x| e^{-y}$$

exponential always +ve

$$|e^{iz}| = e^{-y} < 1 \quad \text{so long } y > 0$$

$$\therefore \left| \int_{C_R} \frac{e^{iz}}{z^2+1} dz \right| < \int_R \frac{e^{-y}}{R^2-1} |dz| \to 0 \text{ as } R \to \infty$$

$$\int_{-\infty}^{\infty} \frac{e^{ix}}{(x^2+1)} dx = \frac{\pi}{e} = \int_{-\infty}^{\infty} \frac{\cos x + i \sin x}{x^2+1} dx$$

equating Real to Real

$$\int_{-\infty}^{\infty} \frac{\cos x \, dx}{x^2+1} = \frac{\pi}{e} \quad \text{i.e.} \quad \int_{0}^{\infty} \frac{\cos x}{x^2+1} = \frac{\pi}{2e}$$

$$\therefore e^{i\theta} = \cos\theta + i\sin\theta$$

$$|e^{i\theta}| = 1 \quad \text{in case of Real } \theta \text{ where } \theta \text{ is Real}$$

$$= e^{-R^2} \int_0^a e^{-y^2} dy < e^{-R^2} e^{-a^2} \int_0^a dy$$

$$= e^{-R^2} e^{-a^2} a \to 0 \quad \text{as } R \to \infty$$

$$\int_{0}^{R} e^{-(x+ia)^2} dx = \int_{0}^{R} e^{-x^2 - 2iax + a^2} dx$$

$$= e^{a^2} \int_{0}^{R} e^{-x^2} (\cos 2ax - i \sin 2ax) dx$$

Hence
$$\int_{0}^{\infty} e^{-x^2} dx - e^{a^2} \int_{0}^{\infty} e^{-x^2} (\cos 2ax - i \sin 2ax) dx$$

$$- \int_{0}^{a} e^{y^2} i \, dy = 0$$

equating the real part to zero

$$\int_{0}^{\infty} e^{-x^2} dx - e^{a^2} \int_{0}^{\infty} e^{-x^2} \cos 2ax \, dx = 0$$

$$\frac{\sqrt{\pi}}{2} - e^{a^2} \int_{0}^{\infty} e^{-x^2} \cos 2ax \, dx = 0$$

$$\therefore \int_{0}^{\infty} e^{-x^2} \cos 2ax \, dx = \frac{\sqrt{\pi}}{2} e^{-a^2}$$

9.7.4. Cauchy's Integral formula

18\12\71

Cauchy's integral formula :-

Let $f(z)$ be a f^n regular in a domain D & let (a) be any pt of that domain then :-

$$f(a) = \frac{1}{2\pi i} \int_{c} \frac{f(z)}{z-a} dz$$ where c is any closed domain surrounding a and lying in D with (a) centre draw circle γ radiusρ lying inside C hence according to cauchy's theory as

$$\int_c \frac{f(z)}{z-a}\,dz = \int_\gamma \frac{f(z)}{z-a}\,dz.$$

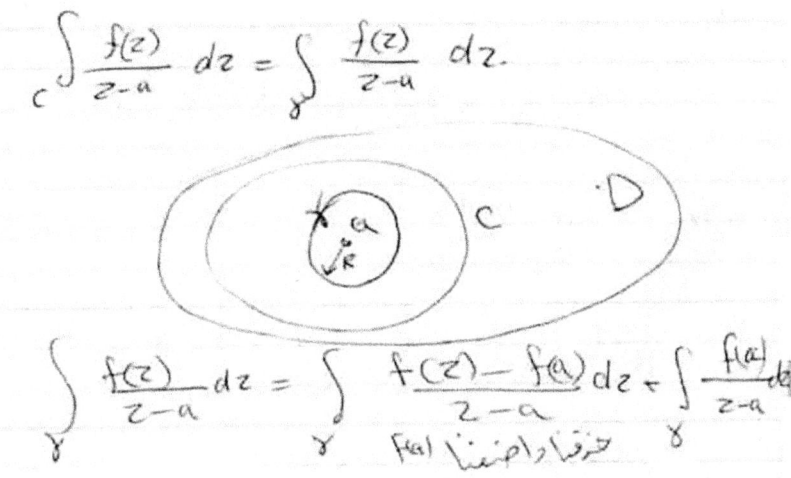

$$\int_\gamma \frac{f(z)}{z-a}\,dz = \int_\gamma \frac{f(z)-f(a)}{z-a}\,dz + \int_\gamma \frac{f(a)}{z-a}\,dz$$

f(a) ثابت بالنسبة

Now a last integral $= f(a)\, 2\pi i = 2\pi i \cdot f(a)$

$f(a)$ Now by hypothesis is Regular at $(z)=a$ hence it has a derivative at $z=a$ & is therefore

continuous at $z = a$ hence according to the Def. of continuity $f(z) \to f(a)$ as $z \to a$ hence given any positive small number ϵ There exists $\delta = \delta(\epsilon)$.

i.e dependent on ϵ such that.

$|f(z) - f(a)| < \epsilon$ for all z satisfying $|z-a| < \delta$. Now take $\rho < \delta$

$$\left| \int_\gamma \frac{f(z)-f(a)}{z-a}\,dz \right| \leq \left| \frac{f(z)-f(a)}{z-a} \right| |dz|$$

$$< \frac{\epsilon}{|\delta|} |dz|$$

$$= \frac{\epsilon}{\delta} 2\pi\rho = 2\pi\epsilon$$

This is however small ε may be. hence the integral on the left is zero z

$$\int_{\gamma} \frac{f(z)}{z-a} dz = \int_{\gamma} \frac{f(z)-f(a)}{\underset{\to 0}{z-a}} dz + \int_{\gamma} \frac{f(a)}{z-a} dz$$

$$\therefore \quad f(a) = \frac{1}{2\pi i} \int_{\gamma} \frac{f(z)}{z-a} dz$$

9.8. Taylor's and Laurent's expansions

$$\bar{\rho} = z - \bar{a}$$

I. Taylor's & Laurent's exp^{ns}

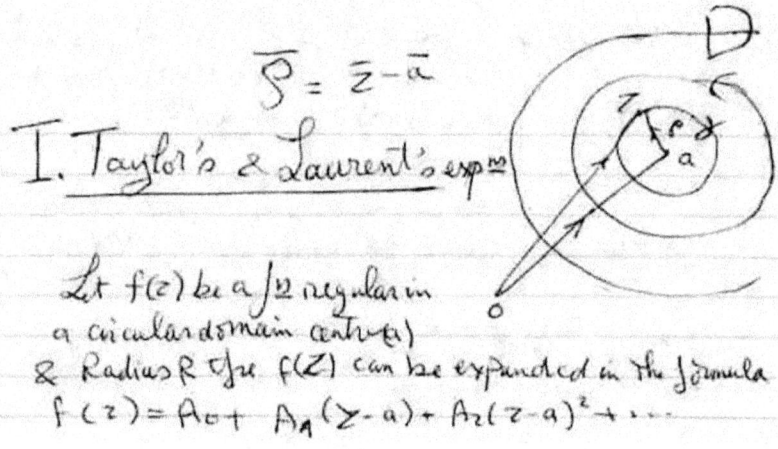

Let $f(z)$ be a f^{ns} regular in a circular domain centre (a) & Radius R then $f(z)$ can be expanded in the formula

$$f(z) = A_0 + A_1 (z-a) + A_2 (z-a)^2 + \cdots$$

i.e. $\displaystyle\sum_{n=0}^{\infty} A_n (z-a)^n$

i.e in Powers of $(z-a)$ in the neighbourhood of $z=a$ where $A_n = \frac{1}{2\pi i} \int_c \frac{f(z) \, dz}{(z-a)^{n+1}}$

or $A_n = \frac{f^{(n)}(a)}{n!}$ & hence

70

$$f(z) = f(a + \overline{z-a}) = f(a) + (z-a) f'(a)$$
$$+ \frac{(z-a)^2 f''(a)}{2!} + \cdots$$
$$\frac{(z-a)^n}{n!} f^n(a)$$

C is any circle centre a whose Radius is less than R (annulus)

Next consider the case of an annulus which is bounded by two circles of radius R_1, R_2. Let $f(z)$ be a $f(z)$ regular in this annulus the according to Laurent's exp^n $f(z)$ can be expanded in the form —

$$f(z) = \sum_{n=-\infty}^{\infty} A_n (z-a)^n$$

$$\left(\text{When } A_n = \frac{1}{2\pi i} \oint \frac{f(z)}{(z-a)^{n+1}} \, dz \right)$$

where C is any circle centre a & whose Radius lies bet. R_1, R_2

CHAPTER 10: INTRODUCTION TO THE THEORY OF PROBABILITY

10.1. Introduction to the Theory of probability
10.1.1. Mathematical definition of probability

29\2\??

<u>Introduction to the Theory of probability</u>

<u>Mathematical def^n of prob.</u>

If an event can happen in (a) ways, and fails to happen in (b) ways, and if all the ways are <u>equally</u> likely to occur, then the prob. of the happening of the event is

$$\text{prob} = \frac{a}{a+b} \text{ of happening}$$

and the prob of failing is

$$\text{prob} = \frac{b}{a+b} \text{ of failing.}$$

ex: The chance (prob) of picking an (ace) ورقة لعب out of the pack of carts is $\frac{4}{52}$ أي 35.

The chance that a number (3) appears in one through of an ordinary (die) داي جي is $\frac{1}{6}$.

An event which is certain to happen is a prob (1) one.
An event will never happen is a prob (0) zero.
the prob. of a doubtful event there is from zero to one

In general the prob of happening of an event is (P) & its failing to happen is (q) then

$$P + q = 1 \quad\quad ①$$

$$\frac{a}{a+b} + \frac{b}{a+b} = 1$$

this is the mathematical def.ⁿ of probs ⸖

10.1.2. Exclusive events

Exclusive events ظل تقبل

Two events are said to be exclusive if the occurance of one event is incompatible غير with occurance of the other event. If several events are exclusive, then the prob of the occurance of one or other of the events is equal to the sum of the separate probabilities

72

e.g. The chance that one (1) or (3) may appear in one through of a die op: is $\frac{1}{6} + \frac{1}{6} = \frac{1}{3}$

Should be noticed that prob$\underline{s}$ are only add$\underline{d}$ if the events are exclusive. $P_{.} = P + P'$

10.1.3. Independent events

Independent events

Two events are said to be independent if the occurrence of one event does not affect the occurrence of the other. We shall now show that if (P) is the prob. of the occurrence of one event & (P') is the prob of the occurrence of the 2$\underline{nd}$ independent event then the prob that $\underline{both}$ events will happen is $P_{.} = PP'$

we illustrate by the following example.

e.g. A bag contains two red balls (R_1, R_2) & one below ball (B_1). And a second bag contains 4- red balls (r_1, r_2, r_3, r_4) & three blue balls (b_1, b_2, b_3).

73

A ball is drawn at random from the 1st bag & another is drawn from the 2nd. What is the chance that the two Balls will both red?

Soln

$$R_1 , R_2 , B_1$$

$$r_1 , r_2 , r_3 , r_4, b_1, b_2, b_3$$

No. of methods = 3 × 7 = 21

successful methods for picking red balls = 2 × 4 = 8

Here the events are independent the total no. of draws that can be made from picking a ball from 1st bag and a ball from the 2nd bag is 3 × 7. out of this draws there are 2 × 4 draws which give a red ball from the 1st bag & 2nd bag

From the 1st bag & a hence according to the definition of prob, the chance that two balls are red is $\frac{2 \times 4}{3 \times 7}$. put the chance of picking a red ball from the 1st bag is $\frac{2}{3}$ & from the 2nd bag is $\frac{4}{7}$

& since $\frac{2}{3} \cdot \frac{4}{7} = \frac{2 \times 4}{3 \times 7}$ then it is a numerative example show the probable of occurance of 2 independent events is equal to the product of the prob. that each event will happen separately. hence we have

10.1.4.. Rules governing probability of independent events

74

the following fundamental results.

(13) If A & B are two _independent_ & if P is the prob. that A happens & P' is the prob. that B happens then

i) $PP' =$ Prob that both A & B happen.

ii) $P(1-P') =$ Prob. that A happens & B fails to happen.

iii) $P'(1-P) =$ Prob that B happens & A fails to happen.

iv) $(1-P)(1-P') =$ Prob. that A & B _both_ fail to happen

v) $1 - (1-P)(1-P') =$ prob that at _least_ A & B possibly both may happen)

ex. 1 — Two submarines attacked simultaneously a ship. If the chance that the two submarines will separately hit the ship are $\frac{1}{4}, \frac{1}{5}$ find.

i) the chance that the two submarines may hit the ship.
ii) the chance that " " will fail to hit the ship.
iii) the " " at _least_ one submarine may hit the ship

Soln

i) $PP' = \frac{1}{4} \times \frac{1}{5} = \frac{1}{20}$ for every 20-trial the two hit the ship.

ii) $(P-1)(P-P') = \frac{3}{4} \times \frac{4}{5} = \frac{3}{5}$ failure of 1st & 2nd.

iii) $1 - (P-1)(1-P') = 1 - \frac{3}{5} = \frac{2}{5}$

ex. 2   what is the chance that 4 (four) may appear at least once in six throws of an ordinary die.

sol.

The chance that (4) will not appear in the 1st throw is $(\frac{5}{6})$ similarly for the other throws, & since all throws are independent. Then the chance that (4) will not appear at all in six throws is $(\frac{5}{6})^6$ hence the chance that 4 may appear at least once is

$$1 - (\frac{5}{6})^6 \simeq 0.668$$

10.1.5. Dependent events

Dependent events

This means that the occurance of one event affects the occurance of the other events.

ex:

One ball was drawn at random from bag containing six (6) white & 4 black balls. Find the chance that a ball now drawn at random from the bag will be white? in 2nd draw

sol.   the chance that the 2nd draw is a white ball is the sum of the chances of 2 exclusive events i) white & then white. chance $(\frac{6}{10} \times \frac{5}{9})$
ii) black & then white. chance $(\frac{4}{10} \times \frac{6}{9})$

76

$$\text{Sum} = \frac{6}{10} \times \frac{5}{9} + \frac{4}{10} \times \frac{6}{9} = \frac{3}{5}$$

OR

i) if 1st ball is white & 2nd is white.

∴ prob of 1st $P_1 = \frac{6}{10}$

" " 2nd $P_2 = \frac{5}{9}$

$P_e = P_1 P_2 = \frac{6}{10} \times \frac{5}{9}$ two exclusive

ii) if 1st ball is black & 2nd is white

∴ prob of 1st $P_1' = \frac{4}{10}$

" " 2nd $P_2' = \frac{6}{9}$

$P' = P_1' P_2' = \frac{4}{10} \times \frac{6}{9}$ Two exclusive

10.2. Binomial law of probability

11⋅3⋅1972

Binomial law of probability

If the prob of the occurrence of an event in one trial is "P" to find the prob of the occurrence of the event exactly (r) in (n) trials. we will illustrate by the following ex.

e.g:

A die is thrown five times (5) & it is required to find the prob that the ace (1) will appear exactly three times. in a five trials.

soln.

We 1st assume that the (ace) appeared in the 1st, 2nd & 3rd thrown.
To appear in the 1st thrown the chance is $\frac{1}{6}$ & so in the 2nd & 3rd throws.
The chance that it will not appear in the 4th thrown is $\frac{5}{6}$ & so for 5th thrown

77

Since these events are independent the chance that the ace will appear in the 1st 3 throws & fails to appear in the 4th & 5th throws is $\left(\frac{1}{6}\right)^3\left(\frac{5}{6}\right)^2$

Now in order that the ace may appear three times it is not necessary that this should occur in the 1st or 2nd & 3rd trial. it may appear for example in the 1st, 2nd & 4th trials these are

$$5C_3 \text{ in number & they are all exclusive}$$

	trials :	1	2	3	4	5
i)	prob of Occurance :	$\frac{1}{6}$	$\frac{1}{6}$	$\frac{1}{6}$	$\frac{1}{6}$	$\frac{1}{6}$
ii)	prob of failure :	$\frac{5}{6}$	$\frac{5}{6}$	$\frac{5}{6}$	$\frac{5}{6}$	$\frac{5}{6}$

iii) prob of Occurance in 5 trials = $\left(\frac{1}{6}\right)^5$

iv) $\left\{\begin{array}{l} \sim \quad \sim \quad \sim \quad 3 \text{ times from} \\ 5 \text{ trials in the 1st, 2nd, 3rd} \\ \text{throws} \end{array}\right\} = \left(\frac{1}{6}\right)^3\left(\frac{5}{6}\right)^2$

v) $\left.\begin{array}{l} \text{prob of occurance in any} \\ \underline{\text{three}} \text{ trials from } \underline{5 \text{ trial}} \end{array}\right\} = 5C_3\left(\frac{1}{6}\right)^3\left(\frac{5}{6}\right)^2$

w) $5C_3 = \dfrac{5!}{3!(5-3)!} = \dfrac{5!}{3! \times 2!}$

hence the chance that the (ace) will appear 3 times in 5 throws is $5C_3\left(\frac{1}{6}\right)^3\left(\frac{5}{6}\right)^2$ hence have the following fundamental law. known as the binomial law of probabilities.

If the prob. of the occurance of an event in one trial is (P) &
its failing to occure is (q). then the prob of the occurence
of the event _exactly_ (r) times in (n) trials is

$$\boxed{n\,C_r\;P^r q^{n-r}}$$ Binomial law of prob.

ex.   The chance the a market may hit a certain target in
one trial is $\frac{1}{10}$. find the chance that it hits the
Target _exactly_ 4 times in 6 trials.

So/n
$$Chance = 6\,C_4 \left(\frac{1}{10}\right)^4 \left(\frac{9}{10}\right)^9 =$$
combination four 4 from six 6

Inverse prob. We illustrate by the following example
e.g.
        It is known that a _black_ ball was drawn from
one or other of two bags, one of them containes
2 black balles & 7 white balls. & another contains
5 black N & 4 N   N & it is required to
determine the prob that the black ball was drawn
from the 1st bag.

So/n $B_1\; B_2\;,\; W_1, W_2, W_3, W_4, W_5, W_6, W_7$
                         $b_1, b_2, b_3, b_4, b_5,\; w_1, w_2, w_3, w_4$

Suppose we make a large numbers of drawings 2N since
we draw a ball from eith bag at random. then in
the long run then, N drawings are made from the 1st
bag & N drawings are made from the 2nd bag.
Out of the N drawings from the 1st bag :—

$\frac{2}{9}$ N are black drawings from the 1st bag. — ①

And out of the N drawings from the 2nd bag.
$\frac{5}{9} N$ are black balls. —————— ②

Hence out of the _total_ number of black drawings
$$\frac{2}{9} N + \frac{5}{9} N$$

& $\frac{2}{9} N$ are drawing of black balls from the 1st bag. hence the prob that the black balls is drawn from the 1st bag is

$$\frac{\frac{2}{9} N}{\frac{2}{9} N + \frac{5}{9} N} = \frac{2}{7}$$

10.3. Random variables

Random experiment & random variable :-

by a random experiment we mean an experiment in which the result cannot be predicted f i.e. however care is taken to repeat the experiment under exactly the same conditions. A random variable is the outcome sv of a random exp.

e.g.

Suppose that we throw a pair of (dice) & we are interesting in the sum of the two numbers that will appear. the experiment is a random exp. & the outcome of the exp which is the sum of the two numbers is the random variable. the random variable may be for example discrete i.e. takes isolated values as in the above example. Since the sum of the two numbers must be an integer. The random variable may be continuous as for example a measuring the stature of a man or the time of duration of a telephone call.

10.4. Discrete distribution

Discrete distribution :— we illustrate by the following example
e.g.   Let a coin be thrown three times & consider the No of heads that may appear as the random variable.   no

$\frac{1}{8}$ = chance that head will appear is $\left(\frac{1}{2}\right)^3$

$\frac{3}{8}$ ~   ~ one ~ ~ ~ ~ $3C_1\left(\frac{1}{2}\right)\left(\frac{1}{2}\right)^2$

$\frac{3}{8}$ ~   ~ two ~ ≤ ~ ~ ~ $3C_2\left(\frac{1}{2}\right)^2\left(\frac{1}{2}\right)$

$\frac{1}{8}$ ~   ~ three ~ ≥ ~ ~   ~ $\left(\frac{1}{2}\right)^3$

the sum of prob≥ is unity since one of the above events must happen

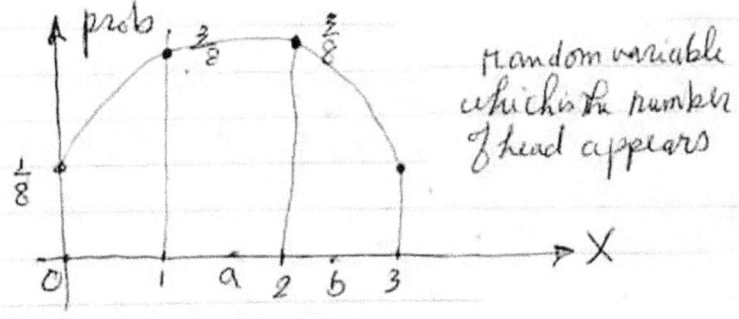

random variable which is the number of head appears

We now defined what is known as a distribution function $F(x)$ at any pt (x) is the prob that the random variable X may take any value less than x or equal x i.e

$$F(x) = P(X \leq x)$$

this means that $F(x)$ at any pt x is equal to the sum of prob^s to the lift of x and including (x) itself

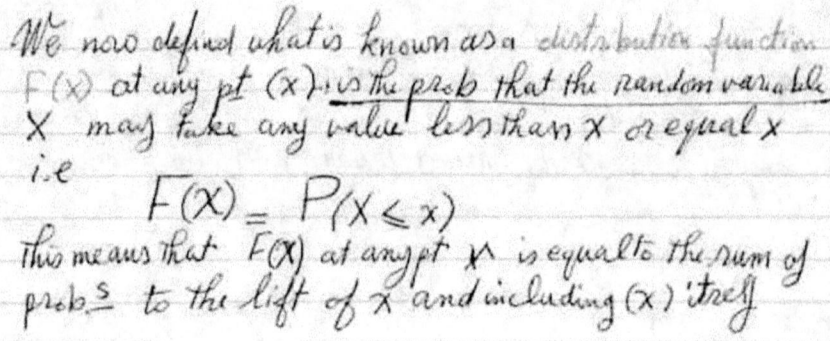

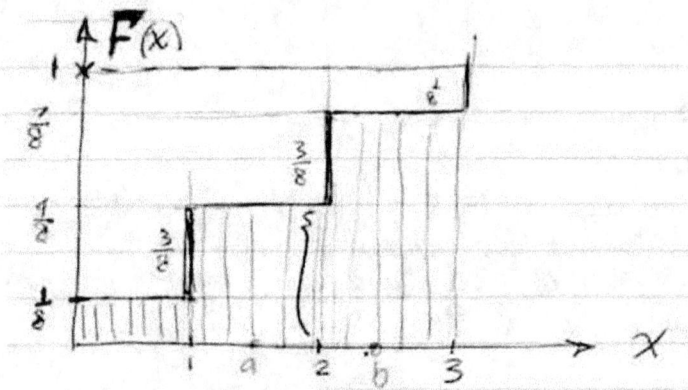

the distributes f^n has the following continuous properties:— not continuous

i) since the prob is never negative then a the distribut^n function $F(x)$ is a non decreasing f^n. if $a < b$.

ii) ∴ $F(b) - F(a) = P(a < x \leq b)$

iii) $\lim_{x \to \infty} F(x) = 1$ Right

iv) $\lim_{x \to -\infty} F(x) = 0$ Lft

v) $f^n F(x)$ is continuous to the Right at any pt (x)

vi) $F(x)$ is discontinuous at any pt (x) at which there is a +ve prob. when we approch x from the lft. since there is a finite jump at this pt.

10.5. Continuous distribution

82

9\3\72

Continuous Distribution

The horizontal axis is the random variable. the verticle axis is represented what is known as the frequency function $f(x)$ or the probability density

Here $f(x)dx$ represents the prob that the random variable may lie bet^n x, $x+dx$.

Note $F(b) - F(a) = f(x)dx$

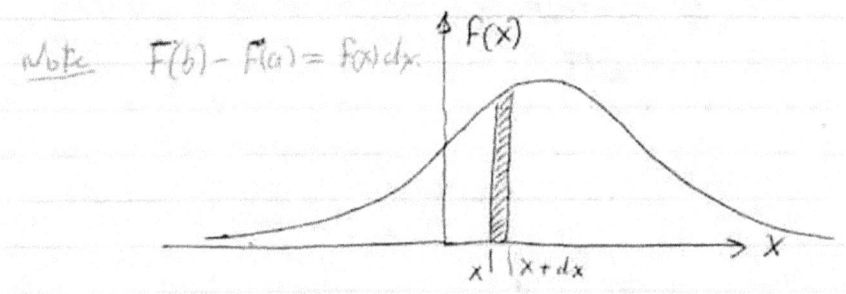

i.e $f(x)dx = P(x < X \leqslant x+dx)$.

this means that the area under the curve bet^n x & $x+dx$ represents the prob. that the random variable should lie bet^n x & $x+dx$. Similarly

$$\int_a^b f(x)\,dx = P(a < x \leqslant b)$$

i.e the area under the curve bet^n $x = a$ & $x = b$ represents the prob that the random variable should lie bet^n a & b. from this it follows that.

$$\int_{-\infty}^{\infty} f(x)dx = 1$$

since it is certain that the random variable will take some value between $(-\infty \, \& +\infty)$ The distribution function $F(x)$ is defined such that

$$F(x) = \int_{-\infty}^{x} f(x)\,dx$$

i.e : it is equal to the area under the curve betⁿ $(-\infty) \, \& \, (x)$ from this it follows that $F'(x) = f(x)$

<u>Note</u>

$$F(x) \, (capital) = \text{Distribution function}$$

$$f(x)\,dx \, (small) = \text{probability Density}$$
$$f(x) \qquad = (\text{frequency function})$$

$$f(x-)\,dx = F(x)_1 - F(x)_2 = d\,F(x)$$
or
$$F(x) = \int F(x)\,dx.$$

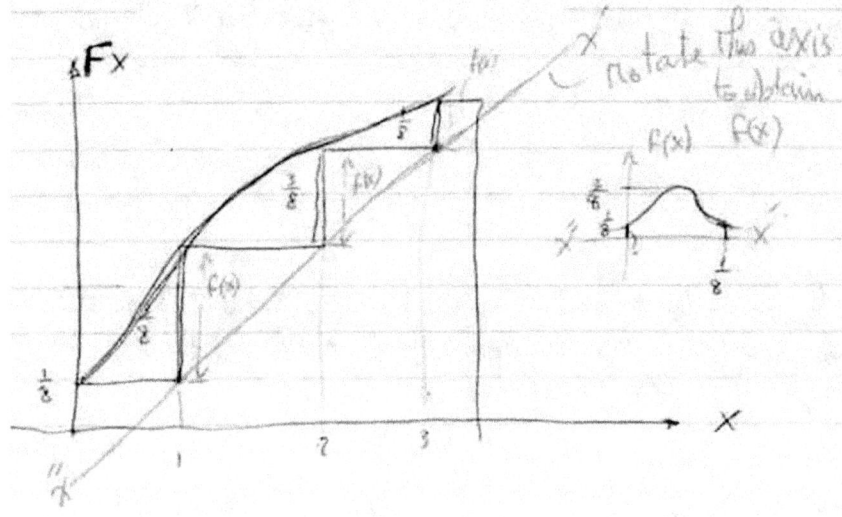

i.e. The derivative of Distribution f_{12} is equal to the frequency f_{11}.

$$f(x) = \frac{1}{\pi(1+x^2)} \quad \text{for all real value of } x \text{ is}$$

known as the Cauchy Distribution. Find the distribution fn $F(x)$ & the prob that the random variable lies bet^{11} 0 & 1

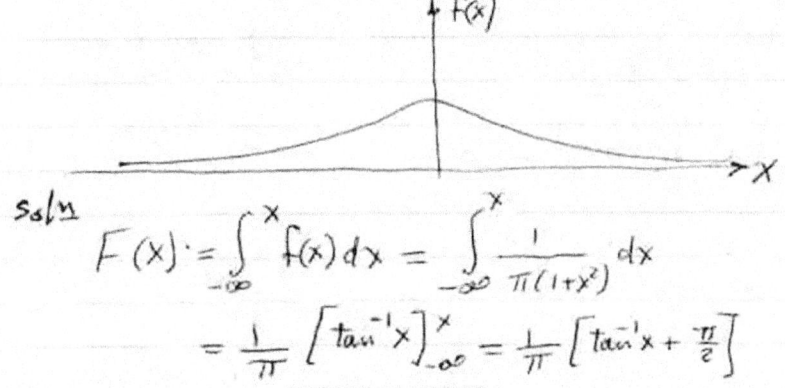

Sol^n

$$F(x) = \int_{-\infty}^{x} f(x)\,dx = \int_{-\infty}^{x} \frac{1}{\pi(1+x^2)}\,dx$$

$$= \frac{1}{\pi}\left[\tan^{-1}x\right]_{-\infty}^{x} = \frac{1}{\pi}\left[\tan^{-1}x + \frac{\pi}{2}\right]$$

The prob that the random variable should lie bet^2 zero & (1) is

$$= \frac{1}{\pi}\left[\tan^{-1}x\right]_{0}^{1} = \frac{1}{\pi}\left[\tan^{-1}\right]$$

$$= \frac{1}{\pi}\cdot\frac{\pi}{4} = \frac{1}{4}$$

10.6. Joint distribution of several variables

Joint Distribution of several variables

In many random experiments we take measurements of more than (1) random variable. Thus for example. we may measure the stature of a man and his weight or we may measure the hardness & percentage of carbon in a sample of steel

If we denote the two variables (random) by (X, Y) we may locate the pt^2 (X, Y) on a plane. it described distributn the pt^2 are scattered on the plane & associated with every point is the prob at that pt. & the sum of prob^2 is unity.

at the pt (x,y) let the frequency fn be $f(x,y)$ Then $f(x,y)dxdy$ is the prob that pt (X,Y) belongs to the infinitesimal rectangle shown.

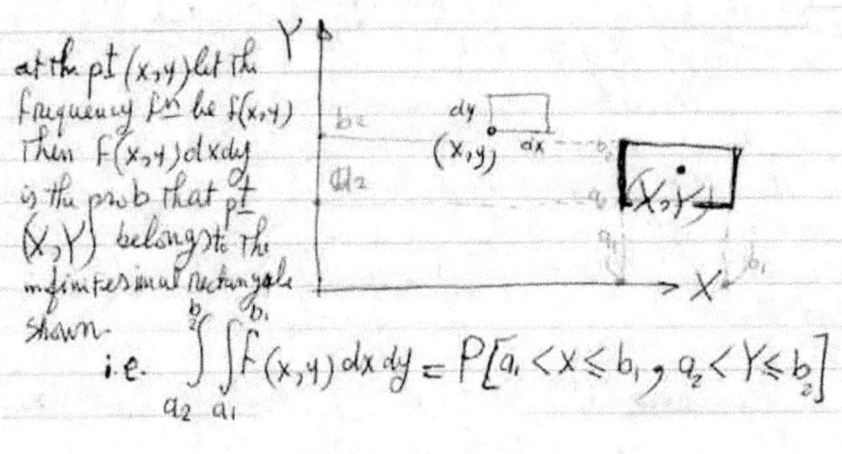

i.e. $$\int_{a_2}^{b_2}\int_{a_1}^{b_1} f(x,y)\,dx\,dy = P\left[a_1 < x \leqslant b_1 , \; a_2 < Y \leqslant b_2\right]$$

i.e $$F(x,y)\,dxdy = P\left(x < X \leqslant x+dx , \; y < Y \leqslant y+dy\right]$$

i.e the prob. that pt (X,Y) belongs to the rectangle shown

$$\int_{-\infty}^{\infty}\int_{-\infty}^{\infty} f(x,y)\,dx\,dy = 1$$

10.7. Expectation

Expectation

If a person has a sum of money x and if his chance to win that sum is p that his expectation is defined as Px. similarly y the chances for a person to win the sums $x_1, x_2, \ldots x_n$ are $p_1, p_2, \ldots p_n$ respectively then there expectation is $p_1 x_1 + p_2 x_2 + \cdots + p_n x_n$.

i.e $$\sum_{r=1}^{n} p_r x_r$$

86

10.8. Parameters of a frequency distribution
10.8.1. Mean

Parameters of a frequency Distribution

(I) The arithmetic mean (الوسط الحسابي) ...

The arithmetic mean (A.M) of a set of numbers $x_1, x_2, \ldots x_n$ is

$$A.M = \frac{x_1 + x_2 + \cdots + x_n}{n}$$

now suppose we have the following frequency distribution:-

Random variable:- $\quad x_1, \quad x_2 \quad \cdots \quad x_n$

frequency :- $\quad f_1, \quad f_2 \quad \cdots \quad f_n$

$$\text{mean value} = \frac{f_1 x_1 + f_2 x_2 + \cdots + f_n x_n}{f_1 + f_2 + \cdots f_n} = \frac{\sum f_r x_r}{\sum f_r}$$

now $\frac{f_r}{\sum f_r}$ is a relative frequency.

i.e The prob which we denote by P_r. hence mean value

$$\text{mean value} = \sum_{r=1} P_r x_r$$

which is the same expression as expectation. we denote the mean value by $E(x)$ expectation.

i.e $\quad E(x) = \sum_{r=1}^{n} P_r x_r.$

The continuous distribution $f(x) dx$ takes place of P_r and the summation is replaced by integral.

$$E(x) = \int_{-\infty}^{\infty} x \, f(x) \, dx$$

87

Now if we have two random variables X, Y then the mean of the product X, Y is given by.

$$E(XY) = \int\int_{-\infty}^{\infty} xy\, f(x,y)\, dx\, dy.$$

now if the random variables X, Y are independent that

$$f(x,y) = f_1(x) \cdot f_2(y)$$

$$\therefore \quad E(XY) = \int\int_{-\infty}^{\infty} xy\, f_1(x)\, f_2(y)\, dx\, dy$$

$$= \int_{-\infty}^{\infty} x\, f_1(x)\, dx \int_{-\infty}^{\infty} y\, f_2(y)\, dy$$

$$\boxed{E(XY) = E(X) \cdot E(Y)}$$

Similarly $\boxed{E(X+Y) = E(X) + E(Y).}$

10.8.2. Median

(II) The Median. — to obtain the median of a set of numbers we 1st arrange these numbers either in ascending i.e order of magnitude or in a descending i.e order. the numbers in the middle is the median & if there are two numbers in the middle we take the arithmetic mean (A.M) of these two.

e.g.   5   3   9   7   4
         3   4   ⑤   7   9

median = 5

88

from a frequency distribut^n the median is the value of the variable corresponding to the ordinate $\frac{1}{2}$ in the frequency function curve

$F(x)$

$F(x)$ Distributo fn

1 —○

$\frac{1}{2}$ —○ a   b

Median

1 —○

$\frac{1}{2}$ —○

median

مُنحني دالة التوزيع إذا كان عنده الاحداثي $\frac{1}{2}$ فإن أي قيمة متغير تقابله عند median

$f(x)$ Frequency fn

① ②

Median
area = area
①   ②

$F(x)$

$\frac{1}{2}$

Median

Distribution function

10.8.3. Mode

$15 \backslash 3 \backslash 72$

Ⅲ <u>The Mode</u>: The mode is a value of a variable at which the prob. density is <u>max</u>. For descrete distribution we used the emperical formula.

Mean − Mode = 3 (Mean − median)

which proved to give excellent results.

$f(x)$

mode

10.8.4. Standard Deviation and Variance

Ⅳ) The <u>standard Deviation</u> الانحراف المعياري.

Def. This is defined as the <u>square root</u> of the mean of the squares of the deviations from the arithmetic mean.

ويعتبر أهم مقياس للتشتت واكثرها استعمالاً.

89

S.D

$$\sigma = \sqrt{\frac{\sum f_r (x_r - \bar{x})^2}{\sum f_r}}$$

since $\dfrac{f_r}{\sum f_r}$ is relative frequency i.e the prob. P_r then

$$\sigma^2 = \sum P_r (x_r - \bar{x})^2$$

is called _variance_. i.e: σ^2 is the mean value of $(x_r - \bar{x})^2$

$$\therefore \quad \sigma^2 = E(x-\bar{x})^2 = E(x^2 - 2x\bar{x} + \bar{x}^2)$$
$$= E(x^2) - 2\bar{x}E(x) + \bar{x}^2$$
$$= E(x^2) - 2\bar{x}^2 + \bar{x}^2$$
$$= E(x^2) - \bar{x}^2$$
$$= E(x^2) - E^2(x)$$

$$\boxed{\sigma^2 = E(x^2) - E^2(x)}$$

mean of   Square of mean
Square
variable

For continueous distribution $\sigma^2 = \displaystyle\int_{-\infty}^{\infty} (x - \bar{x})^2 f(x)\,dx$

The variance of the sum of two
Independent random variable

$$\sigma^2_{x+y} = E(x+y)^2 - E^2(x+y)$$
$$= E(x^2 + 2xy + y^2) - [E(x) + E(y)]^2$$
$$= E(x^2) + 2E(xy) + E(y^2) - E^2(x) - 2E(x)E(y) - \text{...}$$
$$= E(x^2) - E^2(x) + E(y^2) - E^2(y)$$

$$\boxed{\sigma^2_{x+y} = \sigma^2_x + \sigma^2_y}$$ i.e the variance of the sum of
2 independent random

variables is equal to the sum of the seperate variances

10.8.5. Dispersion

(V) Dispersion

we illustrate by the following example

ex. the temp's in two places A & B are measured & found as follows

Place A	38°	39°	40°	41°	42°
place B	14°	25°	48°	50°	63°

we notice that the mean temp is the same for places A & B mainly 40°C. but I is evident that it erroneous lip to say that the two places have the same weather. since for place A the temp's are clustered around the mean.

whereas in place B they are scattered or dispersed over a longe range this suggests that we made take the range as a measure for the dispersion but this actually

is the worst measure since it depends only on the ends & takes no consideration of order outer mediate values.

The end pts themselves may be expesinal pts. The Standard deviation (σ) is used as a measure of dispersion & to use a dimensionless No to compare the dispersion of two phenomena of different units we divide the standard deviation σ by the mean $\bar{x}$

$\frac{\sigma}{\bar{x}}$ is the measure of relative dispersion $= \frac{S.D}{A.M}$

10.8.6. Skewness

(VI) Skewness:- In Symmetrical distribution the mean, mode & median coincide if they do not coincide the distribution is said to by asymmetric or skew

A measure of skewness is $\gamma_1 = \frac{M_3}{\sigma^3}$ where $M_3 = E(X-\bar{x})^3$ the dist division by σ^3 is to obtain a dimensionless number which can then be used for comparison of two phenomena of different unit.

10.8.7. Peakness

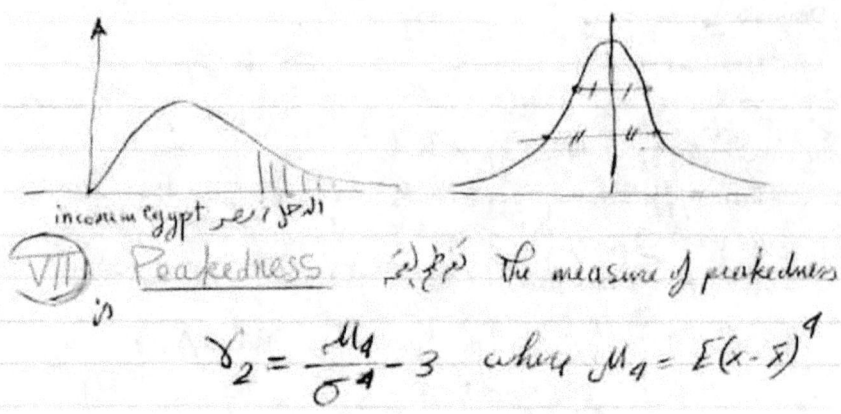

income in egypt الدخل في مصر

(VII) Peakedness نسبة التفرطح the measure of peakedness is

$$\gamma_2 = \frac{\mu_4}{\sigma^4} - 3 \quad \text{where} \quad \mu_4 = E(x - \bar{x})^4$$

10.8.8. Relationships between the mean and the standard deviation

Simplificat^{n} in the calculat^{ns} of the mean & S.D

Let α be the hypothetical mean يعني that is a Number near to
the actual mean. defined a new variable X_r such that

$X_r = x_r - \alpha$ multiply both sides by f_r ∴ $f_r X_r = f_r x_r - \alpha f_r$

يعني & take the sum $\sum f_r X_r = \sum f_r x_r - \alpha \sum f_r$

Divide by the total N^o of the observat^{ns} of measurements.

$$\frac{1}{N} \sum f_r X_r = \frac{1}{N} \sum f_r x_r - \frac{\alpha}{N} \sum f_r$$

$$\bar{X} \quad = \quad \bar{x} - \alpha$$

∴ $\bar{x} = \bar{X} + \alpha$

This simplifies to a great degree the calculat^n of the A.M
The actual mean in a normal phenomenon (phenomena يعني)
is not general a round number يعني If we use this number for the
calculat^n of S.D the calculat^n are tedrous لذا for
this purpose we replace $\bar{X}$ by a hypothetical mean α
which we chose to be a round N^o we then calculate σ^2
& this from the. we obtain the actual variance

(from this notation we understand that α replaced x)

$$\sigma^2 = \frac{1}{N} \sum f_r (x_r - \bar{x})^2 = \frac{1}{N} \sum f_r \left(x_r^2 \right) - \frac{2\bar{x}}{N} \sum f_r x_r + \frac{\bar{x}^2}{N} \sum f_r$$

$$\sigma^2 = \frac{1}{N} \sum f_r x_r^2 - 2\bar{x}^2 + \bar{x}^2$$

$$\sigma_a^2 = \frac{1}{N} \sum f_r x_r^2 - \bar{x}^2 \underline{\qquad} ①$$

92

$$\sigma_\alpha^2 = \frac{1}{N}\sum f_r (x_r-\alpha)^2 = \frac{1}{N}\sum f_r x_r^2 - \frac{2\alpha}{N}\sum f_r x_r + \frac{\alpha^2}{N}\sum f_r$$

$$= \frac{1}{N}\sum f_r x_r^2 - 2\alpha\bar{x} + \alpha^2 \quad\text{——②}$$

Substracting ② from ①

$$\sigma^2 - \sigma_\alpha^2 = -\bar{x}^2 + 2\alpha\bar{x} - \alpha^2 = -(\bar{x}-\alpha)^2$$

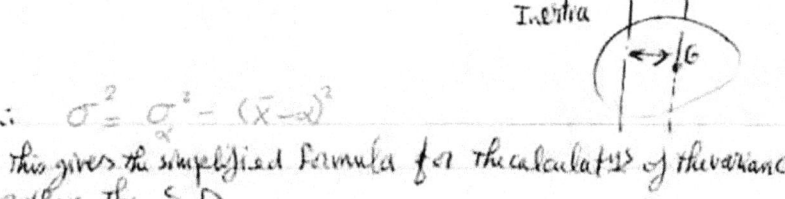
Inertia

$$\therefore \quad \sigma^2 = \sigma_\alpha^2 - (\bar{x}-\alpha)^2$$

this gives the simplified formula for the calculatis of the variance & then the S. D.

10.8.8.1. Example of the mean and s.d.

23\3\1972

example:-

The following table is a frequency table for the weights of hundred 100 adopts ... ï (adolescent, 8%). Calculate the Arithmatic math & Standard Deviation.

centre of interval x lbs	frequency	$X = x - 145$	fX	f X^2
115	2	−30	−60	1800
125	12	−20	−240	4800
135	12	−10	−120	1200
(145)	25	0	0	0
155	27	10	270	2700
165	10	20	200	4000
175	9	30	270	8100
185	3	40	120	4800
the Total	100	$X = x_r - \alpha$	440	27400
	$f_1 + f_2 + f_3 \dots$ $= \sum_{r=1}^{n} f_r$		$\sum f_r X_r$	$\sum f_r X_r^2$

93

$$\bar{X} = \frac{\sum F_r X_r}{\sum F_r} = \frac{440}{100} = 4.4 = \sum P_r X_r$$

$$\sigma' = \sqrt{\frac{\sum f_r (X - \bar{x})^2}{\sum f_r}} \qquad \sigma_\alpha^2 = \frac{\sum_F (X_r - \alpha)^2}{\sum f_r} = 274$$

$x =$ original variable
$X =$ the new variable
$= x - \alpha$

$\underline{\text{Note} >}$

$$\sigma^2 = \frac{\sum f_r (x_r - \bar{x})^2}{\sum f_r}$$

$$\sigma_\alpha^2 = \frac{\sum f_r (x_r - \alpha)^2}{\sum f_r} \qquad \text{Let } X = x_r - \alpha)$$

$$\sigma_\alpha^2 = \frac{\sum f_r X^2}{\sum f_r}$$

$$\sigma^2 = \sigma_\alpha^2 - (\bar{x} - \alpha)^2$$

$$\sigma_\alpha^2 = \sigma^2 + (\bar{x} - \alpha)^2$$

centre of gravity

$$\bar{x} = \alpha + \bar{X} = 145 + 4.4 = 149.4.$$

$$\sigma^2 = \sigma_\alpha^2 - (\bar{x} - \alpha)^2 = 274 - (149.4 - 145)^2$$
$$= 274 - 4.4^2$$
$$= 254.64$$

$$\sigma = 15.96 \text{ lbs}$$

$$\boxed{\sigma^2 = \sigma_\alpha^2 - \bar{X}^2}$$

10.9. Sheppard's correction.

94

Sheppard's correction:-

the association of the frequency with the centre of interval leads to mistakes in both the mean & variance. the error in the mean can be neglected. Since every estimate are liable ε's by underestimations this is not the case for in the variance. Sheppard introduced the following corrects.

$$\sigma_c^2 = \sigma^2 - \frac{c^2}{12}$$

where σ_c^2 is the correction variance. σ^2 is the calculated variance & c is the width of interval. Thus in the above example

$$\sigma_c^2 = \sqrt{254.64 - \frac{(10)^2}{12}} = 15.69 \quad 16.5$$

10.10. Binomial frequency distribution

Binomial Frequency Distribution

We already know that if the prob of the occurance of an event in one trial is (P) & its failing to happen is (q) where $P + q = 1$ then the prob of its occurance exactly r times in n trials is $nC_r\, P^r q^{n-r}$. Now construct the following table

The sum of prob is unity since it is certain that one of the above events must happen. we notice also that the probs are the terms of the binomial expansion $(q+P)^n = 1 = q^n + nC_1 P q^{n-1} + \cdots + P^n$ Differentiat partially w.r.t (P)

N≤ of successes r (occurance)	prob.
0	$nC_0 P^0 q^n = q^n$
1	$nC_1 P q^{n-1}$
2	$nC_2 P^2 q^{n-2}$
r	$nC_r P^r q^{n-r}$
$\vdots$	
n	$nC_n P^n q^0 = P^n$

$$n(q+P)^{n-1} = nC_1 q^{n-1} + 2 nC_2 P q^{n-2} + \cdots + n P^{n-1}$$

multiply both sides by (P)

$$n P(q+P)^{n-1} = nC_1 P q^{n-1} + 2 nC_2 P^2 q^{n-2} + \cdots + n P^n = \sum r P_r$$

i.e the mean of the variable r (r cxpress) = $E(r)$

$$\therefore \quad E(r) = nP$$

i.e the mean of a binomial distribution $= nP$.

Again differentiate partially w.r.t (P)

$$n P^2 (n-1)(P+q)^{n-2} + n(q+P)^{n-1} = nC_1 q^{n-1} + 2^2 \cdot nC_2 P q^{n-2} + n^2 P^{n-1}$$

Again multiply by (P)

$$n P^2 (n-1)(P+q)^{n-2} + n P(q+P)^{n-1} = \sum r^2 P = E(r^2)$$

$$\therefore E(r^2) = n(n-1) P^2 + n P$$

$$= n^2 P^2 - n P^2 + n P$$

$$\therefore \sigma^2 = E(r^2) - E^2(r) = n^2 P^2 - n P^2 + n P - n^2 P^2$$

$$\sigma^2 = n P(1-P) = n P q$$

This gives the variance of a binomial distribution. Hence we have the following results.

$$\text{mean} = n P$$

$$S.D = \sigma = \sqrt{n P q}$$

Thus for example in a binomial distribution in which $P = q = \frac{1}{2}$ which is the case of tossing the coin

$$\sigma = \sqrt{\frac{n}{4}}$$

10.11. Stirling's formula for the approximation of factorials of large numbers

Stirling formula for the approximation of factorials of large numbers.

$$n! \sim \sqrt{2\pi n}\ n^{n}\, e^{-n}$$

the sign $\sim$ means that the ratio of the two sides tends to unity as n tends to infinite.

10.12. Chekychef's Theorem

Chekychef's theorem

let $\bar{x}$ be the mean of a frequency distribution σ be its S.D. let k be a positive constant the theorem states that:-

"The prob that the random variable may lie outside the range $\bar{x} \pm k\sigma$ is less than $\frac{1}{k^2}$"

96

Soln (Proof)

$$\sigma^2 = \int_{-\infty}^{\infty} (x - \bar{x})^2 f(x)\, dx$$

$$\sigma^2 > \int_{-\infty}^{\bar{x} - k\sigma} (x - \bar{x})^2 f(x) + \int_{\bar{x} + k\sigma}^{\infty} (x - \bar{x})^2 f(x)\, dx$$

$$> k^2\sigma^2 \left\{ \int_{-\infty}^{\bar{x} - k\sigma} f(x)\, dx + \int_{\bar{x} + k\sigma}^{\infty} f(x)\, dx \right\}$$

The expression betⁿ brackets is the prob that the random variable may lies outside the range $\bar{x} \pm k\sigma$ hence the prob is less them.

$$\text{Prob} < \frac{\sigma^2}{k^2\sigma^2} = \frac{1}{k^2}$$

10.13. Normal frequency distribution or Gaussian distribution

30\3\1972

Normal Frequency Distribution a Gaussian Distribution.

This a limiting case of binomial distribution in which $n \to \infty$, P is remaining the same. Consider the variable rC where C is a constant which takes in succession the values $0, C, \dots, n$ with prob⁸ which are the successive terms of the expression of $(q + p)^n$

The mean of the binomial distribution is now. $npC = E(rc)$ & the variance is $\sigma^2 = npq\, c^2$.

let y be the prob of the occurance of rC & y_1, the prob of the occurance of $(r+1)C$.

$$y = nC_r\, p^r q^{n-r}$$

$$y_1 = nC_{(r+1)}\, p^{r+1} q^{n-r-1}$$

$$\therefore \frac{y_1}{y} = \frac{n!}{(r+1)!\,(n-r-1)!} \cdot \frac{r!\,(n-r)!}{n!} \cdot \frac{p}{q} = \frac{(n-r)}{(r+1)} \cdot \frac{p}{q}$$

$$\therefore \frac{y_1 - y}{y} = \frac{y_1}{y} - 1 = \frac{n-r}{r+1} \cdot \frac{p}{q} - 1 = \frac{nP - rP - rq - q}{qr + q} = \frac{nP - r(P + q) - q}{rq + q}$$

$$= \frac{nP - r - q}{rq + q} = \frac{npC - rC + qC}{rqC + qC}$$

Put $x = rc - nPc$

$qx = qrc - nPqc$

$\therefore qrc = qx + nPqc$

(variable, mean)

$\therefore \dfrac{y_1 - y}{y} = -\dfrac{x - qc}{qx + nPqc + qc} = \dfrac{-xc - qc^2}{qxc + nPqc^2 + qc^2} = -\dfrac{(x+qc)c}{qxc + \sigma^2 + qc^2}$

Put $c = \Delta x$ & $y_1 - y = \Delta y$

$\therefore$

$$\dfrac{\Delta y}{y} = -\dfrac{(x + q\Delta x)\Delta x}{qx(\Delta x) + \sigma^2 + q(\Delta x)^2}$$

$\dfrac{1}{y}\dfrac{\Delta y}{\Delta x} = -\dfrac{[x + q\Delta x]}{qx(\Delta x) + \sigma^2 + q(\Delta x)^2}$ Let $\Delta x \to 0$

$\dfrac{1}{y}\dfrac{dy}{dx} = -\dfrac{x}{\sigma^2}$ $\longrightarrow$ ①

$\therefore \displaystyle\int \dfrac{dy}{y} = \int -\dfrac{x\,dx}{\sigma^2}$

from discrete continuous

$\log y = -\dfrac{x^2}{2\sigma^2} + \log A$ $\therefore y = Ae^{-\frac{x^2}{2\sigma^2}}$ ②

The constant A can be determined from the fact that the total area under the curve is unity from which :-

$\therefore \displaystyle\int_0^\infty e^{-y^2} dy = \dfrac{\sqrt{\pi}}{2}$ $\therefore \displaystyle\int_{-\infty}^\infty e^{-y^2} dy = 2\int_0^\infty e^{-y^2} dy = \sqrt{\pi}$

$\therefore \displaystyle\int_{-\infty}^\infty Ae^{-\frac{x^2}{2\sigma^2}} = A\int_{-\infty}^\infty e^{-\frac{x^2}{2\sigma^2}} = A\sqrt{\pi \cdot 2\sigma^2} = 1$

$\therefore$

$A = \dfrac{1}{\sigma\sqrt{2\pi}} \cdot$ $\longrightarrow$ ③

$\therefore \boxed{y = \dfrac{1}{\sigma\sqrt{2\pi}} e^{-\frac{x^2}{2\sigma^2}}}$

In this eqn x represents the deviation of the variable from the mean. i.e the axis coincides the line of symmetry which cuts the x-axis in a pt which is the

mean, mode, median.

If the line of symmetry makes the x-axis in the mean value $\bar{x}$ Then the eq^n of the curve becomes.

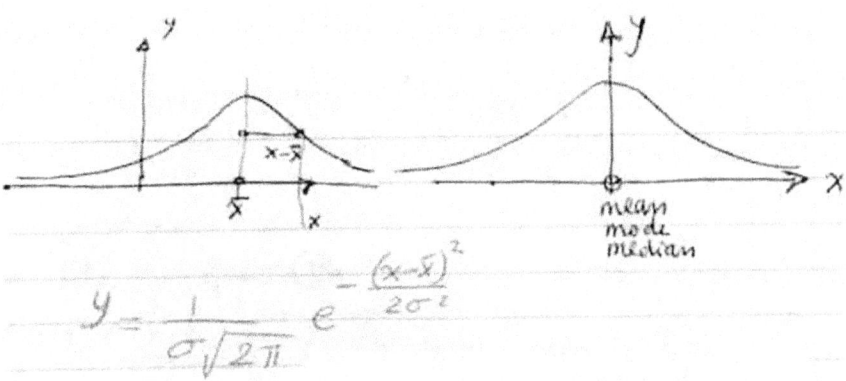

$$y = \frac{1}{\sigma\sqrt{2\pi}} e^{-\frac{(x-\bar{x})^2}{2\sigma^2}}$$

$$y = \frac{1}{\sigma\sqrt{2\pi}} e^{-\frac{(x-\bar{x})^2}{2\sigma^2}} \quad \text{Famous eq}^n$$

which is the eq^n of the normal frequency Distribut^n curve the area bet^n two pts x_1, x_2 on the curve (under the curve) represents the probability that the random variable should lie bet^n x_1, x_2.

There are tables constructive which give the value of the integral for various values of x- from the table we find that.

① 99.73% of the area under the curve lies bet^n $\bar{x} \pm 3\sigma$ the prob that the random variable lies outside the range $\bar{x} \pm 3\sigma$ is 0.27% which is a very small prob. that is why we can neglect the curve beyond $\bar{x} \pm 3\sigma$.

② 95.45% of the area under the curve lies bet^n $\bar{x} \pm 2\sigma$.
③ 68.27% of the area under the curve lies bet^n $\bar{x} \pm \sigma$
④ 50% ~ ~ ~ ~ ~ ~ ~ ~ $\bar{x} \pm 0.6745\sigma$

10.14. Poission's distribution

99

<u>Poisson's Distribution</u> This occurs when the prob in one trial is very small & we shell obtain the distribut² as a limiting case from binomial distributs when $n \to \infty$, $P \to 0$ in such away that the mean $m = nP$ remains constant. the prob of the occurance of an event r times in $\underline{n}$ trials is $nC_r P^r q^{n-r}$

$$nC_r \, P^r q^{n-r} = \frac{n(n-1)(n-2)\cdots(n-r+1)}{r!} \, P^r (1-P)^{n-r}$$

as

$$nC_r = \frac{n!}{r!(n-r)!} = \frac{n(n-1)(n-2)\cdots(n-r+1)(n-r)!}{r!} \, \frac{1}{(n-r)!}$$

$$\therefore nC_r \, P^n q^{n-r} = \frac{nP(nP-P)(nP-2P)\cdots(nP-rP+P)}{r!\,(1-P)^r} \left\{ (1-P)^n \right\}$$

Since $m = nP$

$$\therefore nC_r \, P^n q^{n-r} = \frac{nP(nP-P)(nP-2P)\cdots(nP-rP+P)}{r!\,(1-P)^r} \left\{ (1-P)^{-\frac{1}{P}} \right\}^{-m}$$

Since $e = \lim_{x \to 0} (1+x)^{\frac{1}{x}}$

$\therefore$ as P tends to 0 $\quad nC_r P^r q^{n-r} = \dfrac{m(m-P)(m-2P)\cdots(m-rP+P)\left\{(1-P)^{-\frac{1}{P}}\right\}}{r!\,(1-P)^r}$

$$\boxed{nc_r \, P^r q^{n-r} = \frac{m^r}{r!} e^{-m}}$$

$\lim\limits_{P \to 0} (1-P)^{-\frac{1}{P}}$

hence we can construct the following table.

Number of successes r	prob.
0	e^{-m}
1	$\frac{m}{1!} e^{-m}$
$\vdots$	$\vdots$
r	$\frac{m^r}{r!} e^{-m}$
$\vdots$	$\vdots$
Check	$\vdots$

Sum of prbs

$$= e^{-m}\left(1 + m + \frac{m^2}{2!} + \cdots + \frac{m^r}{r!} + \cdots\right) = e^{-m} \cdot e^{m}$$

$$= 1$$

Limiting case of binomial distribution. mean $m = np$
variance $\sigma^2 = npq$ & since $p \to 0$, $q \to 1$

$\therefore$ $\sigma^2 = np = $ mean.

This means in poisson's distribute the variance is equal to the mean, this could used as a preliminary $\frac{y_c}{c} \delta_-$ test to find whether a given distribut^n conforms, $\dot{\mathscr{e}}$, with the poisson's distribut^n.

ex. If in a big city, two persons die dialy in the way by an accident. what is the prob that in one day 3 person will die.

here $m = 2$ $\qquad r = 3$

$$prob = \frac{2^3}{3!} e^{-2} = \frac{4}{3e^2} = 0.18$$

10.15. Law of large numbers

Law of large numbers
If P is the prob of occurance of an event in one trial & r is the number of successes in (n) trials & ϵ is a positive constant then

$$\lim_{n\to\infty} P\left\{\left|\frac{r}{n}-p\right|<\epsilon\right\} = 1$$

The meaning of this law is that it is certain that the relative frequency $\frac{r}{n}$ will in the limit when n becomes infinite equal to P e.g: if we throw die then the prob that an ace will turn up is $\frac{1}{6}$ this is P, $\frac{1}{6}$ actually means that if we make a large number of throws then in approximately one sixth of these throws the ace will turn up & as we increase the No of throws the relative frequency $\frac{r}{n}$ approaches $\frac{1}{6}$.

10.16. Correlation

Correlation In many physical phenomena there is a certain relation bet corresponding values x, y of two phenomena. large value of x may be accompanied by large value of y & small value of x may be accompanied by small value of y. This is a positive correlation On the other hand _large_ value of x may be accompanied by small value of y & vise versa this is a negative correlation.

An example of the 1st case is that the stature of a man & that of his son.

An example of the 2nd case is the rate of marriage & the number of unemployee. hence in general variations of x are accompanied by variation in y. to measure this variat^n we need some base point from which diviations are measurable this is taken as the arithmetic mean

Hence for one phenomenon we have

$$x_1-\bar{x}, \ x_2-\bar{x}, \ x_3-\bar{x}, \dots, x_s-\bar{x}, \ x_n-\bar{x} \quad \text{& for the}$$

other phenomenon
$$y_1-\bar{y}, \ y_2-\bar{y}, \ y_3-\bar{y}, \dots, y_s-\bar{y}, \ y_n-\bar{y}$$

In order to be able to compare phenomena of different units we use expressions which are dimensionless for this purpose we divide the deviation from the mean by the standard deviat: S.D.

$$\frac{x_1-\bar{x}}{\sigma_x}, \frac{x_2-\bar{x}}{\sigma_x}, \quad \cdots \quad \frac{x_s-\bar{x}}{\sigma_s}, \cdots \frac{x_n-\bar{x}}{\sigma_x}$$

$$\frac{y_1-\bar{y}}{\sigma_y}, \frac{y_2-\bar{y}}{\sigma_y}, \cdots \quad , \frac{y_s-\bar{y}}{\sigma_y}, \cdots \frac{y_n-\bar{y}}{\sigma_y}$$

the product-moment correlation coefficient (r) is defined as follows :-

$$r = \sum_{s=1}^{n} \frac{(x_s-\bar{x})(y_s-\bar{y})}{n\,\sigma_x\,\sigma_y}$$

(r) varies bet: ±1. if there is complete correlation then r = 1 & y is fn of x, on the other hand if r = 0 then there is no correlat: at all bet: x, y.

ex.   find the product moment correlat: coff± for the following two phenomena.

x	1	2	3	4	5
y	2	5	3	8	7

here

$$\bar{x} = \frac{1+2+3+4+5}{5} = \frac{15}{15} = 3$$

$$\bar{y} = \frac{2+5+3+8+7}{5} = \frac{25}{5} = 5$$

we now form the following table.

x	$x-\bar{x}$	$(x-\bar{x})^2$	y	$y-\bar{y}$	$(y-\bar{y})^2$
1	-2	4	2	-3	9
2	-1	1	5	0	0
3	0	0	3	-2	4
4	1	1	8	3	9
5	2	4	7	2	4
		10			26

$$\sigma_x = \sqrt{\frac{\sum f_r (x-\bar{x})^2}{\sum f_r}} = \sqrt{\frac{f_r \sum(x-\bar{x})^2}{\sum f_r}} = \sqrt{\sum (x-\bar{x})^2}\sqrt{\frac{f_r}{\sum f_r}}$$

$$= \sqrt{10}\sqrt{P} = \sqrt{\frac{10}{5}} = 1.414$$

$$\sigma_y = \sqrt{\sum(y-\bar{y})^2}\sqrt{P} = \sqrt{\frac{26}{5}} = 2.28$$

$$\sum (x-\bar{x})(y-\bar{y}) = 6+0+0+3+4 = 13$$

$$r = \frac{\sum(x-\bar{x})(y-\bar{y})}{n\,\sigma_x\,\sigma_y} = \frac{13}{5 \times 1.414 \times 2.28} = 0.8 \text{ approx.}$$

10.17. Scatter diagram

Scatter diagram — regression lines

Suppose that x & y are two corresponding values of two phenomena need x, y as a coordinates of a point locate this pts then we have what is as a Scattered Diagram.

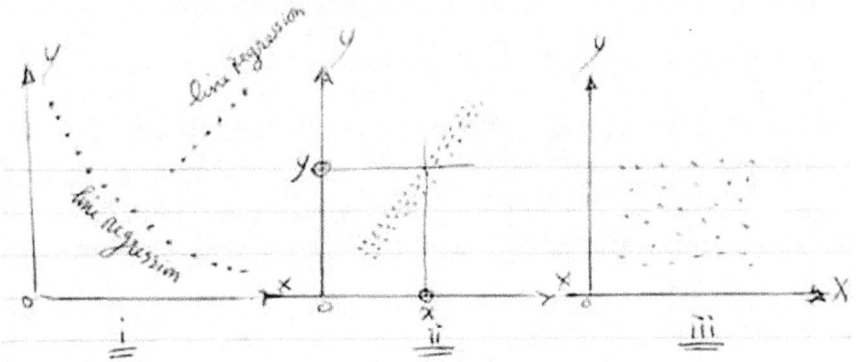

i) In fig. i there is complete correlation betᵘ x, y i.e $r = 1$ & y is $f \Rightarrow f(x)$ The pts lie on a curve which we call a regression line if the pts lie on a straight line then we have a linear regression

ii) In fig ii we have a certain degree of independence.

iii) There is no correlation at all & pts are scattered over the plane.

we now consider linear regression.

(X, y_m) corresponding to any pt (x) there is more than one value of y. let y_m denotes the mean value of this width. locate the pt (X, y_m) & find the best straight line to approximate these pts by using a methods of the least squares that is by minimizing the sum of the squares of the vertical departures from the line in this way are obtain the eqn.

$$y - \bar{y} = \frac{r \sigma_y}{\sigma_x} (x - \bar{x})$$

this is called the regression of (y) on x & the coeff $\frac{r \sigma_y}{\sigma_x}$ is called the coeff of regression b_{xy}

(X_m, y) Again corresponding to any point (y) there is more than one value of (x) let x_m by the mean of these values locate the pts (x_m, y) & find the best straight line to fit these pts by applying the methode of squares which in this case aught to minimizing the sum of squares of the horizontal deviations. in this case we obtain a line with eqᵘ is $\quad x - \bar{x} = r \dfrac{\sigma_x}{\sigma_y}(y - \bar{y})$

which is the regression of x on y. the two line will coincide then there is a complete correlation. i.e $r = 1$

Methodes of calculations

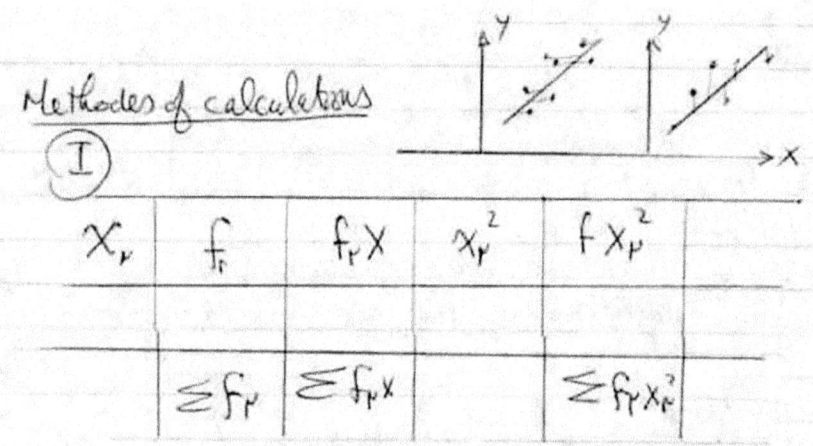

(I)

x_r	f_r	$f_r x$	x_r^2	$f x_r^2$
	$\sum f_r$	$\sum f_r x$		$\sum f_r x_r^2$

$$\bar{x} = \frac{\sum f_r x}{\sum f_r} \quad \textcircled{1} = E(x) \quad \text{by squaring}$$

$$E(x^2) = \frac{\sum f_r x^2}{\sum f_r} \quad \textcircled{2} \quad E(x^2)$$

$$\boxed{\underset{\text{variance}}{\sigma^2} = E(x^2) - E(x)^2}$$

$$\underset{\text{from } \textcircled{2}}{\downarrow} \qquad \underset{\text{from } \textcircled{1}}{\downarrow}$$

10.18. Standard error of the mean SEM

106

Standard error of the mean — sampling distribution of the mean:-
Suppose we have a frequency distribution consisting of n
individuals whose mean is $\bar{x}$ say & whose S.D is σ.
now suppose the n-individuals are nearly a sample of a much
large number. individuals which we may call a Father
population. It is required to find a true mean m

i.e the the mean of the total population. for this purpose
we take a large number of samples each of
n – individuals. & let their means be
$\bar{x}_1, \bar{x}_2 \cdots \cdots \bar{x}_n$.

Now an important theorem in statistics استاتسنك states that:-
the means of the samples $\bar{x}_1, \bar{x}_2, \bar{x}_n$ form a frequency
distribution. which is very nearly a normal frequency.
distribution. whose mean is the true mean (m) &
whose S.D is $\dfrac{\sigma}{\sqrt{n}}$ where σ is the S.D of the

father population

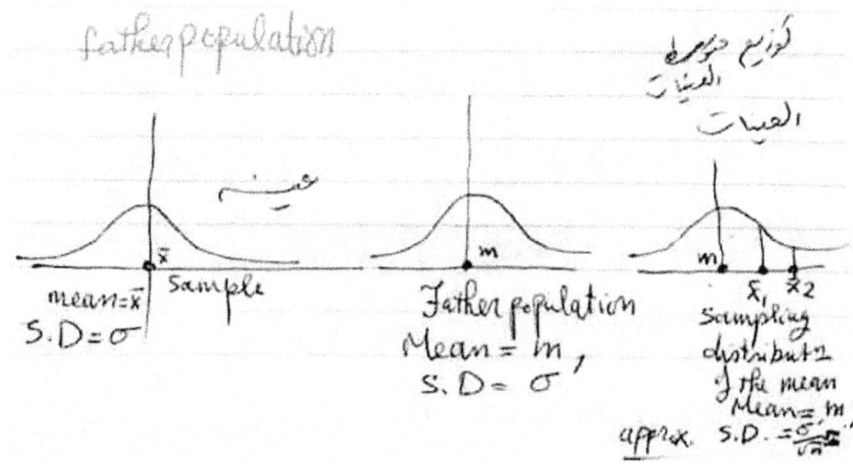

mean=$\bar{x}$ Sample
S.D = σ

Father population
Mean = m,
S.D = σ'

$\bar{x}_1, \bar{x}_2$
Sampling
distribut²
of the mean
Mean= m
approx. S.D = $\dfrac{\sigma'}{\sqrt{n}}$

Since it is difficult to calculate σ' which is the S.D of the father population., we agree to make (σ) to equal approximately σ'. Now from the properties of normal frequency distribution we can say that it is not likely that $\bar{x}$ will differ from m by more than 3 times the S.D. i.e $\dfrac{3\sigma}{\sqrt{n}}$ stated in other words :–

The true mean (m) is not likely to differ from the mean of the sample $\bar{x}$ by more than $\dfrac{3\sigma}{\sqrt{n}}$ is called the standard error of the mean.

hence it is customary when $\bar{x}$ is the mean of a sample to write that the true mean is $\bar{x} \pm \dfrac{\sigma}{\sqrt{n}}$ on the

understanding that the true mean lies bet^n $\bar{x} \pm \dfrac{3\sigma'}{\sqrt{2}}$ In this way we obtained from the mean of the sample a range with which the true mean lies.

Méthode (II)   simplification   here the original mean
$$\boxed{\bar{x} = \bar{X} + \alpha}$$

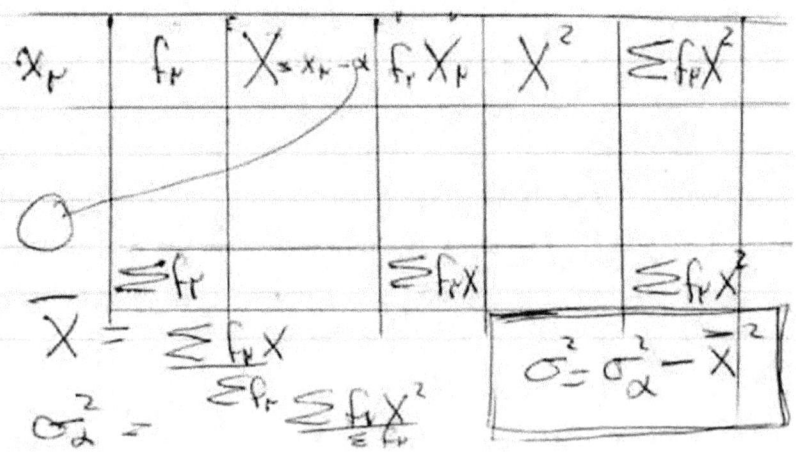

x_p	f_p	$X = x_p - \alpha$	$f_p X_p$	X^2	$\sum f_p X^2$
	$\sum f_p$		$\sum f_p X$		$\sum f_p X^2$

$$\bar{X} = \dfrac{\sum f_p X}{\sum f_p}$$

$$\sigma_\alpha^2 = \dfrac{\sum f_p X^2}{\sum f_p}$$

$$\boxed{\sigma^2 = \sigma_\alpha^2 - \bar{X}^2}$$

CHAPTER 11: LAP LACE TRANSFORMATION

11.1.1. Integral transforms

The integral transform f(p) of a given function f(x) in the range (a,b) is defined as follows.

$$\bar{f}(p) = \int_a^b f(x) \; k(p,x) \; dx$$

where $k(p,x)$ is a known function of p and x known as the kernel of the transform. This is on the assumption that the integral exists. If the limits of integration are 0 and ∞ then
we have the following special cases :—

(i) If $k(p,x) = e^{-px}$, then

$$\bar{f}(p) = \int_0^\infty f(x) \; e^{-px} \; dx$$

and this is known as the Laplace transform of f(x).

(ii) If $k(p,x) = \sin px$, then

$$\bar{f}(p) = \int_0^\infty f(x) \sin px \; dx$$

and this is the infinite Fourier sine transform of f(x).

(iii) If $k(p,x) = \cos px$, then

$$\bar{f}(p) = \int_0^\infty f(x) \cos px \; dx$$

and this is the infinite Fourier cosine transform.

(iv) If $k(p,x) = x \, J_n(px)$ where $J_n(px)$ is the Bessel function of the first kind of order n, then

$$\bar{f}(p) = \int_0^\infty x \, f(x) \, J_n(px) \; dx$$

and this is known as the Hankel transform of f(x).

(v) If $k(p,x) = x^{p-1}$, then

$$\bar{f}(p) = \int_0^\infty f(x) \; x^{p-1} dx$$

and this is known as the Mellia transform.

Such transforms have been widely used to obtain solutions of both ordinary and partial linear differential equations.

The main object of the present work is to discuss in some detail the Laplace transformation and its applications to the solution of differential equations with given initial conditions and in solving boundary — value problems.

11.1.2. Laplace transformation.

Let F(t) be a function defined for all positive values of the real variable t. The Laplace transform of F(t) usually denoted
by

$$L\left\{F(t)\right\} = \int_0^\infty e^{-st}\, F(t)\, dt = f(s)$$

provided that the integral exists. We shall denote the original function by a capital letter and its transform by the corresponding lower case letter. We often call F(t) the object function and f(s) the result function. s may be real or complex but we shall consider first real values of s.

We shall start by finding the Laplace transforms of some of the elementary functions from first principles. Later on simpler methods will be used for the derivation of these transforms and we shall give the limitations on F(t) as well as on the range of s.

I. $L\left\{1\right\},\ t > 0$

$$L\left\{1\right\} = \int_0^\infty e^{-st}\, dt = \left[\frac{e^{-st}}{-s}\right]_0^\infty = \frac{1}{s},\ s > 0$$

$$\therefore\ \ L\left\{1\right\} = \frac{1}{s},\ s > 0$$

II. $L\left\{e^{at}\right\},\ t > 0$

$$L\left\{e^{at}\right\} = \int_0^\infty e^{-st}\cdot e^{at}\, dt = \int_0^\infty e^{-(s-a)t}\, dt$$

$$= \left[\frac{e^{-(s-a)t}}{-(s-a)}\right]_0^\infty = \frac{1}{s-a},\ s > a$$

$$\therefore\ \ L\left\{e^{at}\right\} = \frac{1}{s-a},\ s > a$$

III. $L\left\{t^k\right\}$ where $k > -1$

$$L\left\{t^k\right\} = \int_0^\infty e^{-st}\, t^k\, dt$$

Put $st = x$ where $s > 0$ $\therefore t = \dfrac{x}{s}$, $dt = \dfrac{dx}{s}$

$$\therefore L\left\{t^k\right\} = \int_0^\infty e^{-x}\, \frac{x^k}{s^k}\, \frac{dx}{s} = \frac{1}{s^{k+1}}\int_0^\infty e^{-x}\, x^k\, dx$$

$$= \frac{\Gamma(k+1)}{s^{k+1}}$$

Note that the condition $k > -1$ is necessary for the convergence of the integral.

$$\therefore L\left\{t^k\right\} = \frac{\Gamma(k+1)}{s^{k+1}}, \quad k > -1, \ s > 0$$

Corollary. 1. If k is a positive integer equal to n, then

$$L\left\{t^n\right\} = \frac{\Gamma(n+1)}{s^{n+1}} = \frac{n!}{s^{n+1}}, \quad s > 0$$

Thus $L\left\{t\right\} = \dfrac{1}{s^2}$, $L\left\{t^2\right\} = \dfrac{2!}{s^3}$, $L\left\{t^3\right\} = \dfrac{3!}{s^4}$, ...

Corollary. 2. If $k = -\tfrac{1}{2}$, then

$$L\left\{t^{-\frac{1}{2}}\right\} = \frac{\Gamma(\frac{1}{2})}{s^{\frac{1}{2}}} = \frac{\sqrt{\pi}}{\sqrt{s}} = \sqrt{\frac{\pi}{s}}, \quad s > 0$$

IV. $L\left\{\sin kt\right\} = \int_0^\infty e^{-st}\sin kt\, dt$

$$= \left[-\frac{(s\sin kt + k\cos kt)\, e^{-st}}{s^2 + k^2}\right]_0^\infty = \frac{k}{s^2 + k^2}, \quad s > 0$$

$$L \left\{ \cos kt \right\} = \int_0^\infty e^{-st} \cos kt \, dt$$

$$= \left[\frac{(k \sin kt - s \cos kt) \, e^{-st}}{s^2 + k^2} \right]_0^\infty = \frac{s}{s^2 + k^2}, \; s > 0$$

11.1.3. Laplace transformation of the sum of two functions

Let A and B be two arbitrary constants, then

$$L \left\{ A \, F(t) + B \, G(t) \right\} = \int_0^\infty e^{-st} \left\{ A \, F(t) + B \, G(t) \right\} dt$$

$$= A \int_0^\infty e^{-st} F(t) \, dt + B \int_0^\infty e^{-st} G(t) \, dt$$

$$= A \, L \left\{ F(t) \right\} + B \, L \left\{ G(t) \right\}$$

This means that the Laplace transform of a linear combination of two functions is the same linear combination of the transforms of these functions.

$$\text{e. g.} \quad L \left\{ \cosh kt \right\} = L \left\{ \frac{e^{kt} + e^{-kt}}{2} \right\}$$

$$= \tfrac{1}{2} \left[\frac{1}{s - k} + \frac{1}{s + k} \right] = \frac{s}{s^2 - k^2}, \; s > |k|$$

$$\text{Similarly} \quad L \left\{ \sinh kt \right\} = \frac{k}{s^2 - k^2}, \; s > |k|$$

11.1.4. Sectionally or piecewise continuous functions

A function is said to be sectionally continuous or piecewise continuous in an interval $a \leqslant t \leqslant b$ if this interval can be subdivided into a finite number of intervals in each of which the function is continuous and has finite right and left hand limits. It follows from the definition that sectionally continuous functions include two main classes of functions:
(i) continuous functions,
(ii) discontinuous functions in which the discontinuity consists of finite jumps,

112

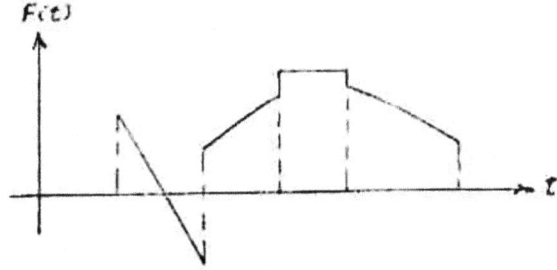

It is worthy of notice that sectionally continuous functions are integrable functions, the integral being the sum of the integrals of the continuous functions over the subintervals.

Example 1. Find the Laplace transform of the function F(t) defined by

$$F(t) = F_0 \qquad 0 < t < t_0$$
$$F(t) = 0 \qquad t > t_0$$

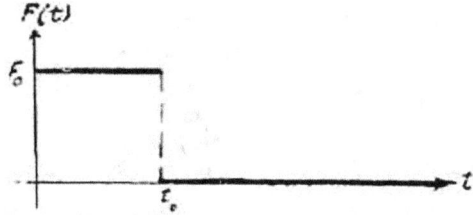

Solution:

$$L\left\{ F(t) \right\} = \int_0^\infty e^{-st} \ F(t) \ dt = \int_0^{t_0} e^{-st} \ F_0 \ dt$$

$$= F_0 \left[\frac{e^{-st}}{-s} \right]_0^{t_0} = \frac{F_0}{s} \left(1 - e^{-t_0 s} \right)$$

Example 2. Find the Laplace transform of H(t) where H(t) = t at 0 < t < 4 and H(t) = 5 at t > 4

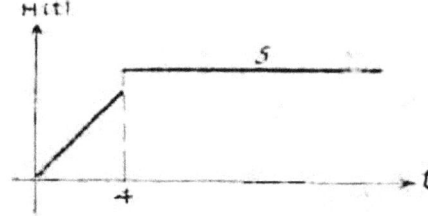

Solution:

$$L\left\{ H(t) \right\} = \int_0^\infty e^{-st} \ H(t) \ dt$$

113

$$= \int_0^4 e^{-st}\, t\, dt + \int_4^\infty e^{-st} \times 5\, dt$$

$$= \left[-\frac{t}{s}\, e^{-st} - \frac{1}{s^2}\, e^{-st} \right]_0^4 + \left[-\frac{5}{s}\, e^{-st} \right]_4^\infty = \frac{1}{s^2} + \frac{e^{-4s}}{s} - \frac{e^{-4s}}{s^2}$$

11.1.5. Functions of exponential order

If two positive constants M and a exist such that for all

$$t > T, \quad \left| e^{-at}\, F(t) \right| < M \quad \text{or} \quad \left| F(t) \right| < M e^{at},$$

then we say that F(t) is a function of exponential order a as $t \longrightarrow \infty$ and write F(t) is $O\left(e^{at}\right)$. This actually means that functions of exponential order cannot grow in absolute value more rapidly than Me^{at} as t increases.

Example 1. Let F(t) = C where C is a constant. Choose a positive constant M such that M> | C |

$\therefore$ $|C| < M e^{at}$ as $t \longrightarrow \infty$ for all $a \geqslant 0$

$\therefore$ C is $O\left(e^{at}\right)$ where $a \geqslant 0$

Similarly for all bounded functions e.g. sin kt or cos kt.
Thus sin kt and cos kt are

$O\left(e^{at}\right)$ where $a \geqslant 0$.

Example 2. Let $F(t) = t^n$ to where n is a positive integer.

$$\frac{t^n}{e^{at}} \longrightarrow 0 \text{ as } t \longrightarrow \infty \text{ for all } a > 0$$

$\therefore$ $\left| \dfrac{t^n}{e^{at}} - 0 \right| < M$ as $t \longrightarrow \infty.$

i.e. $\left| t^n \right| < M e^{at}$ as $t \longrightarrow \infty$ for all $a > 0.$

$\therefore$ t^n is $O\left(e^{at}\right)$ where $a > 0.$

Example 3. Let $F(t) = e^{t^3}$

$$\therefore \quad \left| e^{-at} \cdot e^{t^3} \right| = e^{t^3 - at}$$

can be made larger than any given constant by increasing t. Hence e^{t^3} is not of exponential order.

11.1.6. Sufficient conditions for the existence of the Laplace transform

We shall now show that the Laplace transform of F(t) exists when:

(i) F(t) is sectionally continuous in every finite interval in the range $t \geqslant 0$, (ii) F(t) is $O(e^{at})$ as $t \longrightarrow \infty$,

i.e $| F(t) | < Me^{at}$ for all $t > T$ (say).

$$\text{For} \int_0^\infty e^{-st} F(t) \, dt = \int_0^T e^{-st} F(t) \, dt + \int_T^\infty e^{-st} F(t) \, dt$$

The first integral on the R.H.S. exists since F(t) is sectionally continuous in every finite interval $0 \leqslant t \leqslant T$. The second integral on the right also exists since F(t) is $O(e^{at})$ for $t > T$. This follows from the fact that

$$\left| \int_T^\infty e^{-st} F(t) \, dt \right| \leqslant \int_T^\infty \left| e^{-st} F(t) \right| \, dt$$

$$< \int_T^\infty e^{-st} Me^{at} \, dt$$

$$= M \int_T^\infty e^{-(s-a)t} \, dt = M \left[\frac{e^{-(s-a)t}}{-(s-a)} \right]_T^\infty$$

$$= \frac{Me^{-(s-a)T}}{s-a} \quad , \quad s > a$$

$$< \frac{M}{s-a}$$

Hence the Laplace integral $\int_0^\infty e^{-st} F(t) \, dt$ exists if $s > a$.

115

It should be noticed that the above conditions though sufficient are not necessary. They are however simple to apply in most practical problems. If the above conditions are not satisfied the Laplace transform may or may not exist. Thus for example

$$L\left\{ t^{-\frac{1}{2}} \right\} = \sqrt{\frac{\pi}{s}} \ , \ s > 0$$

as we have already shown in 1.2, but $F(t) = t^{-\frac{1}{2}}$ does not satisfy the above conditions.

11.1.7. Null functions

A null function N(t) is a function of t such that for all t > 0

$$\int_0^t N(\tau) \, d\tau = 0$$

e.g. the function

F(t) = 2, t = 1
F(t) = 1, t = 2
F(t) = -1, t ---, 3
F(t) = 0 otherwise

is a null function. In general any function which is zero at all but a countable set of points is a null function. It is evident that the Laplace transform of a null function is zero.

11.1.8. Inverse Laplace transforms

If L { F(t)} = f(s) , then the inverse Laplace transform is often written

$$L^{-1}\left\{ f(s) \right\} = F(t)$$

i.e., the inverse Laplace transform of f(s) is F(t). Now the Laplace transform of a null function N (t) is zero. Hence if

L {F(t)} = f(s) then L {F(t) + N(t)} = f(s) .

Thus we can have two different functions with the same Laplace transform. Hence if we allow null functions the inverse transform is not unique, but if we do not allow null functions, then it has been shown by Lerch that if we restrict ourselves to functions of exponential order for t > T and which are sectionally continuous in every finite interval $0 \leqslant t \leqslant T$
Then

$$L^{-1}\left\{ f(s) \right\} = F(t)$$

is unique, We shall now write the previous transforms which we have already obtained in the inverse form.

$$L^{-1}\left\{ \frac{1}{s} \right\} = 1 \ , \quad L^{-1}\left\{ \frac{1}{s^n} \right\} = \frac{t^{n-1}}{(n-1)!} \ (n \ \text{is a} + \text{ve integer})$$

$$L^{-1}\left\{\frac{1}{s-a}\right\} = e^{at}, \quad L^{-1}\left\{\frac{1}{s+a}\right\} = e^{-at}$$

$$L^{-1}\left\{\frac{1}{s^2+k^2}\right\} = \frac{1}{k}\sin kt, \quad L^{-1}\left\{\frac{s}{s^2+k^2}\right\} = \cos kt$$

$$L^{-1}\left\{\frac{1}{s^2-k^2}\right\} = \frac{1}{k}\sinh kt, \quad L^{-1}\left\{\frac{s}{s^2-k^2}\right\} = \cosh kt$$

11.1.9. The inverse Laplace transformation of the sum of two functions

We have shown that

$$L\left\{A\,F(t) + B\,G(t)\right\} = A\,f(s) + B\,g(s)$$

$$\therefore\ L^{-1}\left\{A\,f(s) + B\,g\,(s)\right\} = A\,F(t) + B\,G(t)$$
$$= A\,L^{-1}\left\{f(s)\right\} + B\,L^{-1}\left\{g(s)\right\}$$

Example.

$$L^{-1}\left\{\frac{3}{s+1} + \frac{2s}{s^2+25} - \frac{4}{s^2+9}\right\}$$

$$= L^{-1}\left\{\frac{3}{s+1}\right\} + L^{-1}\left\{\frac{2s}{s^2+25}\right\} - L^{-1}\left\{\frac{4}{s^2+9}\right\}$$

$$= 3e^{-t} + 2\cos 5t - \frac{4}{3}\sin 3t$$

11.1.10. Transforms of derivatives

Let F (t) be a function of order e^{at} as $t \longrightarrow \infty$ and suppose that F(t) is continuous with a sectionally continuous derivative
F'(t) in every finite interval $0 \leqslant t \leqslant T$. Then provided s > a, the Laplace transform of F'(t) exists and is given by

$$L\left\{F'(t)\right\} = s\,L\left\{F(t)\right\} - F(o)$$
$$= s\,f(s) - F(o)$$

where F(o) owing to the continuity of F(t) is the same as F(o+), i.e., the same as the limiting value of F(t) when t approaches 0 through positive values.

117

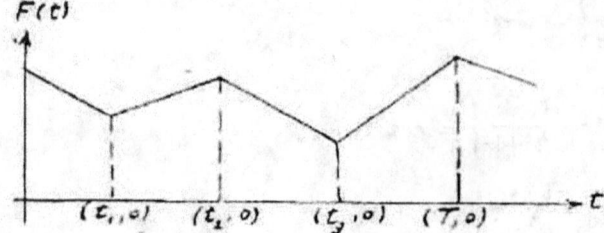

The figure shows a continuous function with a sectionally continuous derivative in the interval $0 \leqslant t \leqslant T$ and we have

$$L\left\{F'(t)\right\} = \lim_{T \to \infty} \int_0^T e^{-st} F'(t)\, dt$$

Now $\displaystyle \int_0^T e^{-st} F'(t)\, dt$

$$= \int_0^{t_1} e^{-st} F'(t)\, dt + \int_{t_1}^{t_2} e^{-st} F'(t)\, dt + .. + \int_{t_n}^{T} e^{-st} F'(t)\, dt$$

$$= \int_0^{t_1} e^{-st}\, d F(t) + \int_{t_1}^{t_2} e^{-st}\, d F(t) + .. + \int_{t_n}^{T} e^{-st}\, d F(t)$$

Integrating by parts we get

$$\int_0^T e^{-st} F'(t)\, dt = \left[e^{-st} F(t) \right]_0^{t_1} + \left[e^{-st} F(t) \right]_{t_1}^{t_2} + \cdots$$

$$+ \left[e^{-st} F(t) \right]_{t_n}^{T} + s \int_0^T e^{-st} F(t)\, dt$$

Now, owing to the continuity of $F(t)$, $F(t_1 - 0) = F(t_1 + 0)$
i.e., the limiting value of F(t) as t---->t1 is the same whether we approach t1 from the right or left and similarly for t 2, t 3 , ..., t n .

$$\therefore \int_0^T e^{-st} F'(t)\, dt = - F(o) + e^{-sT} F(T) + s \int_0^T e^{-st} F(t)\, dt$$

118

Now, since F(t) is $O(e^{at})$, then $|F(t)| < M e^{at}$ for large t and consequently

$$|e^{-sT} F(T)| < M e^{-(s-a)T} \longrightarrow 0 \text{ as } T \longrightarrow \infty \quad (s > a)$$

Also $\int_0^T e^{-st} F(t) \, dt \longrightarrow L \{ F(t) \}$ as $T \longrightarrow \infty$

$$\therefore \int_0^{\infty} e^{-st} F'(t) \, dt = -F(0) + s L \{ F(t) \}$$

i.e. $\quad L \{ F'(t) \} = s f(s) - F(o)$

The transform of the second derivative F"(t) can be obtained in a similar manner by applying the above theorem to F'(t). Assuming F(t) to be continuous and of exponential order and F"(t) to be sectionally continuous, then

$$L \{ F''(t) \} = s L \{ F'(t) \} - F'(o)$$
$$= s [s f(s) - F(o)] - F'(o)$$
$$= s^2 f(s) - s F(o) - F'(o)$$

In general suppose that F(t) together with its derivatives

$F'(t), \ F''(t), \ ..., \ F^{(n-1)}(t)$ be of order e^{at} as $t \longrightarrow \infty$

and that F(t) has a continuous derivative $F^{(n-1)}(t)$ and a sectionally continuous derivative $F^{(n)}t$ in every finite interval $0 \leqslant t \leqslant T$, then

$$L \{ F^{(n)}(t) \} = s^n f(s) - s^{n-1} F(0) - s^{n-2} F'(0) - ... - F^{(n-1)}(0)$$

Example 1.   Find L { sinh kt}

Solution:

Put F(t) = sinh kt, then F'(t) = k cosh kt, F"(t) = k^2 sinh kt
$\qquad$ F(o) = o, F'(o) = k

$$L \{ F''(t) \} = s^2 f(s) - s F(o) - F'(o)$$

119

$\therefore$ $L\left\{k^2\sinh kt\right\}=s^2 f(s)-k$

i.e $k^2 L\left\{\sinh kt\right\}=s^2 L\left\{\sinh kt\right\}-k$

$\therefore$ $L\left\{\sinh kt\right\}=\dfrac{k}{s^2-k^2}$

Example 2. Find $L\left\{t^n\right\}$ where n is a positive integer.

Solution:

Put $F(t)=t^n$, then $F'(t)=n\,t^{n-1}$, $F''(t)=n(n-1)t^{n-2}$...

$F^{(n)}(t)=n!$

$\therefore$ $F(o)=F'(o)=F''(o)=..=F^{(n-1)}(o)=o$

$L\left\{F^{(n)}(t)\right\}=s^n\,f(s)-s^{n-1}F(o)-s^{n-2}F'(o)...-F^{(n-1)}(o)$

$\therefore L\left\{n!\right\}=s^n L\left\{t^n\right\}$

$\therefore$ $\dfrac{n!}{s}=s^n L\left\{t^n\right\}$ i.e $L\left\{t^n\right\}=\dfrac{n!}{s^{n+1}}$

Example 3. For the function

$F(t)=t+1 \quad 0\leqslant t\leqslant 2$
$\quad\quad=3 \quad\quad\quad t>2$

draw the graphs of F(t) and F'(t) and find $L\left\{F'(t)\right\}$

Solution:

$L\left\{F'(t)\right\}=\dfrac{1}{s}\left(1-e^{-2s}\right),s>0$

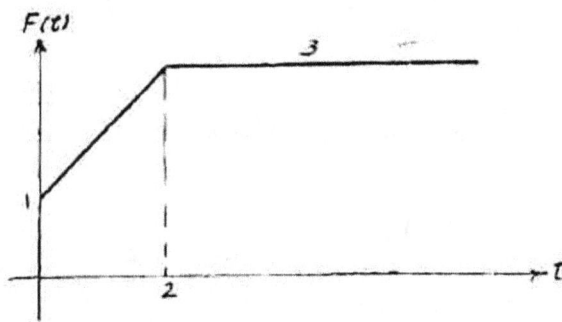

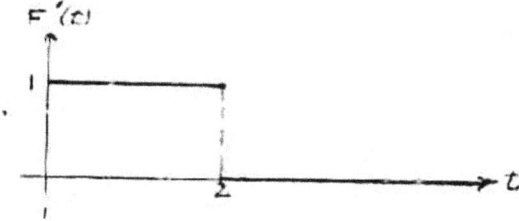

11.1.11. Transforms of integrals.

If F(t) be sectionally continuous and of exponential order then

$$L\left\{\int_0^t F(\tau)\,d\tau\right\} = \frac{1}{s}\,f(s) \text{ where } f(s) = L\left\{F(t)\right\}$$

Put

$$G(t) = \int_0^t E(\tau)\,d\tau$$

then G(t) is continuous and of exponential order and G(o) = o. Since G'(t) = F(t) then transforming both sides we get

$$s\,L\left\{G(t)\right\} - G(o) = f(s)$$

$$\therefore L\left\{G(t)\right\} = \frac{1}{s}\,f(s)$$

$$\text{i.e } L\left\{\int_0^t F(\tau)\,d\tau\right\} = \frac{1}{s}\,f(s)$$

Hence we have the following result:

The division of the transform of a function by s corresponds to the integration of that function between o and t. The rule can be extended to a repeated division by s.

Thus the division of the transform of F(t) by s^n corresponds to a repeated integration of F(t) n times from o to t.

Example 4.   We know that

$$L\left\{\sin kt\right\} = \frac{k}{s^2 + k^2}, \; s > 0$$

$$\therefore L\left\{\int_0^t \sin k\tau\,d\tau\right\} = \frac{k}{s\,(s^2+k^2)}$$

121

$$\therefore \quad L \left\{ \left[- \frac{\cos k\tau}{k} \right]_0^t \right\} = \frac{k}{s(s^2 + k^2)}$$

$$\text{i.e } L \left\{ \frac{1}{k_2} (1 - \cos kt) \right\} = \frac{1}{s(s^2 + k^2)}$$

By integrating again we get:

$$L \left\{ \frac{1}{k^3} (kt - \sin kt) \right\} = \frac{1}{s^2(s^2 + k^2)}$$

Hence we have the following two results:

$$L^{-1} \left\{ \frac{1}{s(s^2 + k^2)} \right\} = \frac{1}{k^3} (1 - \cos kt)$$

$$L^{-1} \left\{ \frac{1}{s^2(s^2 + k^2)} \right\} = \frac{1}{k^3} (kt - \sin kt)$$

11.1.12 The first shift theorem of multiplying the object function by e^{at}

Let F(t) be a sectionally continuous function of order e^{at}. Then

$$\text{Then } L \left\{ F(t) \right\} = \int_0^{\infty} e^{-st} F(t) \, dt = f(s), \; s > a$$

$$\therefore \quad L \left\{ e^{at} F(t) \right\} = \int_0^{\infty} e^{-st} \cdot e^{at} F(t) \, dt$$

$$= \int_0^{\infty} e^{-(s-a)t} F(t) \, dt$$

$$= f(s-a), \; s - a > a \text{ i.e } s > a + a$$

Hence we have the following theorem on substitution:

The change of the variable s in the transform f(s) of F(t) into s - a corresponds to the multiplication of F(t) by e^{at}

Since f(s-a) is a shift of f(s) to the right a distance a, the above theorem is known as the first shift theorem.

$$\text{e.g. } L \left\{ e^{at} \sin kt \right\} = \frac{k}{(s-a)^2 + k^2}, \; s > a,$$

$$L \left\{ e^{-at} \cos kt \right\} = \frac{s + a}{(s+a)^2 + k^2}, \; s > -a,$$

$$L^{-1}\left\{\frac{1}{(s-a)^n}\right\} = e^{at}\frac{t^{n-1}}{(n-1)!}$$

$$L\left\{e^{3t}\cosh 4t\right\} = \frac{s-3}{(s-2)^2-16} = \frac{s-3}{s^2-6s-7}$$

Example 5. Prove that

$$L\left\{e^{at}\ t^{-\frac{1}{2}}\ (1+2at)\right\} = \sqrt{\pi}\ \frac{s}{(s-a)^{3/2}}$$

Solution:

$$L\left\{e^{at}\ t^{-1/2}\ (1+2at)\right\} = L\left\{e^{at}(t^{-1/2}+2at^{1/2})\right\}$$

$$= L\left\{e^{at}\ t^{-1/2}\right\} + 2aL\left\{e^{at}\ t^{1/2}\right\}$$

Since $L\left\{t^k\right\} = \frac{\Gamma(k+1)}{s^{k+1}}$, $k > -1$

then $L\left\{t^{-1/2}\right\} = \sqrt{\frac{\pi}{s}}$,

$$L\left\{t^{1/2}\right\} = \frac{\Gamma(3/2)}{s^{3/2}} = \frac{\frac{1}{2}\Gamma(1/2)}{s^{3/2}} = \frac{\sqrt{\pi}}{2\ s^{3/2}}$$

$$\therefore L\left\{e^{at}\ t^{-\frac{1}{2}}\right\} = \frac{\sqrt{\pi}}{(s-a)^{1/2}}$$

$$L\left\{e^{at}\ t^{\frac{1}{2}}\right\} = \frac{\sqrt{\pi}}{2\ (s-a)^{3/2}}$$

$$\therefore L\left\{e^{at}\ t^{-\frac{1}{2}}\ (1+2at)\right\} = \frac{\sqrt{\pi}}{(s-a)^{\frac{1}{2}}} + \frac{2a\ \sqrt{\pi}}{2\ (s-a)^{3/2}}$$

$$= \frac{\sqrt{\pi}\ s}{(s-a)^{3/2}}$$

11.1.13. Two useful Transforms

(i) We know that

$$L\left\{\cos kt\right\} = \int_0^\infty e^{-st}\cos kt\ dt = \frac{s}{s^2+k^2}\ ,\ s>0$$

Differentiate w.r.t. k

$$\frac{\partial}{\partial k} \int_0^\infty e^{-st} \cos kt \; dt = \int_0^\infty \frac{\partial}{\partial k} (e^{-st} \cos kt) \; dt$$

$$= -\int_0^\infty e^{-st} \; t \sin kt \; dt = \frac{-2ks}{(s^2+k^2)^2}$$

$$\therefore L\left\{ t \sin kt \right\} = \frac{2ks}{(s^2+k^2)^2} \quad , \; s > 0$$

$$\text{or } L^{-1}\left\{ \frac{s}{(s^2+k^2)^2} \right\} = \frac{1}{2k} \; t \sin kt$$

(ii) Dividing the transform by s which corresponds to the integration' of the object function between o and t we get

$$L^{-1}\left\{ \frac{1}{(s^2+k^2)^2} \right\} = \frac{1}{2k} \int_0^t \tau \sin k\tau \; d\tau$$

$$= -\frac{1}{2k^2} \int_0^t \tau \; d \cos k\tau$$

$$= -\frac{1}{2k^2} \left[\tau \cos k\tau - \frac{\sin k\tau}{k} \right]_0^t$$

$$= \frac{1}{2k^3} \left(\sin kt - kt \cos kt \right)$$

11.1.14. The multiplication of the variable by a positive constant

Suppose that $L\{ F(t) \} = f(s)$, for $s > a$, and let a be a positive constant, then

$$L\left\{ F(at) \right\} = \int_0^\infty e^{-st} F(at) \; dt. \text{ Put } at = \tau, \text{ then}$$

$$L\left\{ F(at) \right\} = \frac{1}{a} \int_0^\infty e^{-\frac{s\tau}{a}} F(\tau) \; d\tau = \frac{1}{a} f\left(\frac{s}{a}\right), \; \frac{s}{a} > a$$

$$\therefore L\left\{ F(at) \right\} = \frac{1}{a} f\left(\frac{s}{a}\right) \text{ where } s > a a \; , \; a > 0.$$

This formula can also be written in the form

$$L^{-1}\left\{ f(cs) \right\} = \frac{1}{c} F\left(\frac{t}{c}\right), \; c > 0.$$

e.g $L\left\{ \sin t \right\} = \frac{1}{s^2+1}$

124

$$\therefore \quad L\left\{\sin kt\right\} = \frac{1}{k} \cdot \frac{1}{\dfrac{s^2}{k^2} + 1} = \frac{k}{s^2 + k^2} \ , s > 0.$$

11.1.15. Determination of the inverse transforms by the aid of partial fractions.

A proper fraction $\dfrac{p(s)}{q(s)}$ in which the degree of the numerator is less than that of the denominator of the form

$$\frac{p(s)}{(s-a)\,(s-b)^r\,(s^2+cs+d)\,(s^2+ls+m)^n}$$

can be resolved in a series of partial fractious in the form

$$\frac{p(s)}{q(s)} = \frac{A}{s-a} + \left[\frac{B_1}{(s-b)^r} + \frac{B_2}{(s-b)^{r-1}} + \cdots + \frac{Br}{s-b} \right]$$

$$+ \frac{Cs + D}{s^2+cs+d} + \left[\frac{E_1s+F_1}{(s^2+ls+m)^n} + \frac{E_2s + F_2}{(s^2+ls+m)^{n-1}} \right.$$

$$\left. + \cdots + \frac{E_n s + F_n}{s^2 + ls + m} \right]$$

On clearing fractions on both sides we obtain an identity from which the unknown constants can be obtained either by substituting zeros of the denominator in the identity or equating equal powers on both sides.

The following are examples on finding the inverse transform by resolving the transform into its partial fractions.

Example 1.

Find $L^{-1}\left\{ \dfrac{s^2+3}{(s+1)\,(s-2)\,(s-4)} \right\}$

Solution:

$$\frac{s^2+3}{(s+1)\,(s-2)\,(s-4)} = \frac{4}{15\,(s+1)} - \frac{7}{6\,(s-2)} + \frac{19}{10\,(s-4)}$$

$$\therefore \quad L^{-1}\left\{ \frac{s^2+3}{(s+1)\,(s-2)\,(s-4)} \right\} = \frac{4}{15} e^{-t} - \frac{7}{6} e^{2t} + \frac{19}{10} e^{4t}$$

Example 2.

Find $L^{-1}\left\{ \dfrac{s+2}{s^3\,(s-1)^2} \right\}$

Solution:

$$\frac{s+2}{s^3(s-1)^2} = \frac{2}{s^3} + \frac{5}{s^2} + \frac{8}{s} + \frac{3}{(s-1)^2} - \frac{8}{s-1}$$

$$\therefore \ L^{-1}\left\{\frac{s+2}{s^3(s-1)^2}\right\} = t^3 + 5t + 8 + e^t(3t-8)$$

Example 3. Find $L^{-1}\left\{\dfrac{s^2+1}{(s-1)(s-2)^2}\right\}$

Solution:

Let $\dfrac{s^2+1}{(s-1)(s-2)^2} = \dfrac{A}{s-1} + \dfrac{B}{(s-2)^2} + \dfrac{C}{s-2} \quad \cdots \ (1)$

Multiply both sides by s-1 and let $s \to 1$

$$\lim_{s \to 1} \frac{s^2+1}{(s-2)^2} = A \ \therefore \ A = 2$$

Multiply both sides of (1) by $(s-2)^2$ and let $s \to 2$

$$\lim_{s \to 2} \frac{s^2+1}{s-1} = B \ \therefore \ B = 5$$

To find C multiply both sides of (1) by s and let $s \to \infty$

$$\lim_{s \to \infty} \frac{s^3+s}{(s-1)(s-2)^2} = \lim_{s \to \infty} \frac{As}{s-1} + \lim_{s \to \infty} \frac{Bs}{(s-2)^2}$$

$$+ \lim_{s \to \infty} \frac{Cs}{s-2}$$

$$\therefore \ 1 = A+C = 2+C \ \therefore \ C = -1$$

$$\therefore \ \frac{s^2+1}{(s-1)(s-2)^2} = \frac{2}{s-1} + \frac{5}{(s-2)^2} - \frac{1}{s-2}$$

$$\therefore \ L^{-1}\left\{\frac{s^2+1}{(s-1)(s-2)^2}\right\} = 2e^t + 5t\,e^{2t} - e^{2t}$$

Example 4. Find $L^{-1}\left\{\dfrac{3s^2+9s+16}{(s+1)(s^2+4s+13)}\right\}$

Solution:

$$\frac{3s^2+9s+16}{(s+1)(s^2+4s+13)} = \frac{1}{s+1} + \frac{2s+3}{s^2+4s+13} = \frac{1}{s+1} + \frac{2(s+2)-1}{(s+2)^2+9}$$

126

$$= \frac{1}{s+1} + \frac{2(s+2)}{(s+2)^2 + 9} - \frac{1}{(s+2)^2+9}$$

$$\therefore \ L^{-1} \left\{ \frac{3s^2 + 9s + 16}{(s+1)(s^2+4s+13)} \right\} = e^{-t} + 2e^{-2t} \cos 3t$$

$$- \tfrac{1}{3} e^{-2t} \sin 3t$$

Example 5.   Find $L^{-1} \left\{ \dfrac{2s + 5}{(s^2 + 6s + 25)^2} \right\}$

Solution:

$$\frac{2s + 5}{(s^2 + 6s + 25)^2} = \frac{2(s+3) - 1}{[(s+3)^2 + 16]^2}$$

$$= \frac{2(s + 3)}{[(s+3)^2 + 16]^2} - \frac{1}{[(s+3)^2 + 16]^2}$$

Now $L^{-1} \left\{ \dfrac{s}{(s^2 + k^2)^2} \right\} = \dfrac{1}{2k} t \sin kt$

and $L^{-1} \left\{ \dfrac{1}{(s^2 + k^2)^2} \right\} = \dfrac{1}{2k^3} (\sin kt - kt \cos kt)$

$$\therefore \ L^{-1} \left\{ \frac{2s + 5}{(s^2+6s+25)^2} \right\}$$

$$= \frac{2}{8} e^{-3t} t \sin 4t - \frac{e^{-3t}}{128} (\sin 4t - 4t \cos 4t)$$

$$= \frac{e^{-3t}}{128} (32 t \sin 4t - \sin 4t + 4t \cos 4t)$$

Example 6.   Find $L^{-1} \left\{ \dfrac{1}{(s^2 + a^2)(s^2 + b^2)} \right\}$, $a^2 \neq b^2$

Solution:

$$\frac{1}{(s^2+a^2)(s^2+b^2)} = \frac{1}{b^2-a^2} \left(\frac{1}{s^2+a^2} - \frac{1}{s^2+b^2} \right)$$

$$\therefore \ L^{-1} \left\{ \frac{1}{(s^2+a^2)(s^2+b^2)} \right\} = \frac{1}{b^2-a^2} \left(\frac{1}{a} \sin at - \frac{1}{b} \sin bt \right)$$

Example 7. $L^{-1} \left\{ \dfrac{s}{(s^2+a^2)(s^2+b^2)} \right\} = L^{-1} \left\{ \dfrac{1}{b^2-a^2} \left(\dfrac{s}{s^2 + a^2} - \dfrac{s}{s^2+b^2} \right) \right\}$

$$= \frac{1}{b^2-a^2} (\cos at - \cos bt)$$

Example 8.

$$L^{-1}\left\{\frac{s^3}{(s^2 + a^2)(s^2 + b^2)}\right\} = L^{-1}\left\{\frac{1}{b^2 - a^2}\left(\frac{s^2}{s^2 + a^2} - \frac{s^2}{s^2 + b^2}\right)\right\}$$

$$= L^{-1}\left\{\frac{1}{b^2 - a^2}\left(1 - \frac{a^2}{s^2 + a^2} - 1 + \frac{b^2}{s^2 + b^2}\right)\right\}$$

$$= \frac{1}{b^2 - a^2}(b \sin bt - a \sin at)$$

11.1.16. Laplace's solution of linear differential equations with constant coefficients.

Example 1.   Solve the equation $Y'''(t) - 6 Y''(t) + 11Y'(t) - 6Y(t) = 1$, given that $Y(o) = Y'(0) = Y''(0) = 0$.

Solution:

Transforming the equation i.e., multiplying both sides by e^{-st} and integrating w.r.t. t between o and ∞ p we get :

$$s^3\, y(s) - 6\, s^2\, y(s) + 11\, s\, y(s) - 6\, y(s) = \frac{1}{s}$$

$$\therefore\ y(s) = \frac{1}{s(s^3 - 6s^2 + 11s - 6)}$$

$$= \frac{1}{s(s-1)(s-2)(s-3)}$$

$$= -\frac{1}{6s} + \frac{1}{2(s-1)} - \frac{1}{2(s-2)} + \frac{1}{6(s-3)}$$

Performing the inverse transformation we get

$$Y(t) = -\frac{1}{6} + \frac{1}{2}e^t - \frac{1}{2}e^{2t} + \frac{1}{6}e^{3t}$$

Example 2.   Solve the equation $Y''(t) - 3 Y'(t) + 2 Y(t) = e^t$, given that $Y(0) = Y'(0) = 0$

Solution:

$$(s^2 - 3s + 2)\, y(s) = \frac{1}{s-1}$$

$$\therefore\ y(s) = \frac{1}{(s-1)^2(s-2)}$$

$$= -\frac{1}{(s-1)^2} - \frac{1}{s-1} + \frac{1}{s-2}$$

$$\therefore\ Y(t) = -t\, e^t - e^t + e^{2t}$$

Example 3.   Solve the equation $X''(t) + 2 X'(t) + X(t) = 3\, t\, e^{-t}$
given that $X(0) = 4,\ X'(0) = 2$

Solution:

128

$$s^2 x(s) - s X(o) - X'(o) + 2[s\,x(s) - X(o)] + x(s) = \frac{3}{(s+1)^2}$$

$$\therefore\quad s^2 x(s) - 4s - 2 + 2s\,x(s) - 8 + x(s) = \frac{3}{(s+1)^2}$$

$$(s^2 + 2s + 1)\,x(s) = 4s + 10 + \frac{3}{(s+1)^2}$$

$$\therefore\quad x(s) = \frac{4s}{(s+1)^2} + \frac{10}{(s+1)^2} + \frac{3}{(s+1)^4}$$

$$= \frac{4(s+1-1)}{(s+1)^2} + \frac{10}{(s+1)^2} + \frac{3}{(s+1)^4}$$

$$= \frac{4}{s+1} + \frac{6}{(s+1)^2} + \frac{3}{(s+1)^4}$$

$$\therefore\quad X(t) = 4e^{-t} + 6te^{-t} + 3e^{-t} \cdot \frac{t^3}{3!}$$

$$= e^{-t}\left(4 + 6t + \frac{1}{2}t^3\right)$$

Example 4. Solve the equation $Y''(t) + 6 Y'(t) + 9 Y(t) = 6t^2 e^{-3t}$ given that $Y(0) = Y'(0) = 0$

Solution:

$$(s^2 + 6s + 9)\,y(s) = \frac{6 \times 2}{(s+3)^3}$$

$$\therefore\quad y(s) = \frac{12}{(s+3)^5}$$

$$\therefore\quad Y(t) = 12\,e^{-3t}\frac{t^4}{4!} = \tfrac{1}{2}\,e^{-3t}\,t^4$$

Example 5 Solve the equation $Y''(t) + 2 Y'(t) + Y(t) = t$ given that $Y(0) = 3,\ Y(1) = -1$

Solution:

$$s^2 y(s) - s Y(0) - Y'(0) + 2[sy(s) - Y(0)] + y(s) = \frac{1}{s^2}$$

Put $Y'(0) = B$

$$\therefore\quad s^2 y(s) + 3s - B + 2sy(s) + 6 + y(s) = \frac{1}{s^2}$$

$$(s^2 + 2s + 1)\,y(s) + 3s + 6 - B = \frac{1}{s^2}$$

$$y(s) = \frac{-3s - 6 + B}{(s+1)^2} + \frac{1}{s^2(s+1)^2}$$

$$= \frac{-3(s+1) - S + B}{(s+1)^2} + \frac{1}{s^2 (s+1)^2}$$

$$= -\frac{3}{s+1} + \frac{B-3}{(s+1)^2} + \frac{1}{s^2 (s+1)^2}$$

$$= -\frac{3}{s+1} + \frac{B-3}{(s+1)^2} - \frac{2}{s} + \frac{1}{s^2} + \frac{2}{s+1}$$

$$+ \frac{1}{(s+1)^2}$$

$$= -\frac{1}{s+1} + \frac{B-2}{(s+1)^2} - \frac{2}{s} + \frac{1}{s^2}$$

$$\therefore \quad Y(t) = -e^{-t} + (B-2) \, te^{-t} - 2 + t$$

Since $Y(1) = 1$, then

$$-1 = -e^{-1} + (B-2) \, e^{-1} - 2 + 1$$

$$= (B-3) \, e^{-1} - 1$$

$$\therefore \quad B - 3 = 0 \quad \text{i.e } B = 3$$

$$\therefore \quad Y(t) = (t-1) \, e^{-t} - 2 + t$$

Example 6   Solve the equation $Y'''(t) + Y(t) = 1$
given that $Y(0) = Y'(0) = Y''(0) = 0$

Solution:

$$(s^3 + 1) \, y(s) = \frac{1}{s}$$

$$\therefore \quad y(s) = \frac{1}{s(s^3+1)} = \frac{1}{s(s+1)(s^2-s+1)}$$

$$= \frac{1}{s} - \frac{1}{3(s+1)} - \frac{1}{3} \frac{2s-1}{s^2-s+1}$$

$$= \frac{1}{s} - \frac{1}{3(s+1)} - \frac{2}{3} \frac{s - \frac{1}{2}}{(s-\frac{1}{2})^2 + \frac{3}{4}}$$

$$\therefore \quad Y(t) = 1 - \frac{1}{3} e^{-t} - \frac{2}{3} e^{\frac{1}{2}t} \cos \frac{\sqrt{3}}{2} t$$

Example 7   Solve the equation $Y''(t) + k^2 Y(t) = 0$

Solution:
Put $Y(0) = A$, $Y'(0) = B$

$$s^2 y(s) - sY(0) - Y'(0) + k^2 y(s) = 0$$

$$(s^2 + k^2) \, y(s) = As + B$$

$$y(s) = \frac{As}{s^2 + k^2} + \frac{B}{s^2 + k^2}$$

130

$$\left(\therefore\; Y(t) = A\cos kt + \frac{B}{k}\sin kt\right.$$

This is an example on simple harmonic oscillation, A and B being respectively the initial displacement and velocity.

Example 8
Solve the equation X"(t) + 4 X(t) = 10 sin 3t
given that X(0) = X'(0) = 0

Solution:

$$(s^2+4)\; x(s) = 10 \times \frac{3}{s^2+9}$$

$$\therefore\; x(s) = \frac{30}{(s^2+4)(s^2+9)} = 6\left(\frac{1}{s^2+4} - \frac{1}{s^2+9}\right)$$

$$\therefore\; X(t) = 3\sin 2t - 2\sin 3t$$

This is an example on forced oscillation without damping.

Example 9 Solve the equation Y"(t) + n² Y(t) = a sin nt
given that Y(0) = Y'(0) = 0

Solution:

$$(s^2 + n^2)\; y(s) = \frac{an}{s^2+n^2}$$

$$\therefore\; y(s) = \frac{an}{(s^2 + n^2)^2}$$

$$\text{Now } L^{-1}\left\{\frac{1}{(s^2 + k^2)^2}\right\} = \frac{1}{2k^3}(\sin kt - kt\cos kt)$$

$$\therefore\; Y(t) = \frac{an}{2n^3}(\sin nt - nt\cos nt) = \frac{a}{2n^2}(\sin nt - nt\cos nt)$$

This is an example on resonance when the frequency of the impressed oscillations is equal to the natural frequency.

Example 10 Solve the equation Y"(t) + 4Y'(t) + 13Y(t) = 5 cos 3t
given that Y(0) = ¼, Y'(0) = 2

Solution:

$$[\, s^2 y(s) - sY(o) - Y'(o)\,] + 4\,[\, sy(s) - Y(o)\,] + 13y(s)$$
$$= \frac{5s}{s^2 + 9}$$

$$s^2 y(s) - \frac{1}{4}s - 2 + 4\,[\, sy(s) - \frac{1}{4}\,] + 13y(s) = \frac{5s}{s^2 + 9}$$

$$(s^2 + 4s + 13) \ y(s) - \frac{1}{4} s - 3 = \frac{5s}{s^2 + 9}$$

$$y(s) = \frac{s + 12}{4 \ (s^2 + 4s + 13)} + \frac{5s}{(s^2 + 9) \ (s^2 + 4s + 13)}$$

$$= \frac{s^3 + 12s^2 + 2^?s + 108}{4 \ (s^2 + 9) \ (s^2 + 4s + 13)}$$

$$= \frac{s + 9}{8(s^2 + 9)} + \frac{s + 11}{8(s^2 + 4s + 13)}$$

$$= \frac{1}{8} \frac{s}{s^2 + 9} + \frac{9}{8} \frac{1}{s^2 + 9} + \frac{1}{8} \frac{(s+2) + 9}{(s+2)^2 + 9}$$

$$= \frac{1}{8} \frac{s}{s^2 + 9} + \frac{9}{8} \frac{1}{s^2 + 9} + \frac{1}{8} \frac{s + 2}{(s+2)^2 + 9}$$

$$+ \frac{9}{8} \frac{1}{(s+2)^2 + 9}$$

$$\therefore \ Y(t) = \frac{1}{8} \cos 3t + \frac{3}{8} \sin 3t + \frac{1}{8}e^{-2t} \cos 3t + \frac{3}{8} e^{-2t} \sin 3t$$

$$= \frac{1}{8} \ (1 + e^{-2t}) \ (\cos 3t + 3 \sin 3t)$$

This is an example on forced oscillations with damping.

Example 11.   Solve the simultaneous differential equations

$$\dot{x} - x + 2y = o$$
$$\dot{y} - 5x - 3y = o$$

given that $x(0) = y(0) = 1$

Solution:

$$s\bar{x} - 1 - \bar{x} + 2\bar{y} = o$$
$$s\bar{y} - 1 - 5\bar{x} - 3\bar{y} = o$$
$$\therefore \quad (s - 1) \ \bar{x} + 2\bar{y} = 1$$
$$- 5\bar{x} + (s-3) \ \bar{y} = 1$$

Eliminating $\bar{y}$ we get

$$[\,(s-1)\,(s-3)\,+\,10\,]\,\bar{x} = s - 3 - 2$$

$$(s^2 - 4s + 13)\,\bar{x} = s - 5$$

$$\bar{x} = \frac{s-5}{s^2 - 4s + 13} = \frac{(s-2)-3}{(s-2)^2+9}$$

$$= \frac{s-2}{(s-2)^2+9} - \frac{3}{(s-2)^2+9}$$

$$\therefore \quad x = e^{2t}\cos 3t - \frac{3}{3}e^{2t}\sin 3t$$

i.e. $x = e^{2t}(\cos 3t - \sin 3t)$

Again eliminating $\bar{x}$ we get

$$[\,10 + (s-1)\,(s-3)\,]\,\bar{y} = 5 + s - 1$$

$$(s^2 - 4s + 13)\,\bar{y} = s + 4$$

$$\bar{y} = \frac{s+4}{s^2 - 4s + 13} = \frac{s-2+6}{(s-2)^2+9} = \frac{s-2}{(s-2)^2+9} + \frac{6}{(s-2)^2+9}$$

$$\therefore \quad y = e^{2t}(\cos 3t + 2\sin 3t)$$

11.1.17. Exercises on Laplace transformation

Find the Laplace transform of each of the following functions

(1) $3e^{5t}$ 　　　　　　　　Ans. $\dfrac{3}{s-5}$, $s > 5$

(2) $4e^{-3t}$ 　　　　　　　　Ans. $\dfrac{4}{s+3}$, $s > -3$

(3) $4t - 5$ 　　　　Ans. $\dfrac{4}{s^2} - \dfrac{5}{s}$, $s > 0$

(4) $2\cos 6t$ 　　　　Ans. $\dfrac{2s}{s^2+36}$, $s > 0$

(5) $6\sin 2t - 5\cos 2t$ 　　Ans. $\dfrac{12-5s}{s^2+4}$, $s > 0$

(6) $t^2 - 3t + 5$ 　　　　Ans. $\dfrac{2}{s^3} - \dfrac{3}{s^2} + \dfrac{5}{s}$, $s > 0$

(7) $\cos^2 kt$ Ans. $\dfrac{s^2 + 2k^2}{s(s^2 + 4k^2)}$, $s > o$

(8) $(\sin t - \cos t)^2$ Ans. $\dfrac{s^2 - 2s + 4}{s(s^2 + 4)}$, $s > o$

(9) $\cosh^2 4t$ Ans. $\dfrac{s^2 - 32}{s(s^2 - 64)}$, $s > |8|$

(10) $\sin t + 3 \cos t$ Ans. $\dfrac{1 + 3s}{s^2 + 1}$, $s > o$

(11) $F(t) = 4$ $o < t < 1$ [Ans. $\dfrac{1}{s}(4 - e^{-s})$, $s > o$]

         $= 3$ $t > 1$

(12) $\Phi(t) = \sin 2t$ $o < t < \pi$ [Ans. $\dfrac{2(1 - e^{-\pi s})}{s^2 + 4}$, $s > o$]

         $= o$ $t > \pi$

(13) $t^{5/2}$ [Ans. $\dfrac{15}{8s^3}\left(\dfrac{\pi}{s}\right)^{1/2}$, $s > o$]

(14) $t^3 e^{-2t}$ $\left[\text{Ans. } \dfrac{6}{(s+2)^4}\right]$

(15) $e^{-t} \cos 4t$ $\left[\text{Ans. } \dfrac{s + 1}{s^2 + 2s + 17}\right]$

(16) $e^{2t} \sin t$ $\left[\text{Ans. } \dfrac{1}{s^2 - 4s + 5}\right]$

(17) $e^{-4t} \cosh 2t$ $\left[\text{Ans. } \dfrac{s + 4}{s^2 + 8s + 12}\right]$

<u>Find the inverse transform of:</u>

(18) $\dfrac{15}{s^2+4s+13}$ [Ans. $5e^{-2t} \sin 3t$]

(19) $\dfrac{s+1}{s^2+6s+25}$ [Ans. $e^{-3t}(\cos 4t - \tfrac{1}{2}\sin 4t)$]

(20) $\dfrac{s}{s^2-8s+16}$ [Ans. $e^{4t}(1+4t)$]

(21) $\dfrac{1}{s^2+2s+5}$ [Ans. $\tfrac{1}{2}e^{-t}\sin 2t$]

(22) $\dfrac{s}{s^2-6s+13}$ [Ans. $e^{3t}(\cos 2t + \dfrac{3}{2}\sin 2t)$]

(23) $\dfrac{1}{s^2+8s+16}$ [Ans. te^{-4t}]

(24) $\dfrac{s-5}{s^2+6s+13}$ [Ans. $e^{-3t}(\cos 2t - 4\sin 2t)$]

(25) $\dfrac{3s+1}{(s+1)^4}$ $\left[\text{Ans. } e^{-t}\left(\dfrac{3}{2}t^2 - \dfrac{1}{3}t^3\right)\right]$

(26) $\dfrac{s^2}{(s+2)^3}$ [Ans. $e^{-2t}(1 - 4t + 2t^2)$]

(27) $\dfrac{s+2}{s^2-6s+8}$ [Ans. $3e^{4t} - 2e^{2t}$]

(28) $\dfrac{2s^2+5s-4}{s^3+s^2-2s}$ [Ans. $2 + e^t - e^{-2t}$]

(29) $\dfrac{2s^2+1}{s(s+1)^2}$ [Ans. $1 + e^{-t} - 3te^{-t}$]

(30) $\dfrac{1}{s^3(s^2+1)}$ [Ans. $\tfrac{1}{2}t^2 - 1 + \cos t$]

(31) $\dfrac{5s - 2}{s^2 (s+2)(s-1)}$ [Ans. $t - 2 + e^t + e^{-2t}$]

(32) $\dfrac{s^3 + s - 4}{(s^2-2s+2)(s^2+2s-3)}$ [Ans. $e^t (\cos t + \sin t) - e^{-t} \sinh 2t$]

(33) $\dfrac{s}{(s^2+1)(s^2+3)}$ [Ans. $\frac{1}{2} (\cos t - \cos \sqrt{3}\, t)$]

(34) $\dfrac{5s + 3}{(s-1)(s^2+2s+5)}$ [Ans. $e^t - e^{-t}(\cos 2t - \frac{3}{2} \sin 2t)$]

(35) $\dfrac{s + 1}{(s^2 + 2s + 2)^2}$ [Ans. $\frac{1}{2} t\, e^{-t} \sin t$]

(36) $\dfrac{s + 1}{s (s^2 + s - 6)}$ [Ans. $- \dfrac{1}{6} + \dfrac{3}{10} e^{2t} - \dfrac{2}{15} e^{-3t}$]

Solve the following differential equations

(37) $Y'(t) - Y(t) = 4e^t \cos t$, given that $Y(0) = 3$

[Ans. $Y(t) = 3e^t + 4e^t \sin t$]

(38) $Y''(t) - 5Y'(t) + 6Y(t) = e^{2t}$, given that $Y(0)=1$, $Y'(0)=0$

[Ans. $Y(t) = (2-t) e^{2t} - e^{3t}$]

(39) $Y''(t) + 2Y'(t) + 5Y(t) = e^{-t} \sin t$, given that $Y(0) = 0$,

$Y'(0) = 1$ [Ans. $Y(t) = \frac{1}{3} e^{-t}(\sin t + \sin 2t)$]

(40) $Y^{(4)}(t) + 4Y'''(t) + 4Y''(t) = 0$, given that $Y(0) = 1$,

$Y'(0) = Y''(0) = Y'''(0) = 0$ [Ans. $Y(t) = 1$]

(41) $Y''(t) + Y(t) = t$ given that $Y(0) = 1$, $Y'(0) = -2$

[Ans. $Y(t) = t + \cos t - 3 \sin t$]

(42) $Y''(t) - 3Y'(t) + 2Y(t) = 4e^{2t}$, given that $Y(0) = -3$,

$Y'(0) = 5$ [Ans. $Y(t) = -7e^t + 4e^{2t} + 4te^{2t}$]

(43) $Y''(t) + 9Y(t) = \cos 2t$, given that $Y(0) = 1$, $Y\left(\dfrac{\pi}{2}\right) = -1$

[Ans. $Y(t) = \dfrac{4}{5} \cos 3t + \dfrac{4}{5} \sin 3t + \dfrac{1}{5} \cos 2t$]

(44) $Y''(t) + 2Y'(t) + Y(t) = \sin 2t$, given that $Y(0) = Y'(0)=0$

[Ans. $Y(t) = \dfrac{4}{25} e^{-t} + \dfrac{2}{5} te^{-t} - \dfrac{4}{25} \cos 2t - \dfrac{3}{25} \sin 2t$]

(45) $Y'(t) + Y(t) = t^2 e^{-t}$ given that $Y(0) = Y_0$

[Ans. $Y(t) = 1/_3 t^3 e^{-t} + Y_0 e^{-t}$]

(46) $x''(t) + 4x'(t) + 4x(t) = 4e^{-2t}$ given that $x(0) = -1$,

$x'(0) = 4$ [Ans. $x(t) = e^{-2t} (2t^2 + 2t - 1)$]

(47) $x''(t) + x(t) = 6 \cos 2t$, given that $x(0) = 3$, $x'(0) = 1$

[Ans. $x(t) = 5 \cos t + \sin t - 2 \cos 2t$]

(48) $Y'''(t) - 3Y''(t) + 3Y'(t) - Y(t) = t^3 e^t$, given that
$Y(0) = 1$, $Y'(0) = 0$, $Y''(0) = -2$

[Ans. $Y(t) = (1 - t - \frac{1}{2} t^2 + 1/_{60} t^5) e^t$]

(49) $y''(x) + 9y(x) = 40e^x$, $y(0) = 5$, $y'(0) = -2$

[Ans. $y(x) = 4e^x + \cos 3x - 2 \sin 3x$]

(50) $x''(t) - 4x'(t) + 4x(t) = e^{2t}$, $x'(0) = 0$, $x(1) = 0$

[Ans. $x(t) = \frac{1}{2} (1-t)^2 e^{2t}$]

(51) $x'''(t) - 3x''(t) + 3x'(t) - x(t) = 16e^{3t}$, given that
$x(0) = 0$, $x'(0) = 4$, $x''(0) = 6$

[Ans. $x(t) = 2e^{3t} - (5t^2 + 2) e^t$]

Solve the following simultaneous equations

(52) $3X'(t) + 2X(t) - Y(t) = t$

$2Y'(t) - X(t) + Y(t) = 5e^{-t}$

given that $X(0) = Y(0) = 0$

[Ans. $X(t) = 6e^{-t/6} - (t+1) e^{-t} + t - 5$

$Y(t) = 9e^{-t/6} + (t-2) e^{-t} + t - 7$]

(53) $(D-2) x + 3y = 0$

$2x + (D-1) y = 0$

given that $x(0) = 8$, $y(0) = 3$

[Ans. $x = 5e^{-t} + 3e^{4t}$, $y = 5 e^{-t} - 2e^{4t}$]

(54) $x''(t) - x(t) + 5y'(t) = t$

   $y''(t) - 4y(t) - 2x'(t) = -2$

   given that $x(0) = x'(0) = y(0) = y'(0) = 0$

[Ans. $x(t) = -t + 5 \sin t - 2 \sin 2t$,

   $y(t) = 1 - 2 \cos t + \cos 2t$]

(55) $5\dot{x} - 2\dot{y} + 4x - y = e^{-t}$

   $\dot{x} + 8x - 3y = 5e^{-t}$

   given that $x = y = 0$ when $t = 0$

[Ans. $x = 2e^{-t} + e^{t} - 3e^{-2t}$

   $y = 3e^{-t} + 3e^{t} - 6e^{-2t}$]

11.2. General Theorems on the Laplace Transformation

11.2.1. The unit step function

This is a function which is equal to zero when $t < a$ and is equal to unity when $t > a$, a being a positive constant. Let this function be denoted by $U(t-a)$, then

$U(t-a) = 0 \qquad t < a$

$\qquad\quad = 1 \qquad t > a$

Let us obtain the Laplace transform of this function,

$$L \left\{ U(t-a) \right\} = \int_0^a e^{-st} \, 0 \, dt + \int_a^\infty e^{-st} \, 1 \, dt = \frac{e^{-as}}{s}, \quad s > 0.$$

If $a = 0$ we have the following special case
$L \{ U(t) \} = 1/s$ where $U(t) = 0$, $t < 0$ and $U(t) = 1$ $t > 0$

Any function $F(t)$ multiplied by $U(t-a)$ will have a value zero for $t < a$ and $F(t)$ for $t > a$

i.e $U(t-a) \, F(t) = 0 \qquad 0 < t < a,$

$\qquad\qquad\qquad = F(t) \qquad t > a.$

11.2.2. The second translation or shifting property

We shall now show that, if a is a positive constant and if

138

$$L^{-1} \left\{ f(s) \right\} = F(t) \quad , \quad \text{then}$$

$$L^{-1} \left\{ e^{-as} f(s) \right\} = 0 , \qquad 0 < t < a$$

$$= F(t-a), \qquad t > a$$

This can also be written according to the in the form

$$L^{-1} \left\{ e^{-as} f(s) \right\} = U(t-a) \, F(t-a)$$

For

$$e^{-as} f(s) = e^{-as} \int_0^\infty e^{-st} \, F(t) \, dt = \int_0^\infty e^{-s(t+a)} \, F(t) \, dt$$

Put $t + a = \tau$, then $dt = d\tau$

When $t = 0$, $\tau = a$ and when $t \rightarrow \infty$, $\tau \rightarrow \infty$

$$\therefore \int_0^\infty e^{-s(t+a)} \, F(t) \, dt = \int_a^\infty e^{-s\tau} \, F(\tau-a) \, d\tau$$

$$= \int_0^a e^{-st} 0 \, dt + \int_a^\infty e^{-st} \, F(t-a) \, dt$$

$$= \int_0^\infty e^{-st} \, U(t-a) \, F(t-a) \, dt = L \left\{ U(t-a) \, F(t-a) \right\}$$

Hence if $\quad L^{-1} \left\{ f(s) \right\} = F(t)$, then

$$L^{-1} \left\{ e^{-as} f(s) \right\} = U(t-a) \, F(t-a)$$

i.e $\qquad L^{-1} \left\{ e^{-as} f(s) \right\} = 0 , \quad 0 < t < a$

$$= F(t-a) , \quad t > a$$

We can now give a simple geometrical interpretation to he function U (t - a) F(t - a). Let the definition of F(t) be extended such that F(t) is equal to zero for negative values of t. Then the graph of U(t —a) F(t--a) is actually the graph obtained by shifting the graph of F(t) to the right a distance a.

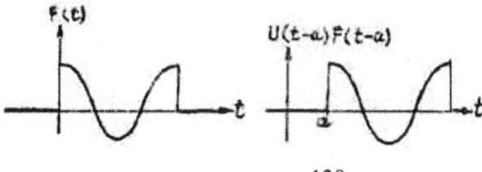

This can also be stated as follows:

If the definition of F(t) be extended such that F(t) is equal to zero for negative values of t, then the effect of shifting the graph of F(t) through a distance a to the right is to multiply the transform f(s) by e^{-as}.

Example 1.

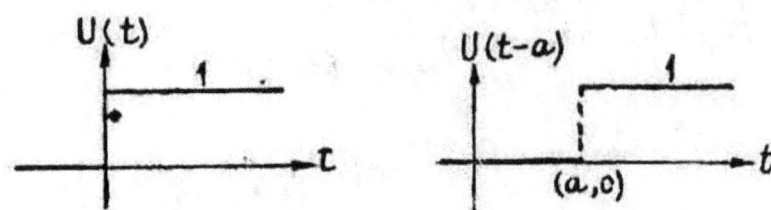

Since the unit step function U(t—a) is actually a shift of U(t) a distance a to the right and since L { U(t) } = 1 / s then

$$L \left\{ U(t-a) \right\} = \frac{e^{-as}}{s}$$

Example 2

Since $L^{-1} \left\{ \dfrac{1}{s^4} \right\} = \dfrac{t^3}{3!}$, then

$$L^{-1} \left\{ \frac{e^{-as}}{s^4} \right\} = U(t-a) \frac{(t-a)^3}{3!}$$

i.e $L^{-1} \left\{ \dfrac{e^{-as}}{s^4} \right\} = 0 \qquad 0 < t < a$

$$= \frac{(t-a)^8}{3!} \qquad t > a$$

Example 3.

Since $L^{-1} \left\{ \dfrac{s}{s^2 + k^2} \right\} = \cos kt$, then

$$L^{-1} \left\{ \frac{e^{-as} s}{s^2 + k^2} \right\} = U(t-a) \cos k(t-a)$$

i.e $\quad L^{-1}\left\{\dfrac{e^{-as}\ s}{s^2 + k^2}\right\} = 0 \qquad o < t < a$

$$= \cos k(t-a) \qquad t > a$$

Example 4

Find

$$L^{-1}\left\{\frac{(s+2)\ e^{-\pi s}}{s^2 + s + 1}\right\}$$

Solution:

$$L^{-1}\left\{\frac{s + 2}{s^2 + s + 1}\right\} = L^{-1}\left\{\frac{(s+1/2) + 3/2}{(s+1/2)^2 + 3/4}\right\}$$

$$= L^{-1}\left\{\frac{s + 1/2}{(s+1/2)^2 + 3/4}\right\} + L^{-1}\left\{\frac{3/2}{(s+1/2)^2 + 3/4}\right\}$$

$$= e^{-\frac{1}{2}t} \cos\frac{\sqrt{3}}{2}t + \sqrt{3}\ e^{-\frac{1}{2}t} \sin\frac{\sqrt{3}}{2}t$$

$$= e^{-\frac{1}{2}t}\left(\cos\frac{\sqrt{3}}{2}t + \sqrt{3} \sin\frac{\sqrt{3}}{2}t\right)$$

$$\therefore\quad L^{-1}\left\{\frac{(s+2)\ e^{-\pi s}}{s^2 + s + 1}\right\} = e^{-\frac{1}{2}(t-\pi)}\left[\cos\frac{\sqrt{3}}{2}(t-\pi)\right.$$

$$\left. + \sqrt{3} \sin\frac{\sqrt{3}}{2}(t - \pi)\right] U(t-\pi)$$

i.e $= 0 \qquad o < t < \pi$

$$= e^{-\frac{1}{2}(t-\pi)}\left[\cos\frac{\sqrt{3}}{2}(t-\pi) + \sqrt{3} \sin\frac{\sqrt{3}}{2}(t - \pi)\right], t > \pi$$

Example 5

If $F(t) = L^{-1}\left\{\dfrac{e^{-3s}}{(s + 1)^3}\right\}$, calculate $F(3/2)$, $F(4)$

$$L^{-1}\left\{\frac{1}{(s + 1)^3}\right\} = \tfrac{1}{2}\ t^2\ e^{-t}$$

$$\therefore\quad F(t) = L^{-1}\left\{\frac{e^{-3s}}{(s + 1)^3}\right\} = \tfrac{1}{2}\ (t-3)^2\ e^{-(t-3)}\ U(t-3)$$

i.e $F(t) = 0 \qquad o < t < 3$

$$F(t) = \tfrac{1}{2}\ (t-3)^2\ e^{-(t-3)} \qquad , t > 3$$

$$\therefore\quad F(3/2) = 0, \quad F(4) = \tfrac{1}{2}\ (4-3)^2\ e^{-(4-3)} = \frac{1}{2e}$$

Example 6. Find and sketch the function F(t) for which,

$$F(t) = L^{-1} \left\{ \frac{3}{s} - \frac{4e^{-s}}{s^2} + \frac{4e^{-3s}}{s^2} \right\}$$

Solution:

Since $L^{-1} \left\{ \frac{1}{s^2} \right\} = t$

then $L^{-t} \left\{ \frac{e^{-s}}{s^2} \right\} = (t-1)\ U(t-1)$

and $L^{-1} \left\{ \frac{e^{-3s}}{s^2} \right\} = (t-3)\ U(t-3)$

$\therefore$ $F(t) = 3 - 4\ (t-1)\ U(t-1) + 4\ (t-3)\ U(t-3)$

Hence for $o < t < 1$, $F(t) = 3$

for $1 < t < 3$, $F(t) = 3 - 4\ (t-1) = 7 - 4t$

and for $t > 3$, $F(t) = 3 - 4(t-1) + 4(t-3) = -5$

Hence the graph of F(t) is as shown in fig. 2-4

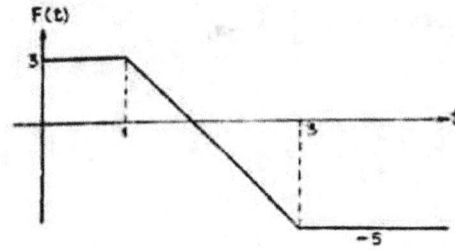

11.2.3. Application of the shift theorem to the solution of difference and differential equations

Example I Solve the first order difference equation

Y(t) = Y(t-h) + 1 at $t > o$

with the condition Y(t) = 0 at. $t < o$

h, being a positive constant.

Solution:

The function Y(t-h) is the same as the translated function U(t-h) Y(t-h). Hence transforming the difference equation using the shifting property we get

$$y(s) = e^{-hs} y(s) + \frac{1}{s}$$

i.e $y(s) = \dfrac{1}{s\ (1-e^{-hs})}$

$$= \frac{1}{s} \left(1 + e^{-hs} + e^{-2hs} + \ldots \right)$$

$$= \frac{1}{s} + \frac{1}{s} e^{-hs} + \frac{1}{s} e^{-2hs} + \ldots$$

$$\therefore \quad Y(t) = U(t) + U(t-h) + U(t-2h) + \ldots$$

i.e $\quad Y(t) = 1 \qquad\qquad o < t < h$

$\qquad\qquad = 1 + 1 = 2 \qquad h < t < 2h$

$\qquad\qquad = 1+1+1 = 3 \qquad 2h < t < 3h$

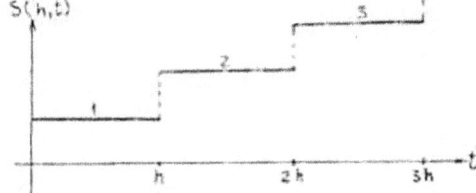

The function $Y(t)$ is known as a <u>staircase function</u> $S(h,t)$ with a positive run h and with a unit rise. Its transform is

$$\frac{1}{s(1-e^{-hs})} = \frac{1}{s} \frac{e^{\frac{hs}{2}}}{e^{\frac{hs}{2}} - e^{-\frac{hs}{2}}}$$

$$= \frac{1}{2s} \frac{\cosh \frac{hs}{2} + \sinh \frac{hs}{2}}{\sinh \frac{hs}{2}}$$

$$= \frac{1}{2s} \left(1 + \coth \frac{hs}{2} \right)$$

Example 2   Find the function $Y(t)$ which satisfies the second order difference equation

$$Y(t) - 4Y(t-h) + 4Y(t-2h) = 1$$

with the condition $Y(t) = 0$ when $t < 0$, h is a positive constant and the right hand side has to be replaced by zero when $t < o$.

Solution:

$$y(s) - 4e^{-hs} y(s) + 4e^{-2hs} y(s) = \frac{1}{s}$$

$$\therefore \quad y(s) [1 - 2e^{-hs}]^2 = \frac{1}{s}$$

143

$$y(s) = \frac{1}{s \,(1-2e^{-hs})^2}$$

$$= \frac{1}{s} \,(1-2e^{-hs})^{-2}$$

$$= \frac{1}{s}\left[1 + 4e^{-hs} + \frac{-2\times-3}{2!}\times 4e^{-2hs} + \cdots \right]$$

$$= \frac{1}{s} + \frac{4}{s}e^{-hs} + \frac{12}{s}e^{-2hs} + \cdots$$

$$\therefore \; Y(t) = U(t) + 4U(t-h) + 12U(t-2h) + \cdots$$

i.e $Y(t) = 1 \qquad\qquad 0 < t < h$

$\qquad\qquad = 1+4=5 \qquad h < t < 2h$

$\qquad\qquad = 1+4+12=17 \quad 2h < t < 3h$

Example 3   Solve the difference — differential equation
$$Y'(t) - 2Y(t-1) = 3$$

with the condition $Y(t) = 0$ when $t \leq 0$ and the constant in the right hand side is to be replaced by zero when $t < 0$.

Solution:

$$sy(s) - 2e^{-s} y(s) = \frac{3}{s}$$

$$(s-2e^{-s}) y(s) = \frac{3}{s}$$

$$\therefore \; y(s) = \frac{3}{s \,(s - 2e^{-s})} = \frac{3}{s^2 \,(1-\dfrac{2e^{-s}}{s})}$$

$$= \frac{3}{s^2}\left(1 + \frac{2e^{-s}}{s} + \frac{4e^{-2s}}{s^2} + \cdots \right)$$

$$= \frac{3}{s^2} + \frac{6e^{-s}}{s^3} + \frac{12e^{-2s}}{s^4} + \cdots$$

$$\therefore \; Y(t) = 3t + \frac{6}{2}(t-1)^2 U(t-1) + \frac{12}{3!}(t-2)^3 U(t-2) + \cdots$$

i.e $Y(t) = 3t \qquad\qquad\qquad 0 < t < 1$

$\qquad\qquad = 3t + 3(t-1)^2 \qquad\quad 1 < t < 2$

$\qquad\qquad = 3t + 3(t-1)^2 + 2(t-2)^3 \quad 2 < t < 3$

Example 4.   Compute $y\left(\dfrac{\pi}{2}\right)$ and $y\left(2+\dfrac{\pi}{2}\right)$

for the function y(x) which satisfies the boundary value problem,

$$y''(x) + y(x) = (x-2)\ U(x-2)\quad y(o) = y'(o) = o$$

$$(x-2)\ U(x-2) = o \qquad\qquad o < x < 2$$

$$= x-2 \qquad\qquad x > 2$$

Transforming the differential equation we get

$$(s^2 + 1)\ \bar{y} = \frac{e^{-2s}}{s^2}$$

$$\therefore\ \bar{y} = \frac{e^{-2s}}{s^2(s^2+1)} = e^{-2s}\left[\frac{1}{s^2} - \frac{1}{s^2+1}\right]$$

$$\therefore\ y(x) = o \qquad\qquad o < x < 2$$

$$= x - 2 - \sin(x-2) \qquad x > 2$$

$$\therefore\ y\left(\frac{x}{2}\right) = o$$

$$y\left(2+\frac{\pi}{2}\right) = 2+\frac{\pi}{2} - 2 - \sin\left(2 + \frac{\pi}{2} - 2\right)$$

$$= \frac{\pi}{2}-1$$

Example 5   Solve the equation

$$x''(t) + 4x(t) = \psi(t),\ x(0) = 1,\ x'(0) = 0\ \text{and where}$$
$$\psi(t)\ \text{is defined by}$$

$$\psi(t) = 4t \qquad 0 \leqslant t \leqslant 1$$

$$= 4 \qquad\quad t > 1$$

We write $\psi(t) = 4t - 4(t-1)\ U(t-1)$

$$\therefore\ L\left\{\psi(t)\right\} = \frac{4}{s^2} - \frac{4}{s^2}e^{-s}$$

$$L\left\{x''(t)\right\} = s^2\bar{x} - s$$

$$\therefore\ s^2\bar{x} - s + 4\bar{x} = \frac{4}{s^2} - \frac{4}{s^2}e^{-s}$$

$$(s^2 + 4)\bar{x} = s + \frac{4}{s^2} - \frac{4}{s^2}e^{-s}$$

$$\therefore\ \bar{x} = \frac{s}{s^2 + 4} + \frac{4}{s^2(s^2 + 4)} - \frac{4e^{-s}}{s^2(s^2+4)}$$

$$= \frac{s}{s^2 + 4} + \frac{1}{s^2} - \frac{1}{s^2 + 4} - \left(\frac{1}{s^2} - \frac{1}{s^2 + 4}\right)e^{-s}$$

$$\therefore\ x(t) = \cos 2t + t - \tfrac{1}{2}\sin 2t - [t - 1$$

$$- \tfrac{1}{2}\sin 2(t-1)]\ U(t-1)$$

145

11.2.4. The unit impulse function

Consider the function F(t) defined by

$$F(t) = \frac{1}{\epsilon} \qquad 0 < t < \epsilon$$

$$= 0 \qquad t > \epsilon$$

The area under the function is unity

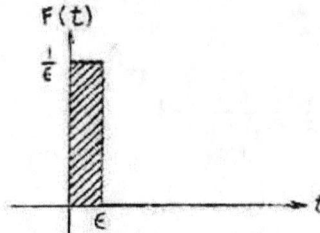

F(t)

Its Laplace transform is

$$\int_0^\epsilon e^{-st} \cdot \frac{1}{\epsilon} \, dt = \frac{1}{\epsilon} \left[\frac{e^{-st}}{-s} \right]_0^\epsilon = \frac{1}{\epsilon s} (1 - e^{-\epsilon s})$$

Now let us take the limiting value of the transform when $\epsilon \longrightarrow 0$

$$\lim_{\epsilon \to 0} \frac{1 - e^{-\epsilon s}}{\epsilon s} = \lim_{\epsilon \to 0} \frac{s e^{-\epsilon s}}{s} = 1$$

The limiting function of F(t) when $\epsilon \longrightarrow 0$ is known as the unit impulse function or the Dirac delta function $\delta(t)$. Thus

$$L\{ \delta(t) \} = 1$$

If the function be shifted a distance T along the positive t axis, then according to the second shift theorem its transform is

$$L\Big\{ \delta(t - T) \Big\} = e^{-Ts}$$

we shall now show that the derivative of the unit step function is the impulse function. For consider the function

$$\frac{1}{\epsilon} U(t) - \frac{1}{\epsilon} U(t - \epsilon) \quad \ldots \qquad (1)$$

Its transform is

$$\frac{1}{\epsilon} \left(\frac{1}{s} - \frac{1}{s} e^{-\epsilon s} \right) = \frac{1}{\epsilon s} \left(1 - e^{-\epsilon s} \right) \quad \ldots \quad (2)$$

and as $\epsilon \to 0$, the expression (1) tends formally to U'(t) and its transform (2) tends to 1.

Thus $L\Big\{ U'(t) \Big\} = L\Big\{ \delta(t) \Big\} = 1$

and $L\Big\{ U'(t - T) \Big\} = L\Big\{ \delta(t - T) \Big\} = e^{-Ts}$

Example 1.

A series circuit of resistance R, inductance. L is connected to a generator which delivers an impulse voltage $V_0\delta(t)$ which is impressed upon the circuit at zero time. Find the current in the circuit.

Solution:

$$L\frac{dI}{dt} + RI = V_0\,\delta(t)$$

$$Lsi(s) + Ri(s) = V_0\,,\quad I(0) = 0$$

$$(Ls+R)i(s) = V_0 \quad\therefore\ i(s) = \frac{V_0}{L}\cdot\frac{1}{s+\frac{R}{L}} \qquad\therefore\quad I(t) = \frac{V_0}{L}\,e^{-\frac{R}{L}t}$$

Note: An impulsive voltage means a large voltage acting for a very short interval of time but the time integral is finite.

11.2.5. The unit doublet

Consider the function F(t) defined by

$$F(t) = \frac{1}{\epsilon^2} \qquad 0<t<\epsilon$$

$$= -\frac{1}{\epsilon^2} \qquad \epsilon<t<2\epsilon$$

$$= 0 \qquad t>2\epsilon$$

$$L\left\{F(t)\right\} = \int_0^\epsilon \frac{1}{\epsilon^2}\,e^{-st}\,dt + \int_\epsilon^{2\epsilon} -\frac{1}{\epsilon^2}\,e^{-st}\,dt$$

$$= \frac{1}{\epsilon^2 s}\left[1 - 2e^{-\epsilon s} + e^{-2\epsilon s}\right]$$

$$= \frac{1}{\epsilon^2 s}\left(1 - e^{-\epsilon s}\right)^2$$

Let us find the limiting value of this transform when $\epsilon \longrightarrow 0$

147

$$\lim_{\epsilon \to 0} \frac{1-2e^{-\epsilon s}+e^{-2\epsilon s}}{\epsilon^2 s} = \lim_{\epsilon \to 0} \frac{-2s^2 e^{-\epsilon s}+4s^2 e^{-2\epsilon s}}{2s}$$

$$= \frac{2s^2}{2s} = s$$

The limiting function of F(t) when $\epsilon \to 0$ is known as the unit doublet and its transform is s. If the function is shifted along the positive t axis a distance T then its transform is
$s\,e^{-Ts}$.

We shall now show that the derivative of the unit impulse function is the unit doublet. For consider the function

$$\frac{1}{\epsilon}\,\delta(t) - \frac{1}{\epsilon}\,\delta(t-\epsilon) \quad . \quad . \quad . \quad (1)$$

whore ϵ may he made as small as we please. The Laplace transform of this is

$$\frac{1}{\epsilon}\,(1-e^{-\epsilon s}) \quad . \quad . \quad . \quad . \quad (2)$$

and as $\epsilon \to 0$ the expression (1) tends formally to $\delta'(t)$ and its transform (2) tends to s .

Thus $\quad L\left\{\delta'(t)\right\} = s$

and $\quad L\left\{\delta'(t-T)\right\} = se^{-Ts}$

In other words

$$L\left\{U''(t)\right\} = s$$

and $\quad L\left\{U''(t-T)\right\} = s\,e^{-Ts}.$

11.2.6. The behavior of f(s) as $s \to \infty$

Let F(t) be a function which is sectionally continuous in every finite interval $0 \le t \le T$ and let it be of the order of $e^{a_0 t}$ as $t \to \infty$, i.e., $\quad |F(t)| < Me^{a_0 t}$.

Take $s \ge a$ where $a \ge a_0$ then

$$\left| e^{-st} F(t) \right| < e^{-at} Me^{a_0 t} = Me^{-(a-a_0)t}$$

$$\therefore \quad \left| f(s) \right| = \left| L\left\{F(t)\right\} \right| = \left| \int_0^\infty e^{-st} F(t)\,dt \right|$$

148

$$\leqslant \int_0^\infty |\ e^{-st}\ F(t)\ |\ dt \qquad < \int_0^\infty M\, e^{-(a-a_0)\,t}\ dt$$

$$= M \left[\frac{e^{-(a-a_0)t}}{-(a-a_0)} \right]_0^\infty \quad = \frac{M}{a-a_0}\ ,\quad s \geqslant a$$

Put $a = s$ $\therefore |\ f(s)\ | < \dfrac{M}{s-a_0} \longrightarrow 0$ as $s \longrightarrow \infty$

$\therefore f(s) \longrightarrow 0$ as $s \longrightarrow \infty$

11.2.7. Initial value theorem

To prove that $\displaystyle \lim_{t \to 0} F(t) = \lim_{s \to \infty} s\, f(s)$

$$L\left\{ F'(t) \right\} = \int_0^\infty e^{-st}\ F'(t)\ dt = s\, f(s) - F(0) \dots \qquad (1)$$

Now if F'(t) is sectionally continuous and of exponential order then according to the theorem in 2.6.

$$\lim_{s \to \infty} \int_0^\infty e^{-st}\ F'(t)\ dt = 0$$

Taking the limit in (1), assuming F(t) to be continuous at t = 0, we get

$$0 = \lim_{s \to \infty} s\, f(s) - F(0)$$

i.e $\displaystyle \lim_{s \to \infty} s\, f(s) = F(0) = \lim_{t \to 0} F(t).$

11.2.8. Final value theorem

To prove that $\displaystyle \lim_{t \to \infty} F(t) = \lim_{s \to 0} s\, f(s)$

$$L\left\{ F'(t) \right\} = \int_0^\infty e^{-st}\ F'(t)\ dt = s\, f(s) - F(0) \ \dots \qquad (1)$$

The limit of the left hand side of (1) as s → 0 is

149

$$\lim_{s \to 0} \int_0^\infty e^{-st} F'(t)dt = \int_0^\infty F'(t)dt = [F(t)]_0^\infty = \lim_{t \to \infty} F(t) - F(o)$$

The limit of right hand side of (1) as s → 0 is

$$\lim_{s \to 0} s\, f(s) - F(o)$$

$$\therefore \lim_{t \to \infty} F(t) - F(o) = \lim_{s \to 0} s\, f(s) - F(o)$$

i.e $\lim_{t \to \infty} F(t) = \lim_{s \to 0} s\, f(s)$

Example 2

Let $F(t) = 2e^{-3t}$, then $f(s) = \dfrac{2}{s+3}$

By the initial value theorem we have

$$\lim_{t \to 0} 2e^{-3t} = \lim_{s \to \infty} \frac{2s}{s+3} = 2$$

and by the final value theorem we have

$$\lim_{t \to \infty} 2e^{-3t} = \lim_{s \to 0} \frac{2s}{s+3} = 0$$

11.2.9. Differentiation of transform

Let F(t) be a sectionally continuous function in every finite interval $o \le t \le T$ and let it be of the order of e^{at} as $t \to \infty$. Then if s>a, we have

$$L\left\{ F(t) \right\} = \int_0^\infty e^{-st} F(t)\, dt = f(s)$$

Differentiate w.r.t. s

$$\frac{\partial}{\partial s} \int_0^\infty e^{-st} F(t)\, dt = \int_0^\infty \frac{\partial}{\partial s} \left\{ e^{-st} F(t) \right\} dt$$

$$= \int_0^\infty - t e^{-st} F(t)dt$$

$$= L\left\{ -t\, F(t) \right\} = f'(s)$$

150

Similarly $\quad L\left\{(-t)^2 \, F(t)\right\} = f''(s)$

and in general $\quad L\left\{(-t)^n \, F(t)\right\} = f^{(n)}(s)$

Hence the differentiation of the transform of a function w. r. t. s corresponds to the multiplication of the function by - t.

$$\text{e.g. } L\left\{\sin kt\right\} = \frac{k}{s^2+k^2}, \quad s > 0$$

$$\therefore \; L\left\{-t \sin kt\right\} = \frac{-2ks}{(s^2+k^2)^2}$$

$$\text{Then } L^{-1}\left\{\frac{s}{(s^2+k^2)^2}\right\} = \frac{1}{2k} t \sin kt$$

$$\text{Similarly } L\left\{\cos kt\right\} = \frac{s}{s^2+k^2}$$

$$\therefore \; L\left\{-t \cos kt\right\} = \frac{s^2+k^2-2s^2}{(s+k^2)^2}$$

$$\therefore \; L\left\{t \cos kt\right\} = \frac{s^2-k^2}{(s^2+k^2)^2} = \frac{s^2+k^2-2k^2}{(s^2+k^2)^2}$$

$$= \frac{1}{s^2+k^2} - \frac{2k^2}{(s^2+k^2)^2}$$

$$= L\left\{\frac{\sin kt}{k}\right\} - \frac{2k^2}{(s^2+k^2)^2}$$

$$\therefore \; L\left\{t \cos kt - \frac{1}{k}\sin kt\right\} = - \frac{2k^2}{(s^2+k^2)^2}$$

$$\text{i.e. } L^{-1}\left\{\frac{1}{(s^2+k^2)^2}\right\} = \frac{1}{2k^3}(\sin kt - kt \cos kt)$$

11.2.10. Application of the differentiation of Laplace transform to the solution of linear differential equations with coefficients as polynomials in t.

Example 1 Find a solution of the differential equation

$$t\frac{d^2x}{dt^2} + t\frac{dx}{dt} + x = 0$$

which satisfies $x = 0$, $dx/dt = 1$ when $t = 0$.

$$-\frac{d}{ds}\left[s^2\bar{x} - sx(0)-x'(0)\right] - \frac{d}{ds}\left[s\bar{x} - x(0)\right] + \bar{x} = 0$$

$$\therefore \; -\frac{d}{ds}\left[s^2\bar{x} - 1\right] - \frac{d}{ds} s\, \bar{x} + \bar{x} = 0$$

$$-s^2\frac{d\bar{x}}{ds} - 2s\bar{x} - s\frac{d\bar{x}}{ds} - \bar{x} + \bar{x} = 0$$

151

$$\therefore \quad (1+s)\frac{d\bar{x}}{ds} + 2\,\bar{x} = 0$$

$$\int \frac{d\bar{x}}{\bar{x}} + 2 \int \frac{ds}{1+s} = 0$$

$$\log \bar{x} + 2 \log (1+s) = \log C$$

$$\therefore \quad \bar{x} = \frac{C}{(1+s)^2}$$

$$x = cte^{-t}$$

The condition $x = 1$ when $t = 0$ is satisfied and the condition $dx/dt = 1$ when $t = 0$ gives $C = 1$

Hence

$$x = te^{-t}$$

Example 2   Find the solution of Bessel's differential equation of zero order

$$t\,Y''(t) + Y'(t)\,t\,Y(t) = 0$$

which has a Laplace transform and which satisfies $Y(0)=1$

Solution:

$$-\frac{d}{ds}[s^2 y(s) - s - Y'(0)] + [\,sy(s) - 1\,] - \frac{d}{ds}\,y(s) = 0$$

$$\therefore \quad -2sy - s^2\frac{dy}{ds} + 1 + sy - 1 - \frac{dy}{ds} = 0$$

$$\text{giving} \quad (s^2 + 1)\frac{dy}{ds} + sy = 0$$

$$\therefore \quad \frac{dy}{y} = \frac{-sds}{s^2 + 1}$$

$$\log y = -\tfrac{1}{2}\log (s^2+1) + \log C$$

$$\therefore \quad y(s) = \frac{C}{\sqrt{s^2+1}}$$

Now by the initial value theorem

$$\lim_{t \to 0} Y(t) = \lim_{s \to \infty} sy(s)$$

$$\therefore \quad 1 = \lim_{s \to \infty} \frac{Cs}{\sqrt{s^2+1}} = C$$

also C can be determined as follows:

$$y(s) = \frac{C}{s}\left(1 + \frac{1}{s^2}\right)^{-\frac{1}{2}}$$

$$= \frac{C}{s}\left(1 - \tfrac{1}{2}\cdot\frac{1}{s^2} - \tfrac{1}{2}\times -\tfrac{3}{2}\times\frac{1}{2!}\cdot\frac{1}{s^4}\cdots\right)$$

152

$$= \frac{C}{s}\left(1 - \frac{1}{2}\frac{1}{s^2} + \frac{1.3}{2^2.2!} \cdot \frac{1}{s^4} - \cdots\right)$$

$$= C\left(\frac{1}{s} - \frac{1}{2} \cdot \frac{1}{s^3} + \frac{1.3}{2^2.2!} \cdot \frac{1}{s^5} - \cdots\right)$$

$$\therefore \ Y(t) = C\left[1 - \frac{1}{2}\frac{t^2}{2!} + \frac{1.3}{2^2.2!} \cdot \frac{t^4}{4!} - \cdots\right]$$

and since $Y(o) = 1 \ \therefore \ C = 1$ and we get Bessel function of zero order of the first kind namely

$$J_0(t) = 1 - \frac{t^2}{2^2} + \frac{t^4}{2^2.4^2} - \frac{t^6}{2^2.4^2.6^2} + \cdots$$

Corollary 1. $\qquad L\left\{J_0(t)\right\} = \dfrac{1}{\sqrt{s^2+1}} \quad . \quad s > o$

Corollary 2. $\qquad L\left\{J_0(at)\right\} = \dfrac{1}{a} \cdot \dfrac{1}{\sqrt{\left(\frac{s}{a}\right)^2+1}} = \dfrac{1}{\sqrt{s^2+a^2}}$

Corollary 3.

Since $J_0'(t) = -J_1(t)$ $\qquad$ then $L\left\{J_0'(t)\right\} = L\left\{-J_1(t)\right\}$

$$s\,L\left\{J_0(t)\right\} - J_0(o) = L\left\{-J_1(t)\right\}$$

$$\frac{s}{\sqrt{s^2+1}} - 1 = -L\left\{J_1(t)\right\}$$

$$\therefore \ L\left\{J_1(t)\right\} = 1 - \frac{s}{\sqrt{s^2+1}} = \frac{\sqrt{s^2+1}-s}{\sqrt{s^2+1}} = \frac{1}{\sqrt{s^2+1}\,(s+\sqrt{s^2+1})}$$

Example 3 $\qquad$ Find the Laplace transform of $\sin\sqrt{t}$

Solution:

Put $Y(t) = \sin\sqrt{t}$

By differentiating twice we get

$4t\,Y''(t) + 2\,Y(t) + Y(t) = 0$

Transforming we get

$$- 4 \frac{d}{ds}[s^2 y - sY(0) - Y'(0)] + 2[sy - Y(0)] + y = 0$$

$$\therefore 4s^2 \frac{dy}{ds} + (6s-1)y = 0$$

$$y = \frac{C}{s^{3/2}} e^{-\frac{1}{4s}}$$

For Small values of t, $\sin \sqrt{t} = \sqrt{t}$ approx.

and $L\left\{\sqrt{t}\right\} = \frac{\sqrt{\pi}}{2s^{3/2}}$ ✳

For large s, $y = \frac{C}{s^{3/2}}$ approx Hence by comparison $C = \frac{\sqrt{\pi}}{2}$

Alternative method

$$\sin \sqrt{t} = \sqrt{t} - \frac{(\sqrt{t})^3}{3!} + \frac{(\sqrt{t})^5}{5!} - \cdots$$

$$= t^{1/2} - \frac{t^{3/2}}{3!} + \frac{t^{5/2}}{5!} - \cdots$$

$$\therefore L\left\{\sin \sqrt{t}\right\} = \frac{\Gamma(3/2)}{s^{3/2}} - \frac{\Gamma(5/2)}{3! \, s^{5/2}} + \frac{\Gamma(7/2)}{5! \, s^{7/2}} - \cdots$$

$$= \frac{\sqrt{\pi}}{2s^{3/2}}\left[1 - (1/2^2 s) + \frac{(1/2^2 s)^2}{2!} - \cdots\right]$$

$$= \frac{\sqrt{\pi}}{2s^{3/2}} e^{-\frac{1}{2^2 s}} = \frac{\sqrt{\pi}}{2s^{3/2}} e^{-\frac{1}{4s}}$$

11.2.11. Integration of transforms

Let $F(t)$ be sectionally continuous of the order of e^{at} as $t \longrightarrow \infty$ and let $x > a$, then

$$f(x) = \int_0^\infty e^{-xt} F(t) \, dt$$

Now let $s > a$ and suppose that $\lim_{t \rightarrow +0} \frac{F(t)}{t}$ exists, then

$$\int_s^\infty f(x) \, dx = \int_s^\infty \left\{\int_0^\infty e^{-xt} F(t) \, dt\right\} dx$$

Assuming that we can invert the order of integration which is true in this case, we have

$$\int_s^\infty f(x) \, dx = \int_0^\infty F(t) \left\{\int_s^\infty e^{-xt} \, dx\right\} dt$$

154

$$= \int_0^\infty F(t) \left[\frac{e^{-xt}}{-t} \right]_s^\infty dt = \int_0^\infty \frac{F(t)}{t} e^{-st} dt$$

$$= \int_0^\infty e^{-st} \left\{ \frac{F(t)}{t} \right\} dt = L \left\{ \frac{F(t)}{t} \right\}$$

$$\therefore \quad \text{If} \quad L\left\{ F(t) \right\} = f(s) \qquad \text{then} \quad L\left\{ \frac{F(t)}{t} \right\} = \int_s^\infty f(x)\, dx$$

Hence the division of the object function by t corresponds to the integration of the transform of that function from s to ∞.

Example 1.   Prove that $L\left\{ \dfrac{1 - \cos kt}{t} \right\} = \frac{1}{2} \log \left(1 + \dfrac{k^2}{s^2}\right), \ s > k > 0.$

Solution:

$$L\left\{ 1 - \cos kt \right\} = \frac{1}{s} - \frac{s}{s^2 + k^2}$$

$$\therefore L\left\{ \frac{1 - \cos kt}{t} \right\} = \int_s^\infty \left(\frac{1}{x} - \frac{x}{x^2 + k^2} \right) dx$$

$$= \left[\log x - \frac{1}{2} \log (x^2 + k^2) \right]_s^\infty$$

$$= \lim_{x \to \infty} \log \frac{x}{\sqrt{x^2 + k^2}} - \left[\log \frac{x}{\sqrt{x^2 + k^2}} \right]_{x=s}$$

$$= 0 - \log \frac{s}{\sqrt{s^2 + k^2}} = \frac{1}{2} \log \frac{s^2 + k^2}{s^2} = \frac{1}{2} \log \left(1 + \frac{k^2}{s^2} \right)$$

Example 2. $L\left\{ e^{-at} - e^{-bt} \right\} = \dfrac{1}{s+a} - \dfrac{1}{s+b}$

155

$$\therefore L\left\{\frac{e^{-at}-e^{-bt}}{t}\right\} = \int_s^\infty \left[\frac{1}{x+a} - \frac{1}{x+b}\right] dx$$

$$= \left[\log\frac{x+a}{x+b}\right]_s^\infty = -\log\frac{s+a}{s+b} = \log\frac{s+b}{s+a}$$

$$\therefore L\left\{\frac{e^{-at}-e^{-bt}}{t}\right\} = \log\frac{s+b}{s+a}, \ s>-a, s>-b$$

Corollary   If $a = 0$ and $b = 1$ we get

$$L\left\{\frac{1-e^{-t}}{t}\right\} = \log\frac{s+1}{s} \quad \therefore L\left\{\frac{1-e^{-t}}{t}\right\} = \log\left(1+\frac{1}{s}\right), s>0$$

Example 3   Prove that $\displaystyle\int_0^\infty \frac{F(t)}{t} dt = \int_0^\infty f(x) dx$ provided that the

integrals converge and hence deduce that

(i) $\displaystyle\int_0^\infty \frac{\sin t}{t} dt = \frac{\pi}{2}$,   (ii) $\displaystyle\int_0^\infty \frac{e^{-t}-e^{-3t}}{t} dt = \log 3$

Solution:

Since $\displaystyle L\left\{\frac{F(t)}{t}\right\} = \int_s^\infty f(x) dx$, then

$$\int_0^\infty e^{-st} \frac{F(t)}{t} dt = \int_s^\infty f(x) dx$$

Let $s \rightarrow 0 +$ and assuming that the integrals converge, then

$$\int_0^\infty \frac{F(t)}{t} dt = \int_0^\infty f(x) dx$$

(i) Take $F(t) = \sin t$, then $f(s) = \frac{1}{s^2+1}$ and we get

$$\int_0^\infty \frac{\sin t}{t} dt = \int_0^\infty \frac{dx}{x^2+1} = \left[\tan^{-1}x\right]_0^\infty = \frac{\pi}{2}$$

156

(ii)   Take $F(t) = e^{-t} - e^{-3t}$, then $f(s) = \dfrac{1}{s+1} - \dfrac{1}{s+3}$

$$\therefore \; L\left\{\frac{e^{-t} - e^{-3t}}{t}\right\} = \int_s^\infty \left[\frac{1}{x+1} - \frac{1}{x+3}\right] dx$$

$$\therefore \; \int_0^\infty e^{-st}\left(\frac{e^{-t} - e^{-3t}}{t}\right) dt = \left[\log \frac{x+1}{x+3}\right]_s^\infty = \log \frac{s+3}{s+1}$$

Taking the limit as s → 0 we get

$$\int_0^\infty \frac{e^{-t} - e^{-3t}}{t} \, dt = \log 3$$

11.2.12. Transforms of periodic functions

If F(t) is a periodic function of period a, i.e., $F(t+a) = F(t)$ and if F(t) is sectionally continuous over a period, then

$$f(s) = \frac{\displaystyle\int_0^a e^{-st} F(t)\, dt}{1 - e^{-as}}$$

$$f(s) = \int_0^\infty e^{-st} F(t)\, dt = \int_0^a e^{-st} F(t)\, dt + \int_a^{2a} e^{-st} F(t)\, dt + \dots$$

$$= \sum_{n=0}^\infty \int_{na}^{(n+1)a} e^{-st} F(t)\, dt$$

Put $t = \tau + na$, then $F(\tau + na) = F(\tau)$

157

$$f(s) = \sum_{n=0}^{\infty} \int_0^a e^{-s(\tau+na)} F(\tau+na)\, d\tau$$

$$= \sum_{n=0}^{\infty} e^{-nas} \int_0^a e^{-s\tau} F(\tau)\, d\tau = \sum_{n=0}^{\infty} e^{-nas} \int_0^a e^{-st} F(t)\, dt$$

$$= (1 + e^{-as} + e^{-2as} + e^{-2as} + \ldots) \int_0^a e^{-st} F(t)\, dt = \frac{\displaystyle\int_0^a e^{-st} F(t)\, dt}{1 - e^{-as}}$$

Example 1   Find the Laplace transform of the square wave or the <u>Meander function</u> $M(c,t)$ defined by

$$M(c,t) = 1 \qquad 0 < t < c$$
$$= -1 \qquad c < t < 2c$$
$$M(c,t+2c) = M(c,t)$$

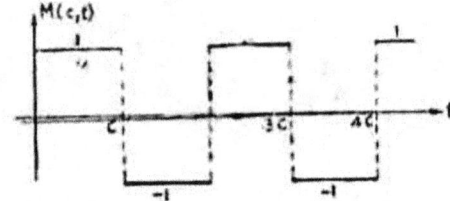

Since the period is 2c, then

$$L\left\{M(c,t)\right\} = \frac{\displaystyle\int_0^{2c} e^{-st} M(c,t)\, dt}{1 - e^{-2cs}}$$

Now $\displaystyle\int_0^{2c} e^{-st} M(c,t)\, dt = \int_0^c e^{-st}\, dt - \int_c^{2c} e^{-st}\, dt$

158

$$= \left[\frac{e^{-st}}{-s} \right]_0^c + \left[\frac{e^{-st}}{s} \right]_c^{2c} = \frac{1}{s} (1 - e^{-cs})^2$$

$$\therefore \ L\left\{ M(c,t) \right\} = \frac{1}{s} \frac{(1 - e^{-cs})^2}{1 - e^{-2cs}} = \frac{1}{s} \frac{1 - e^{-cs}}{1 + e^{-cs}}$$

$$= \frac{1}{s} \frac{e^{\frac{cs}{2}} - e^{-\frac{cs}{2}}}{e^{\frac{cs}{2}} + e^{-\frac{cs}{2}}} = \frac{1}{s} \tanh \frac{cs}{2}, \quad s > 0$$

Example 2. Find the Laplace transform of the triangular wave or the function H(c,t) defined as follows :

$$\mathbf{H(o,t)} = t \qquad 0 < t < c$$
$$= 2c\text{-}t \qquad c < t < 2c$$
$$\mathbf{H(c,t+2c)} = \mathbf{H(c,t)}$$

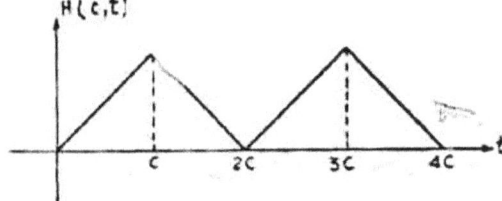

It is evident that H(c,t) at any t is the integral of the function M(c,t) from o to t.

$$\therefore \ L\left\{ \mathbf{H(c,t)} \right\} = \frac{1}{s^2} \tanh \frac{cs}{2}$$

Example 3   Find the Laplace transform of the periodic function F(t) defined by

$$\mathbf{F(t)} = t \qquad 0 < t < 1$$
$$\mathbf{F(t+1)} = \mathbf{F(t)}$$

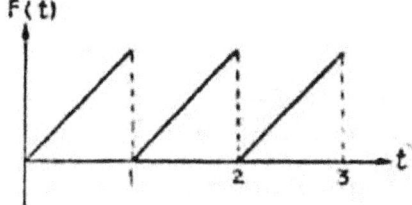

$$L\left\{F(t)\right\} = \frac{\displaystyle\int_0^1 te^{-st}\,dt}{1-e^{-s}}$$

$$\int_0^1 te^{-st}\,dt = -\frac{1}{s}\int_0^1 t\,de^{-st} = -\frac{1}{s}\left[\,te^{-st} + \frac{e^{-st}}{s}\,\right]_0^1$$

$$= -\frac{1}{s}\left[e^{-s} + \frac{e^{-s}}{s} - \frac{1}{s}\right] = \frac{1}{s^2} - \frac{s+1}{s^2}\,e^{-s}$$

$$= \frac{1}{s^2}(1 - e^{-s}) - \frac{1}{s}\,e^{-s}$$

$$\therefore\ L\left\{F(t)\right\} = \frac{1}{s^2} - \frac{e^{-s}}{s(1-e^{-s})}$$

Example 4   Find the Laplace transform of the intermittent sine wave defined by

$$F(t) = \sin t \qquad 0<t<\pi$$
$$= 0 \qquad \pi<t<2\pi$$
$$F(t+2\pi) = F(t)$$

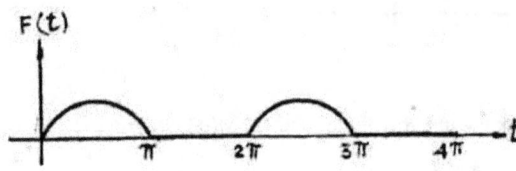

$$L\left\{F(t)\right\} = \frac{\displaystyle\int_0^{2\pi} e^{-st}\,F(t)\,dt}{1 - e^{-2\pi s}} = \frac{\displaystyle\int_0^{\pi} e^{-st}\sin t\,dt}{1 - e^{-2\pi s}}$$

Now $\displaystyle\int_0^{\pi} e^{-st}\sin t\,dt = \frac{1 + e^{-\pi s}}{s^2 + 1}$

$$\therefore\ L\left\{F(t)\right\} = \frac{1 + e^{-\pi s}}{(s^2+1)(1-e^{-2\pi s})} = \frac{1}{(s^2+1)(1-e^{-\pi s})}$$

160

Example 5   Find the Laplace transform of the full-wave rectification | sin t | of the sine function.

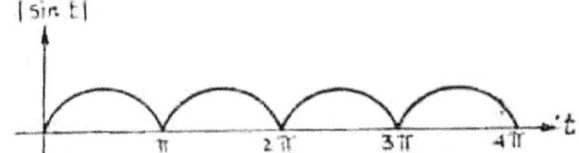

It is evident that in the full wave rectification the ordinates are the sum of the ordinates of F(t) of example 4 + ordinates of F(t) displaced to the right a distance π.

$$\text{Hence } L\{\,|\sin t|\,\} = \frac{1}{(s^2+1)(1-e^{-\pi s})} + \frac{e^{-\pi s}}{(s^2+1)(1-e^{-\pi s})}$$

$$= \frac{1+e^{-\pi s}}{(s^2+1)(1-e^{-\pi s})}$$

$$= \frac{1}{s^3+1} \cdot \frac{e^{\frac{\pi s}{2}} + e^{-\frac{\pi s}{2}}}{e^{\frac{\pi s}{2}} - e^{-\frac{\pi s}{2}}} = \frac{1}{s^2+1} \coth \frac{\pi s}{2}$$

11.2.13. The product theorem—Convolution

This theorem aims at finding the inverse transform of the product of two transforms. Let $F_1(t)$ and $F_2(t)$ be two sectionally continuous functions in every finite interval $0 \leq t \leq T$ and of the order of e^{at} as $t \longrightarrow \infty$. Take $s > a$ and let

$$L\{F_1(t)\} = f_1(s), \quad L\{F_2(t)\} = f_2(s)$$

then

$$L^{-1}\{f_1(s)\,f_2(s)\} = F_1(t)^* F_2(t) \quad \text{where}$$

$$F_1(t)^* F_2(t) = \int_0^t F_1(t-\lambda)\,F_2(\lambda)\,d\lambda$$

F1(t)* F2(t) as defined above is called the convolution of the two functions and the last integral is known as a convolution integral or Faltung integral,

Proof :

161

$$f_1(s) = \int_0^\infty e^{-sx} F_1(x)\ dx,$$

$$f_2(s) = \int_0^\infty e^{-sy} F_2(y)\ dy,$$

$$f_1(s)\ f_2(s) = \int_0^\infty e^{-sx} F_1(x)\ dx \int_0^\infty e^{-sy} F_2(y)\ dy$$

$$= \int_0^\infty \int_0^\infty e^{-s\,(x+y)} F_1(x)\, F_2(y)\ dx\ dy \quad .. \quad (1)$$

FIELD OF INTEGRATION

x-y plane   t-λ plane

the integration *being* carried over the positive quadrant
$x > 0, y > 0$.

Now let us make the transformation of coordinates

$$\left.\begin{array}{l} x = t - \lambda \\ y = \lambda \end{array}\right\} \quad i.e \quad \left.\begin{array}{l} t = x + y \\ \lambda = y \end{array}\right\}$$

Since $x \geqslant o$, then $t - \lambda \geqslant o$ i.e $t \geqslant \lambda$. Hence the field of integration in the $x - y$ plane transforms in the $t - \lambda$ plane into the field between the positive t — axis and the straight line $\lambda = t$.

Now the element of area dx dy transforms into

$$\begin{vmatrix} \dfrac{\partial x}{\partial t} & \dfrac{\partial y}{\partial t} \\[2mm] \dfrac{\partial x}{\partial \lambda} & \dfrac{\partial y}{\partial \lambda} \end{vmatrix} dt\ d\lambda = \begin{vmatrix} 1 & 0 \\ -1 & 1 \end{vmatrix} dt\ d\lambda = dt\ d\lambda$$

Hence the double integral (1) transforms into

$$f_1(s)\ f_2(s) = \int_0^\infty \int_0^t e^{-st} F_1(t-\lambda)\ F_2(\lambda)\ dt\ d\lambda$$

162

$$= \int_0^\infty e^{-st} \left[\int_0^t F_1(t-\lambda)\, F_2(\lambda)\, d\lambda \right] dt$$

$$= L \left\{ \int_0^t F_1(t-\lambda)\, F_2(\lambda)\, d\lambda \right\} = L \left\{ F_1 \cdot F_2 \right\} = L \left\{ F_2 \cdot F_1 \right\} \text{ by symmetry}$$

Hence we have the following result:

$$\text{If} \quad L^{-1} \left\{ f_1(s) \right\} = F_1(t)$$

$$\text{and} \quad L^{-1} \left\{ f_2(s) \right\} = F_2(t)$$

$$\text{then} \quad L^{-1} \left\{ f_1(s)\, f_2(s) \right\} = F_1(t)*F_2(t) = F_2(t)*F_1(t)$$

$$= \int_0^t F_1(t-\lambda)\, F_2(\lambda)\, d\lambda = \int_0^t F_2(t-\lambda)\, F_1(\lambda)\, d\lambda$$

This formula is known in operational calculus as the Borel formula.

Example 1
$$L^{-1} \left\{ \frac{1}{s^2-a^2} \right\} = L^{-1} \left\{ \frac{1}{s-a} \cdot \frac{1}{s+a} \right\} = e^{at} \cdot e^{-at}$$

$$= \int_0^t e^{a(t-\lambda)}\, e^{-a\lambda}\, d\lambda = e^{at} \int_0^t e^{-2a\lambda}\, d\lambda = \frac{e^{at}}{-2a} \left[e^{-2a\lambda} \right]_0^t$$

$$= -\frac{e^{at}}{2a} \left[e^{-2at} - 1 \right] = \frac{1}{2a} \left[e^{at} - e^{-at} \right] = \frac{1}{a} \sinh at$$

Example 2
$$L^{-1} \left\{ \frac{s}{(s^2+a^2)^2} \right\} = L^{-1} \left\{ \frac{s}{s^2+a^2} \cdot \frac{1}{s^2+a^2} \right\} = \cos at \cdot \frac{1}{a} \sin at$$

$$= \frac{1}{a} \int_0^t \cos a(t-\lambda) \sin a\lambda\, d\lambda = \frac{1}{2a} \int_0^t [\sin at - \sin a(t-2\lambda)]\, d\lambda$$

$$= \frac{1}{2a} \left[\lambda \sin at + \frac{\cos a(t-2\lambda)}{-2a} \right]_0^t = \frac{1}{2a}\, t \sin at$$

11.2.14. Application of the product theorem to the solution of differential and integral equations

163

Example 1   Solve the differential equation $Y''(t) + k^2 Y(t) = F(t)$, where k is a constant.

Solution:

$$s^2 y(s) - sY(o) - Y'(o) + k^2 y(s) = f(s)$$

Put $Y(o) = A$ and $Y'(o) = B$

$$(s^2 + k^2)\ y(s) = As + B + f(s)$$

$$\therefore\ y(s) = \frac{As}{s^2 + k^2} + \frac{B}{s^2 + k^2} + \frac{1}{s^2 + k^2} \cdot f(s)$$

$$\therefore\ Y(t) = A \cos kt + \frac{B}{k} \sin kt + \frac{1}{k} \sin kt^* F(t)$$

$$= A \cos kt + \frac{B}{k} \sin kt + \frac{1}{k} \int_0^t \sin k(t-\lambda)\, F(\lambda)\, d\lambda$$

This method is particularly useful in the case when the transform of F(t) is difficult to obtain as in the case of an impressed voltage given for example by an oscillogram. The convolution integral can, however, be calculated numerically.

Example 2   Solve the integral equation

$$Y(t) = 4 \sin t + \int_0^t \sin(t-\lambda)\ Y(\lambda)\ d\lambda \quad \cdots \quad (1)$$

By an integral equation we mean an equation in which the unknown function lies under the integral sign. Here the integral is of the convolution type. Integrals of the convolution type appear in many practical problems. Equation (1) can he written in the form

$Y(t) = 4 \sin t + \sin t* Y(t)$

Transforming we get

$$y(s) = \frac{4}{s^2 + 1} + \frac{1}{s^2 + 1}\ y(s)$$

$$\therefore\ y(s)\left[1 - \frac{1}{s^2 + 1} \right] = \frac{4}{s^2 + 1}$$

$$\text{i.e}\ \ y(s) = \frac{4}{s^2} \qquad \therefore\ Y(t) = 4t$$

Example 3 Solve the integral equation $Y(t) = \frac{1}{2} t^2 - \int_0^t (t-\lambda) \, Y(\lambda) \, d\lambda$

$\therefore \quad Y(t) = \frac{1}{2} t^3 - t^* \, Y(t)$

$y(s) = \frac{1}{s^3} - \frac{1}{s^2} \, y(s)$

$y(s) \left[1 + \frac{1}{s^2} \right] = \frac{1}{s^3}$

$\therefore \quad y(s) \cdot \frac{s^2+1}{s^2} = \frac{1}{s^3}$

i.e $y(s) = \frac{1}{s(s^2+1)} = \frac{1}{s} - \frac{s}{s^2+1}$

$\therefore \quad Y(t) = 1 - \cos t$

Example 4   Solve the integro- differential equation

$$Y'(t) = t + \int_0^t Y(t-\lambda) \, \cos \lambda \, d\lambda$$

given that $Y(0) = 6$.

Solution:

$Y'(t) = t + Y(t) \cdot \cos t$

$sy(s) - 6 = \frac{1}{s^2} + y(s) \cdot \frac{s}{s^2+1}$

$y(s) \left[s - \frac{s}{s^2+1} \right] = 6 + \frac{1}{s^2}$

$y(s) \, \frac{s^3}{s^2+1} = 6 + \frac{1}{s^2}$

$y(s) = \frac{6(s^2+1)}{s^3} + \frac{s^2+1}{s^5} = \frac{6}{s} + \frac{7}{s^3} + \frac{1}{s^5}$

$\therefore \quad Y(t) = 6 + \frac{7}{2} t^2 + \frac{1}{24} t^4$

11.2.15. Power series method for the determination of transforms and inverse transforms

In many problems expansion into infinite series is helpful in finding Laplace transforms and inverse transforms. The following are additional examples on the same principle. A list of well known expansions is given here for reference.

$$(1) \quad \frac{1}{1-x} = \sum_{n=0}^{\infty} x^n \qquad |x| < 1$$

$$(2) \quad \frac{1}{1+x} = \sum_{n=0}^{\infty} (-1)^n x^n \qquad |x| < 1$$

$$(3) \quad e^x = \sum_{n=0}^{\infty} \frac{x^n}{n!} \qquad \text{all } x$$

$$(4) \quad \cos x = \sum_{n=0}^{\infty} \frac{(-1)^n x^{2n}}{(2n)!} \qquad \text{all } x$$

$$(5) \quad \sin x = \sum_{n=0}^{\infty} \frac{(-1)^n x^{2n+1}}{(2n+1)!} \qquad \text{all } x$$

$$(6) \quad \cosh x = \sum_{n=0}^{\infty} \frac{x^{2n}}{(2n)!} \qquad \text{all } x$$

$$(7) \quad \sinh x = \sum_{n=0}^{\infty} \frac{x^{2n+1}}{(2n+1)!} \qquad \text{all } x$$

$$(8) \quad \tan^{-1}x = \sum_{n=0}^{\infty} \frac{(-1)^n x^{2n+1}}{2n+1} \qquad |x| \leqslant 1$$

$$(9) \quad \log(1+x) = \sum_{n=0}^{\infty} \frac{(-1)^n x^{n+1}}{n+1} \qquad -1 < x \leqslant 1$$

$$(10) \quad \log\frac{1+x}{1-x} = 2\sum_{n=0}^{\infty} \frac{x^{2n+1}}{2n+1} \qquad |x| < 1$$

Example 1   Evaluate $F(t) = L^{-1}\left\{\dfrac{1}{s^3 \cosh 2s}\right\}$ and compute $F(12)$

Solution:

$$\frac{1}{s^3 \cosh 2s} = \frac{2}{s^3 \left(e^{2s} + e^{-2s}\right)} = \frac{2e^{-2s}}{s^3 \left(1 + e^{-4s}\right)}$$

$$= \sum_{n=0}^{\infty} \frac{2}{s^3} e^{-2s} (-1)^n e^{-4ns}$$

$$= \sum_{n=0}^{\infty} \frac{2}{s^3} (-1)^n e^{(-4n-2))s}$$

$$L^{-1}\left\{\frac{1}{s^3 \cosh 2s}\right\} = \sum_{n=0}^{\infty} (-1)^n (t-4n-2)^2 \, U(t-4n-2)$$

$$\therefore F(t) = (t-2)^2 \, U(t-2) - (t-6)^2 \, U(t-6) + (t-10)^2 \, U(t-10)$$
$$- (t-14)^2 \, U(t-14) + \ldots$$

Notice that the series in the right hand side terminates once the argument of the function U is negative. Hence for t =12 we have

$$F(t) = (t-2)^2 \, U(t-2) - (t-6)^2 \, U(t-6) + (t-10)^2 \, U(t-10)$$

$$\therefore \quad F(12) = 100 - 36 + 4 = 68$$

Example 2.   Evaluate $L^{-1}\left\{\log \dfrac{s+1}{s-1}\right\}$

Solution:

$$\log \frac{s+1}{s-1} = \log \frac{1+\dfrac{1}{s}}{1-\dfrac{1}{s}} = 2 \sum_{n=0}^{\infty} \frac{1}{(2n+1)\, s^{2n+1}}$$

(from series 10)

Now $L^{-1}\left\{\dfrac{1}{s^{2n+1}}\right\} = \dfrac{t^{2n}}{(2n)!}$,   Hence

$$L^{-1}\left\{\log \frac{s+1}{s-1}\right\} = 2 \sum_{n=0}^{\infty} \frac{t^{2n}}{(2n+1)!} = \frac{2}{t} \sum_{n=0}^{\infty} \frac{t^{2n+1}}{(2n+1)!}$$

$$= \frac{2}{t} \sinh t \text{ (from series 7)}$$

Example 3   Find $L\left\{\displaystyle\int_0^t \frac{\sin u}{u} \, du\right\}$

Solution:

$$\int_0^t \frac{\sin u}{u} \, du = \int_0^t \frac{1}{u}\left[u - \frac{u^3}{3!} + \frac{u^5}{5!} - \ldots\right] du$$

$$= t - \frac{t^3}{3 \cdot 3!} + \frac{t^5}{5 \cdot 5!} - \ldots$$

$$\therefore L\left\{\int_0^t \frac{\sin u}{u} \, du\right\} = \frac{1}{s^2} - \frac{1}{3 \cdot 3!} \cdot \frac{3!}{s^4} + \frac{1}{5 \cdot 5!} \cdot \frac{5!}{s^6} - \ldots$$

$$= \frac{1}{s^2} - \frac{1}{3 s^4} + \frac{1}{5 s^6} - \ldots$$

167

$$= \frac{1}{s} \left\{ \frac{(1/s)}{1} - \frac{(1/s)^3}{3} + \frac{(1/s)^5}{5} \right\} - \cdots = \frac{1}{s} \tan^{-1} \frac{1}{s}$$

Example 4

If $s > 0$ and $n > 1$ prove that

$$L \left\{ \frac{t^{n-1}}{1-e^{-t}} \right\} = \Gamma(n) \left\{ \frac{1}{s^n} + \frac{1}{(s+1)^n} + \frac{1}{(s+2)^n} + \cdots \right\}$$

Solution:

$$\frac{t^{n-1}}{1-e^{-t}} = t^{n-1} (1 + e^{-t} + e^{-2t} + \cdots)$$

$$= t^{n-1} + t^{n-1} e^{-t} + t^{n-1} e^{-2t} + \cdots$$

$$\therefore L \left\{ \frac{t^{n-1}}{1-e^{-t}} \right\} = \frac{\Gamma(n)}{s^n} + \frac{\Gamma(n)}{(s+1)^n} + \frac{\Gamma(n)}{(s+2)^n} + \cdots$$

$$= \Gamma(n) \left\{ \frac{1}{s^n} + \frac{1}{(s+1)^n} + \frac{1}{(s+2)^n} + \cdots \right\}$$

Example 5 Find $L^{-1} \left\{ \frac{e^{-1/s}}{s} \right\}$

Solution:

$$\frac{1}{s} e^{-1/s} = \frac{1}{s} \left\{ 1 - \frac{1}{s} + \frac{1}{2!s^2} - \frac{1}{3!s^3} + \cdots \right\}$$

$$= \frac{1}{s} - \frac{1}{s^2} + \frac{1}{2!s^3} - \frac{1}{3 s^4} + \cdots$$

$$L^{-1} \left\{ \frac{1}{s} e^{-1/s} \right\} = 1 - t + \frac{t^2}{(2!)^2} - \frac{t^3}{(3!)^2} + \cdots$$

$$= 1 - t + \frac{t^2}{1^2.2^2} - \frac{t^3}{1^2.2^2.3^2} + \cdots$$

$$= 1 - \frac{(2t^{\frac{1}{2}})^2}{2^2} + \frac{(2t^{\frac{1}{2}})^4}{2^2.4^2} - \frac{(2t^{\frac{1}{2}})^6}{2^2.4^2.6^2} + \cdots$$

$$= J_0(2\sqrt{t})$$

11.2.16. The error function or probability integral

This is a tabulated function denoted by erf(x) and is defined as follows

168

$$\text{erf}(x) = \frac{2}{\sqrt{\pi}} \int_0^x e^{-u^2} du$$

$$= \frac{2}{\sqrt{\pi}} \times \text{shaded area}$$

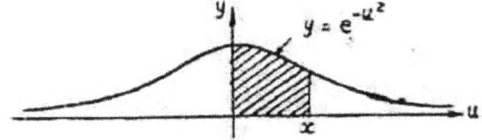

Expanding e^{-u^2} in ascending powers of u and integrating term by term we get

$$\text{erf}(x) = \frac{2}{\sqrt{\pi}} \int_0^x \left(1 - u^2 + \frac{u^4}{2!} - \frac{u^6}{3!} + .. \right) du$$

$$= \frac{2}{\sqrt{\pi}} \left[x - \frac{x^3}{3} + \frac{x^5}{2!5} - \frac{x^7}{3!7} + .. \right]$$

Now since $\int_0^\infty e^{-u^2} du = \frac{\sqrt{\pi}}{2}$, it follows that $\text{erf}(x) \to 1$

as $x \to \infty$.

The complementary error function erfc (x) is defined as follows

$$\text{erfc}(x) = 1 - \text{erf}(x) = \frac{2}{\sqrt{\pi}} \int_0^\infty e^{-u^2} du - \frac{2}{\sqrt{\pi}} \int_0^x e^{-u^2} du$$

$$= \frac{2}{\sqrt{\pi}} \int_x^\infty e^{-u^2} du$$

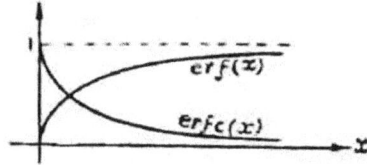

Example 1.   Find $L^{-1} \left\{ \frac{1}{(s-1)\sqrt{s}} \right\}$

Solution:
We have shown that

$$L \left\{ t^{-\frac{1}{2}} \right\} = \sqrt{\frac{\pi}{s}}$$

Hence $L^{-1}\left\{\dfrac{1}{\sqrt{s}}\right\} = \dfrac{1}{\sqrt{\pi t}}$

$$\therefore\ L^{-1}\left\{\dfrac{1}{(s-1)\sqrt{s}}\right\} = e^t \cdot \dfrac{1}{\sqrt{\pi t}} = \dfrac{1}{\sqrt{\pi}}\int_0^t e^{t-\lambda}\dfrac{1}{\sqrt{\lambda}}\,d\lambda = \dfrac{e^t}{\sqrt{\pi}}\int_0^t e^{-\lambda}\dfrac{1}{\sqrt{\lambda}}\,d\lambda$$

Put $\lambda = u^2$, then $\displaystyle\int_0^t e^{-\lambda}\dfrac{1}{\sqrt{\lambda}}\,d\lambda = \int_0^{\sqrt{t}}\dfrac{e^{-u^2}}{u}\cdot 2u\,du = 2\int_0^{\sqrt{t}} e^{-u^2}\,du$

$$\therefore\ L^{-1}\left\{\dfrac{1}{(s-1)\sqrt{s}}\right\} = e^t \times \dfrac{2}{\sqrt{\pi}}\int_0^{\sqrt{t}} e^{-u^2}\,du = e^t\,\mathrm{erf}(\sqrt{t})$$

If we change s into s +1 we get

$$L^{-1}\left\{\dfrac{1}{s\sqrt{s+1}}\right\} = e^{-t}\cdot e^t\,\mathrm{erf}(\sqrt{t}) = \mathrm{erf}(\sqrt{t})$$

Example 2　　Find $L\left\{e^{-2t}\,\mathrm{erf}(\sqrt{t})\right\}$

Solution:

Since $L\left\{\mathrm{erf}(\sqrt{t})\right\} = \dfrac{1}{s\sqrt{s+1}}$, then $L\left\{e^{2t}\,\mathrm{erf}(\sqrt{t})\right\} = \dfrac{1}{(s-2)\sqrt{s-1}}$

Example 3　　Find $L\left\{t\,\mathrm{erf}(2\sqrt{t})\right\}$

Solution:

Since $L\left\{\mathrm{erf}(\sqrt{t})\right\} = \dfrac{1}{s\sqrt{s+1}}$, then

$$L\left\{\mathrm{erf}(2\sqrt{t})\right\} = L\left\{\mathrm{erf}(\sqrt{4t})\right\} = \dfrac{1}{4}\cdot\dfrac{1}{\dfrac{s}{4}\sqrt{\dfrac{s}{4}+1}} = \dfrac{2}{s\sqrt{s+4}}$$

$$\therefore\ L\left\{t\,\mathrm{erf}(2\sqrt{t})\right\} = -\dfrac{d}{ds}\dfrac{2}{s\sqrt{s+4}} = \dfrac{3s+8}{s^2(s+4)^{3/2}}$$

Example 4   Find $L\left\{erfc(\sqrt{t})\right\}$

Solution:

Since $erfc(\sqrt{t}) = 1 - erf(\sqrt{t})$, then

$$L\left\{erfc(\sqrt{t})\right\} = \frac{1}{s} - \frac{1}{s\sqrt{s+1}} = \frac{\sqrt{s+1}-1}{s\sqrt{s+1}}$$

$$= \frac{1}{\sqrt{s+1}\,[\sqrt{s+1}+1]}$$

Example 5   Find $L^{-1}\left\{\frac{1}{1+\sqrt{1+s}}\right\}$

Solution:

$$\frac{1}{1+\sqrt{1+s}} = \frac{1-\sqrt{1+s}}{1-(1+s)} = \frac{1-\sqrt{1+s}}{-s}$$

$$= -\frac{1}{s} + \frac{\sqrt{1+s}}{s} = -\frac{1}{s} + \frac{1+s}{s\sqrt{1+s}}$$

$$= -\frac{1}{s} + \frac{1}{s\sqrt{s+1}} + \frac{1}{\sqrt{s+1}}$$

$$\therefore L^{-1}\left\{\frac{1}{1+\sqrt{1+s}}\right\} = -1 + erf(\sqrt{t}) + \frac{e^{-t}}{\sqrt{\pi t}}$$

$$= \frac{e^{-t}}{\sqrt{\pi t}} - erfc(\sqrt{t})$$

Example 6   Find $L\left\{\int_0^t erf(\sqrt{u})\,du\right\}$

Solution:
Since the integration of the object function between the limits 0 and t corresponds to the division of the transform by s, then

$$L\left\{\int_0^t erf(\sqrt{u})\,du\right\} = \frac{1}{s} \cdot \frac{1}{s\sqrt{s+1}} = \frac{1}{s^2\sqrt{s+1}}$$

11.2.17. The sine-integral function Si(t)

This is a tabulated function defined by

171

$$Si(t) = \int_0^t \frac{\sin u}{u} \, du$$

It is represented by the shaded area under the curve $y = \dfrac{\sin u}{u}$ between o and t.

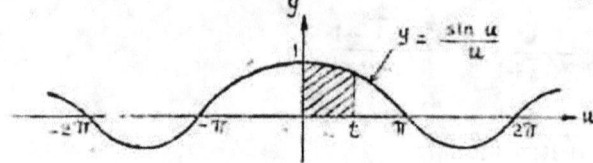

The function $(\sin u)/u$ is an even function which $\to 0$ as $u \to \infty$. It intersects the u — axis where
$$u = \pm\pi, \pm 2\pi, \pm 3\pi, \dots$$

Now $L\left\{\sin t\right\} = \dfrac{1}{s^2 + 1}$

$$\therefore \; L\left\{\frac{\sin t}{t}\right\} = \int_s^\infty \frac{dx}{x^2+1} = \left[\tan^{-1} x\right]_s^\infty$$

$$= \frac{\pi}{2} - \tan^{-1} s = \cot^{-1} s$$

$$\therefore \; L\left\{\int_0^t \frac{\sin u}{u} \, du\right\} = \frac{1}{s} \cot^{-1} s$$

i.e $L\left\{Si(t)\right\} = \dfrac{1}{s} \cot^{-1} s, \qquad s > 0,$

The following is an alternative method for finding $L\{Si(t)\}$

Put $F(t) = \int_0^t \dfrac{\sin u}{u} \, du$

then $F'(t) = \dfrac{\sin t}{t}$ or $t\,F'(t) = \sin t$

$$\therefore \; L\left\{t\,F'(t)\right\} = L\left\{\sin t\right\}$$

$$\therefore \; -\frac{d}{ds}\left[s\,f(s) - F(o)\right] = \frac{1}{s^2+1}$$

$$\therefore \; \frac{d}{ds}\left[s\,f(s)\right] = -\frac{1}{s^2+1}$$

Integrating
$$s\,f(s) = -\tan^{-1} s + C$$

According to the initial value theorem

172

$$\lim_{t \to 0} F(t) = \lim_{s \to \infty} s\, f(s)$$

and since $\lim_{t \to 0} F(t) = F(o) = o$, then $\lim_{s \to \infty} s\, f(s) = o$

$$\therefore \quad -\frac{\pi}{2} + C = o \quad \text{i.e } C = \frac{\pi}{2} \text{ and hence}$$

$$s\, f(s) = \frac{\pi}{2} - \tan^{-1}s = \cot^{-1}s$$

$$\therefore \quad f(s) = \frac{1}{s}\cot^{-1}s \quad \text{i.e } L\left\{ Si\,(t) \right\} = \frac{1}{s}\cot^{-1}s$$

11.2.18. Cosine -integral function Ci(t)

This is a tabulated function defined by

$$Ci(t) = -\int_{t}^{\infty} \frac{\cos u}{u}\, du , \quad t > o .$$

change the variable u into a new variable x by the relation u = xt where t is a constant, then

$$Ci(t) = -\int_{1}^{\infty} \frac{\cos xt}{xt}\, t dx = -\int_{1}^{\infty} \frac{\cos xt}{x}\, dx$$

$$\therefore \quad L\left\{ Ci(t) \right\} = -\int_{0}^{\infty} e^{-st} \left\{ \int_{1}^{\infty} \frac{\cos xt}{x}\, dx \right\} dt$$

$$= -\int_{1}^{\infty} \frac{1}{x} \left\{ \int_{0}^{\infty} e^{-st} \cos xt\, dt \right\} dx$$

(by interchanging the order of integration)

$$= -\int_{1}^{\infty} \frac{1}{x} \left[\frac{s}{s^2 + x^2} \right] dx = -\frac{s}{s^2} \int_{1}^{\infty} \left[\frac{1}{x} - \frac{x}{s^2 + x^2} \right] dx$$

$$= -\frac{1}{s} \left[\log x - \tfrac{1}{2} \log (s^2 + x^2) \right]_{1}^{\infty} = -\frac{1}{s} \left[\log \frac{x}{\sqrt{s^2 + x^2}} \right]_{1}^{\infty}$$

$$= \frac{1}{s} \log \frac{1}{\sqrt{s^2 + 1}} = -\frac{1}{2s} \log (s^2 + 1)$$

$$\therefore \quad L\left\{ Ci(t) \right\} = -\frac{1}{2s} \log (s^2 + 1) , \quad s > o$$

The transform can also be obtained in the following *way:*

$$\text{Put } F(t) = -\int_{t}^{\infty} \frac{\cos u}{u}\, du , \quad \text{then}$$

173

$$F'(t) = \frac{\cos t}{t} \qquad \text{i.e} \quad t\, F'(t) = \cos t$$

$$-\frac{d}{ds}\left[s\, f(s) - F(o)\right] = \frac{s}{s^2+1} \qquad \therefore \frac{d}{ds}\left[s\, f(s)\right] = -\frac{s}{s^2+1}$$

Integrating

$$s\, f(s) = -\tfrac{1}{2} \log (s^2 + 1) + C$$

And according to the final value theorem

$$\lim_{t \to \infty} F(t) = \lim_{s \to 0} s\, f(s)$$

$$\therefore \; o = \lim_{s \to 0} s\, f(s) = C \qquad \therefore \; s\, f(s) = -\tfrac{1}{2} \log (s^2 + 1)$$

$$\text{i.e} \quad f(s) = -\frac{1}{2s} \log (s^2 + 1) \qquad \therefore \; L\left\{Ci(t)\right\} = -\frac{1}{2s} \log (s^2 + 1)$$

11.2.19. Exponential integral function

This is defined by

$$Ei(t) = \int_{-\infty}^{t} \frac{e^u}{u}\, du \qquad (t < o)$$

This can be written

$$- Ei(-t) = \int_{t}^{\infty} \frac{e^{-y}}{y}\, dy = \int_{1}^{\infty} \frac{e^{-tx}}{x}\, dx \qquad (t > o)$$

the last integral being obtained by the substitution $y = tx$ where t is a constant.

$$L\left\{\int_{1}^{\infty} \frac{e^{-tx}}{x}\, dx\right\} = \int_{0}^{\infty} e^{-st}\left\{\int_{1}^{\infty} \frac{e^{-tx}}{x}\, dx\right\} dt$$

$$= \int_{1}^{\infty} \frac{1}{x}\left\{\int_{0}^{\infty} e^{-st}\, e^{-tx}\, dt\right\} dx = \int_{1}^{\infty} \frac{1}{x} \cdot \frac{1}{s+x}\, dx$$

$$= \frac{1}{s}\int_{1}^{\infty}\left[\frac{1}{x} - \frac{1}{s+x}\right] dx = \frac{1}{s}\left[\log \frac{x}{s+x}\right]_{1}^{\infty}$$

$$= -\frac{1}{s} \log \frac{1}{s+1} = \frac{1}{s} \log (s+1) \qquad \therefore \; L\left\{- Ei(-t)\right\} = \frac{1}{s} \log (s+1) ,$$

$$s > o$$

174

11.2.20. Evaluation of definite integrals using the Laplace transformation

Example 1   **Evaluate** $F(t) = \int_0^\infty \dfrac{\sin tx}{x(x^2 + 1)}\, dx$, $t > a$

Solution:

$$L\left\{F(t)\right\} = \int_0^\infty e^{-st}\left\{\int_0^\infty \frac{\sin tx}{x(x^2 + 1)}\, dx\right\} dt$$

$$= \int_0^\infty \frac{1}{x(x^3+1)}\left\{\int_0^\infty e^{-st}\sin tx\, dt\right\} dx$$

(by interchanging the order of integration)

$$= \int_0^\infty \frac{1}{x(x^3+1)} \cdot \frac{x}{s^3+x^2}\, dx = \int_0^\infty \frac{1}{(x^3+1)(x^2+s^2)}\, dx$$

$$= \frac{1}{s^2-1}\int_0^\infty\left[\frac{1}{x^2+1} - \frac{1}{x^2+s^2}\right] dx$$

$$= \frac{1}{s^2-1}\left[\tan^{-1} x - \frac{1}{s}\tan^{-1}\frac{x}{s}\right]_0^\infty$$

$$= \frac{1}{s^2-1}\left(\frac{\pi}{2} - \frac{1}{s}\frac{\pi}{2}\right) = \frac{\pi}{2}\cdot\frac{1}{s^2-1}\left(1 - \frac{1}{s}\right)$$

$$= \frac{\pi}{2}\cdot\frac{1}{s(s+1)} = \frac{\pi}{2}\left[\frac{1}{s} - \frac{1}{s+1}\right]$$

$$\therefore\ F(t) = \frac{\pi}{2}\left[1 - e^{-t}\right].$$

Example 2   **Evaluate** $\int_0^\infty e^{-x^2}dx$

Solution:

Consider $F(t) = \int_0^\infty e^{-tx^2}\, dx$, $t > 0$

$$L\left\{F(t)\right\} = \int_0^\infty e^{-st}\left\{\int_0^\infty e^{-tx^2}dx\right\} dt$$

$$= \int_0^\infty \left\{\int_0^\infty e^{-st}\cdot e^{-tx^2}\, dt\right\} dx$$

175

$$= \int_0^\infty \frac{1}{s+x^2} \, dx = \left[\frac{1}{\sqrt{s}} \tan^{-1} \frac{x}{\sqrt{s}}\right]_0^\infty = \frac{1}{\sqrt{s}} \cdot \frac{\pi}{2}$$

$$\therefore \quad F(t) = \frac{\pi}{2} L^{-1}\left\{\frac{1}{\sqrt{s}}\right\} = \frac{\pi}{2} \cdot \frac{1}{\sqrt{\pi t}} \qquad (1.2)$$

$$= \frac{\sqrt{\pi}}{2} \cdot \frac{1}{\sqrt{t}}$$

$$\because \int_0^\infty e^{-tx^2} dx = \frac{\sqrt{\pi}}{2} \cdot \frac{1}{\sqrt{t}}$$

Putting t = 1 *we* get

$$\int_0^\infty e^{-x^2} \, dx = \frac{\sqrt{\pi}}{2}$$

Example 3   Evaluate $\int_0^t J_o(t-\lambda) \, J_o(\lambda) \, d\lambda$

Solution:

The integral is the convolution $J_o(t)^* J_o(t)$ and its transform is

$$\frac{1}{\sqrt{s^2+1}} \cdot \frac{1}{\sqrt{s^2+1}} = \frac{1}{s^2+1}$$

Taking the inverse transform we get

$$\int_0^t J_o(t-\lambda) \, J_o(\lambda) \, d\lambda = \sin t$$

Example 4   Evaluate $F(t) = \int_0^\infty \frac{\cos tx}{\sqrt{x}} \, dx$, t > 0

Solution:

$$L\left\{F(t)\right\} = \int_0^\infty e^{-st}\left\{\int_0^\infty \frac{\cos tx}{\sqrt{x}} \, dx\right\} dt$$

$$= \int_0^\infty \frac{1}{\sqrt{x}}\left\{\int_0^\infty e^{-st} \cos tx \, dt\right\} dx = \int_0^\infty \frac{1}{\sqrt{x}} \cdot \frac{s}{s^2+x^2} \, dx$$

176

Put $x = s \tan \Theta$ $\therefore$ $dx = s \sec^2\Theta \ d\Theta$

$$\therefore \quad f(s) = \int_0^{\frac{\pi}{2}} \frac{1}{\sqrt{s \tan\Theta}} \cdot \frac{s}{s^2\sec^2\Theta} \cdot s \sec^2\Theta \ d\Theta$$

$$= \frac{1}{2\sqrt{s}} \int_0^{\frac{\pi}{2}} 2 \cos^{\frac{1}{2}}\Theta \sin^{-\frac{1}{2}}\Theta \ d\Theta$$

$$= \frac{1}{2\sqrt{s}} \beta\left(\frac{1}{4}, \frac{3}{4}\right) = \frac{1}{2\sqrt{s}} \frac{\Gamma(^1/_4) \ \Gamma(^3/_4)}{\Gamma(1)}$$

$$= \frac{1}{2\sqrt{s}} \frac{\pi}{\sin\frac{\pi}{4}}, \text{ since } \Gamma(m) \ \Gamma(1-m) = \frac{\pi}{\sin m\pi} \ (0 < m < 1)$$

$$= \frac{\pi}{\sqrt{2}} \cdot \frac{1}{\sqrt{s}}$$

$$\therefore \quad F(t) = \frac{\pi}{\sqrt{2}} \cdot \frac{1}{\sqrt{\pi t}} = \sqrt{\frac{\pi}{2t}}$$

$$\therefore \quad \int_0^{\infty} \frac{\cos tx}{\sqrt{x}} dx = \sqrt{\frac{\pi}{2t}}$$

Example 5   Evaluate $\displaystyle\int_0^{\infty} \sin x^2 \ dx$

Solution:

Let $F(t) = \displaystyle\int_0^{\infty} \sin tx^2 \ dx$, $t > 0$

$$\therefore \quad L\left\{F(t)\right\} = \int_0^{\infty} e^{-st} \left\{\int_0^{\infty} \sin tx^2 \ dx\right\} dt$$

$$= \int_0^{\infty} \left\{\int_0^{\infty} e^{-st} \sin tx^2 \ dt\right\} dx$$

$$= \int_0^{\infty} \frac{x^2}{s^2 + x^4} \ dx$$

Put $x^2 = s \tan \theta$ $\therefore$ $2xdx = s \sec^2\theta \ d\theta$

when $x = 0$, $\theta = 0$ and when $x \to \infty$, $\theta \to \dfrac{\pi}{2}$

177

$$\text{Integral} = \frac{1}{2}\int_0^{\frac{\pi}{2}} \frac{\sqrt{s\,\tan\theta}\; s\,\sec^2\theta\; d\theta}{s^2\,\sec^2\theta}$$

$$= \frac{1}{2\sqrt{s}}\int_0^{\frac{\pi}{2}} \sqrt{\tan\theta}\; d\theta = \frac{1}{2\sqrt{s}}\int_0^{\frac{\pi}{2}} \sin^{\frac{1}{2}}\theta\; \cos^{-\frac{1}{2}}\theta\; d\theta$$

$$= \frac{1}{4\sqrt{s}}\int_0^{\frac{\pi}{2}} 2\,\sin^{\frac{1}{2}}\theta\; \cos^{-\frac{1}{2}}\theta\; d\theta$$

$$= \frac{1}{4\sqrt{s}}\, \beta\,(\tfrac{1}{4},\tfrac{3}{4}) = \frac{1}{4\sqrt{s}}\, \frac{\Gamma(\tfrac{1}{4})\,\Gamma(\tfrac{3}{4})}{\Gamma(1)}$$

$$= \frac{1}{4\sqrt{s}}\, \frac{\pi}{\sin\frac{\pi}{4}} = \frac{\pi}{2\sqrt{2}\sqrt{s}}$$

inverting we get

$$F(t) = \int_0^{\infty} \sin tx^2\, dx = \frac{\pi}{2\sqrt{2}}\cdot\frac{1}{\sqrt{\pi t}} = \frac{\sqrt{\pi}}{2\sqrt{2}\sqrt{t}}$$

Put $t = 1$

$$\int_0^{\infty} \sin x^2\, dx = \frac{1}{2}\sqrt{\frac{\pi}{2}}$$

11.2.21. Heaviside's expansion formulae

Let $f(s) = p(s)/q(s)$ where $p(s)$ and $q(s)$ are polynomials in s, $p(s)$ being of lower degree than $q(s)$.

<u>Case I</u> : Consider the case in which all factors of $q(s)$ are linear and distinct.

Let $q(s) = (s-a_1)(s-a_2) \dots (s-a_n)$, the a's being all distinct.

Then $f(s) = \dfrac{p(s)}{q(s)} = \dfrac{A_1}{s-a_1} + \dfrac{A_2}{s-a_2} + \dots + \dfrac{A_r}{s-a_r}$

$$+ \dots + \frac{A_n}{s-a_n}$$

Clearing fractions we get

$$p(s) = A_1\,(s-a_2)\,(s-a_3) \dots (s-a_n)$$
$$+ A_2\,(s-a_1)\,(s-a_3) \dots (s-a_n)$$
$$+ \cdot \; \cdot \; \cdot \; \cdot \; \cdot \; \cdot \; \cdot \; \cdot \; \cdot$$
$$+ A_n\,(s-a_1)\,(s-a_2) \dots (s-a_{n-1})$$

Put $s = a_1$, then

$$p(a_1) = A_1 (a_1-a_2) (a_1-a_3) \dots (a_1-a_n)$$

$$\therefore A_1 = \frac{p(a_1)}{(a_1-a_2)(a_1-a_3) \dots (a_1-a_n)}$$

Hence A_1 is obtained by substituting a_1 in the numerator and in all factors of the denominator except the factor $s-a_1$ itself. Similarly for $A_2, A_3 \dots, A_n$.

Now
$$\begin{aligned} q'(s) &= (s-a_2)(s-a_3) \dots (s-a_n) \\ &+ (s-a_1)(s-a_3) \dots (s-a_n) \\ &+ \dots \dots \dots \dots \\ &+ (s-a_1)(s-a_2) \dots (s-a_{n-1}) \end{aligned}$$

$$\therefore q'(a_1) = (a_1-a_2)(a_1-a_3) \dots (a_1-a_n)$$

Hence we can write $A_1 = \dfrac{p(a_1)}{q'(a_1)}$ and in general $A_r = \dfrac{p(a_r)}{q'(a_r)}$

$$\therefore f(s) = \frac{p(s)}{q(s)} = \sum_{r=1}^{n} \frac{p(a_r)}{q'(a_r)} \cdot \frac{1}{s-a_r}$$

$$\therefore F(t) = L^{-1} \left\{ \frac{p(s)}{q(s)} \right\} = \sum_{r=1}^{n} \frac{p(a_r)}{q'(a_r)} e^{a_r t}$$

Case II: g(s) has repeated linear factors.

Suppose that the denominator q(s) has a repeated linear factor $(s-a)^n$ and that

$$\frac{p(s)}{q(s)} = \frac{\phi(s)}{(s-a)^n}$$

Where $\phi(s)$ consists of the numerator p(s) and all factors of q(s) except $(s-a)^n$

$$\begin{aligned} \phi(s) &= \phi(a+\overline{s-a}) = \phi(a)+(s-a) \phi'(a) + \frac{(s-a)^2}{2!} \phi''(a)+\dots \\ &+ \frac{(s-a)^{n-1}}{(n-1)!} \phi^{(n-1)}(a) + (s-a)^n h(s). \end{aligned}$$

$$\begin{aligned} \therefore \frac{p(s)}{q(s)} &= \frac{\phi(s)}{(s-a)^n} = \frac{\phi(a)}{(s-a)^n} + \frac{\phi'(a)}{(s-a)^{n-1}} + \frac{\phi''(a)}{2!(s-a)^{n-2}} + \dots \\ &+ \frac{\phi^{(n-1)}(a)}{(n-1)!} \frac{1}{s-a} + h(s). \end{aligned}$$

$$\therefore L^{-1}\left\{\frac{p(s)}{q(s)}\right\} = e^{at}\left[\frac{t^{n-1}}{(n-1)!}\phi(a) + \frac{t^{n-2}}{(n-1)!}\phi'(a)\right.$$

$$+ \frac{t^{n-3}}{(n-3)!}\frac{\phi''(a)}{!} + \dots + \left.\frac{\phi^{(n-1)}(a)}{(n-1)!}\right] + H(t)$$

$$= e^{at}\sum_{r=1}^{n}\frac{t^{n-r}}{(n-r)!}\frac{\phi^{(r-1)}(a)}{(r-1)!}$$

Case III: q(s) has quadratic factors.

let $\dfrac{p(s)}{q(s)} = \dfrac{\phi(s)}{(s-a)^2+b^2}$ where $\phi(s)$ consists of p(s)

and all factors of q (s) except the factor $(s-a)^2+b^2$ itself.

Let $\dfrac{\phi(s)}{(s-a)^2+b^2} = \dfrac{As+B}{(s-a)^2+b^2} + h(s)$

$\therefore \phi(s) \equiv As + B + [(s-a)^2+b^2]\, h(s)$

Put s = a + ib

$\therefore \phi(a+ib) = A(a+ib) + B$

Let $\phi(a+ib) = \phi_1 + i\phi_2$

$\therefore \phi_1 + i\phi_2 = Aa + B + iAb$

$\therefore \phi_1 = Aa + B,\ \phi_2 = Ab$

$\therefore A = \dfrac{\phi_2}{b},\ \phi_1 = \dfrac{a}{b}\phi_2 + B$ i.e $B = \dfrac{b\phi_1 - a\phi_2}{b}$

$$\frac{\phi(s)}{(s-a)^2+b^2} = \frac{\dfrac{\phi_2}{b}s + \dfrac{b\phi_1 - a\phi_2}{b}}{(s-a)^2 + b^2} + h(s)$$

$$= \frac{1}{b}\frac{\phi_2 s + b\phi_1 - a\phi_2}{(s-a)^2 + b^2} + h(s)$$

$$= \frac{1}{b}\frac{\phi_2(s-a) + b\phi_1}{(s-a)^2 + b^2} + h(s)$$

$$\therefore L^{-1}\left\{\frac{p(s)}{q(s)}\right\} = \frac{1}{b}e^{at}[\phi_2\cos bt + \phi_1\sin bt] + H(t)$$

Case IV: q(s) has repeated quadratic factors.

Consider for example the particular case when

$$\frac{p(s)}{q(s)} = \frac{\phi(s)}{[(s-a)^2+b^2]^2} = \frac{As+B}{[(s-a)^2+b^2]^2} + \frac{Cs+D}{(s-a)^2+b^2} + h(s)$$

then $\phi(s) \equiv As+B+(Cs+D)[(s-a)^2+b^2] + h(s)[(s-a)^2+b^2]^2 \dots (1)$

The four unknowns A,B,C,D can be obtained by substituting $s = a + i\,b$ in (1) and in the equation derived by differentiating (I) w.r.t. s and equating real to real and imaginary to imaginary on both sides in each case.

11.2.22. The inversion integral

The inversion integral is a powerful as well as a direct mean for finding inverse Laplace transforms. We have so far assumed s to be real. We shall now consider s to be complex.
Let F(t) be a real function of the positive, real variable t, sectionally continuous in each finite interval $0 \le t \le T$ and of exponential order as $t \to \infty$

$$f(s) = \int_0^\infty e^{-st} F(t)\ dt \quad \text{where} \quad s = x + iy$$

then the inversion integral formula is given by

$$F(t) = \frac{1}{2\pi i} \lim_{\beta \to \infty} \int_{\gamma - i\beta}^{\gamma + i\beta} e^{st} f(s)\ ds \quad \cdots \quad (1)$$

This formula is also known as <u>Bromwich's integral formula,</u> and it is a modified form of the Fourier integral formula. The integration is carried in the complex plane of s along the line $x = y$ which is taken far enough to

the right such that all poles, branch points or essential singularities of $e^{-st} f(s)$ lie to the left of it. In practice, the integral (1) is evaluated by considering first the contour integral

$$\frac{1}{2\pi i} \int_C e^{st} f(s)\ ds$$

where C is the contour consisting of the line AB whose equation is $x = y$ and the circular arc BDA which we

denote by Γ of a circle centre O and radius R. R is taken large enough such that all poles of $e^{-st} f(s)$ lie inside C, and in the limit R is made infinite.

Now since $\beta = \sqrt{R^3 - \gamma^2}$, hence when $\beta \to \infty$, $R \to \infty$ and according to the inversion formula

$$F(t) = \lim_{R \to \infty} \frac{1}{2\pi i} \int_{\gamma - i\beta}^{\gamma + i\beta} e^{st} f(s)\ ds$$

$$= \lim_{R \to \infty} \left\{ \frac{1}{2\pi i} \int_C e^{st} f(s)\,ds - \frac{1}{2\pi i} \int_\Gamma e^{st} f(s)\ ds \right\}$$

181

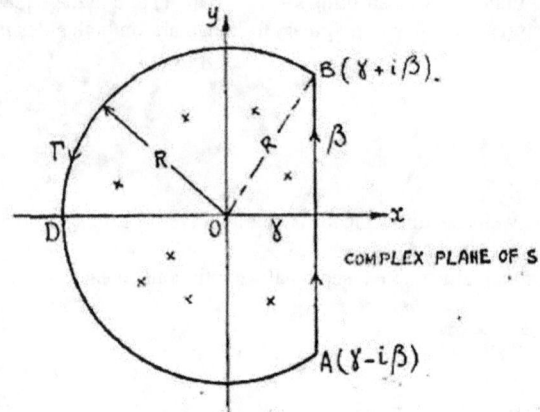

Assuming that all singularities of f(s) are poles and that

$$\int_{\Gamma} e^{st} f(s) \, ds \longrightarrow 0 \text{ as } R \longrightarrow \infty,$$

then, since by Cauchy's theorem of residues

$$\int_{C} e^{st} f(s) \, ds = 2\pi i$$

(sum of residues of $e^{st} f(s)$ at its poles inside C) it follows that

F(t) = sum of residues of $e^{st} f(s)$ at poles of f(s).

This result was obtained on the assumption that

$$\int_{\Gamma} e^{st} f(s) \, ds \longrightarrow 0 \text{ as } R \longrightarrow \infty.$$

This assumption is satisfied in almost all physical problems. In fact a sufficient condition for the ultimate vanishing of the above integral is the following.

If we can find two positive constants M and k such that on Γ where $s = Re^{i\theta}$

$$| f(s) | < \frac{M}{R^k}$$

then the integral around Γ of $e^{st} f(s) \longrightarrow 0$ as $R \longrightarrow \infty$

$$\text{i.e } \lim_{R \longrightarrow \infty} \int_{\Gamma} e^{st} f(s) \, ds = 0$$

11.2.23. Formulae for residues

I. Simple poles.

Let

$$\frac{p(s)}{q(s)} = \frac{\phi(s)}{s-s_0}$$

where $\phi(s)$ consists of p(s) and all factors of q(s) except the factor $s-s_0$ itself. Here, we have a simple pole at $s-s_0$.

Now $\phi(s) = \phi(s_0+s-s_0) = \phi(s_0) + (s-s_0)\phi'(s_0) + \frac{(s-s_0)^2}{2!}\phi''(s_0) + \cdots$

$\therefore \quad \frac{\phi(s)}{s-s_0} = \frac{\phi(s_0)}{s-s_0} + \phi'(s_0) + \frac{s-s_0}{2!}\phi'(s_0) + \cdots$

$\therefore$ **Residue at $s=s_0$ is** $\phi(s_0) = \lim_{s \to s_0}(s-s_0)\frac{\phi(s)}{s-s_0}$

$$= \lim_{s \to s_0}(s-s_0)\frac{p(s)}{q(s)}$$

II. Multiple poles.

Let $\dfrac{p(s)}{q(s)} = \dfrac{\phi(s)}{(s-s_0)^m}$

$\phi(s) = \phi(s_0) + (s-s_0)\phi'(s_0) + \frac{(s-s_0)^2}{2!}\phi''(s_0) + \cdots$

$$+ \frac{(s-s_0)^{m-1}}{(m-1)!}\phi^{(m-1)}(s_0) + \cdots$$

$\therefore \quad \dfrac{\phi(s)}{(s-s_0)^m} = \dfrac{\phi(s_0)}{(s-s_0)^m} + \dfrac{\phi'(s_0)}{(s-s_0)^{m-1}} + \cdots$

$$+ \frac{1}{s-s_0}\frac{\phi^{(m-1)}(s_0)}{(m-1)!} + \cdots$$

$\therefore$ Residue at the pole $s=s_0$ of multiplicity m is given by

$$\frac{1}{(m-1)}\phi^{(m-1)}(s_0) = \frac{1}{(m-1)!}\left[\frac{d^{m-1}}{ds^{m-1}}\phi(s)\right]_{s=s_0}$$

Example 1

Find $L^{-1}\left\{\dfrac{t}{(s-1)(s-2)(s-3)}\right\}$

183

Solution:
This is equal to the sum of residues of

$$\frac{e^{st}}{(s-1)\ (s-2)\ (s-3)}$$

at its poles.

s = 1

$$\text{Residue} = \lim_{s \to 1} (s-1) \frac{e^{st}}{(s-1)(s-2)(s-3)}$$

$$= \lim_{s \to 1} \frac{e^{st}}{(s-2)\ (s-3)} = \frac{e^t}{2}$$

s = 2

$$\text{Residue} = \lim_{s \to 2} \frac{e^{st}}{(s-1)(s-3)} = \frac{e^{2t}}{-1} = -e^{2t}$$

s = 3

$$\text{Residue} = \lim_{s \to 3} \frac{e^{st}}{(s-1)(s-2)} = \tfrac{1}{2} e^{3t}$$

$$\therefore \quad L^{-1}\left\{ \frac{1}{(s-1)\ (s-2)\ (s-3)} \right\} = \tfrac{1}{2} e^t - e^{2t} + \tfrac{1}{2} e^{3t}$$

Example 2

$$\text{Find} \quad L^{-1}\left\{ \frac{1}{s^2(s+1)} \right\}$$

Solution:
This is equal to the sum of residues of

$$\frac{e^{st}}{s^2(s+1)}$$

at its poles.
The residue at the double pole at s = 0 is

$$\left\{ \frac{d}{ds} \frac{e^{st}}{s+1} \right\}_{s=0} = \left\{ \frac{(s+1)\ te^{st} - e^{st}}{(s+1)^2} \right\}_{s=0} = t - 1$$

The residue at s = -1 is

$$\lim_{s \to -1} \frac{e^{st}}{s^2} = e^{-t}$$

$$\therefore \quad L^{-1}\left\{ \frac{1}{s^2(s+1)} \right\} = e^{-t} + t - 1$$

Example 3   Find $L^{-1}\left\{ \dfrac{1}{s^2(s^2+\omega^2)} \right\}$

Solution:

This is equal to the sum of residues of

$$\frac{e^{st}}{s^2(s-i\omega)(s+i\omega)}$$

at its poles.

Residue at the double pole at $s = 0$ is

$$\left\{ \frac{d}{ds}\frac{e^{st}}{s^2+\omega^2} \right\}_{s=0} = \left\{ \frac{(s^2+\omega^2)\,te^{st} - 2se^{st}}{(s^2+\omega^2)^2} \right\}_{s=0} = \frac{t}{\omega^2}$$

Sum of residues at $s = \pm i\omega$ is

$$\left\{ \frac{e^{st}}{s^2(s+i\omega)} \right\}_{s=i\omega} + \left\{ \frac{e^{st}}{s^2(s-i\omega)} \right\}_{s=-i\omega}$$

$$= -\frac{1}{\omega^3}\left[\frac{e^{i\omega t} - e^{-i\omega t}}{2i} \right] = -\frac{\sin \omega t}{\omega^3}$$

$$\therefore \quad L^{-1}\left\{ \frac{1}{s^2(s^2+\omega^2)} \right\} = \frac{1}{\omega^3}(\omega t - \sin \omega t)$$

11.2.24. Inversion in the case of branch points

The path of integration of the inversion integral has to be modified if the integrand has a branch point. This is illustrated by the following example.

Suppose it is required to find the inverse transform of $s^{-\frac{1}{2}}$. Here the function $s^{-\frac{1}{2}}$ has a branch point at the origin. We therefore cut the s-plane along the negative x-axis and take the path shown in this sketch.

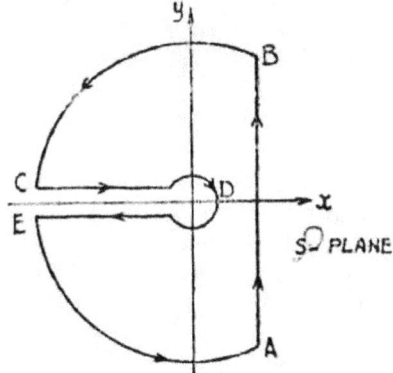

Since $\dfrac{e^{st}}{\sqrt{s}}$ has no singularities within this closed path, hence according to Cauchy's theorem we have:

$$\int_{AB} \frac{e^{st}\,ds}{\sqrt{s}} + \int_{BC+EA} \frac{e^{st}\,ds}{\sqrt{s}} + \int_{CDE} \frac{e^{st}\,ds}{\sqrt{s}} = 0$$

185

It is easy to show that the integral along the circular path of infinite radius is zero and hence

$$\int_{AB} \frac{e^{st}\, ds}{\sqrt{s}} = -\int_{CDE} \frac{e^{st}\, ds}{\sqrt{s}} = \int_{EDC} \frac{e^{st}\, ds}{\sqrt{s}} \quad \text{in the limit,}$$

when the radius of the circular path is infinite.

Now $\displaystyle\int_{EDC}$ consists of three puts, one around the small circle at D and the others along the straight paths ED and DC. We shall now evaluate each of the three integrals. Consider first the integral around the small circle at D. Taking the radius of this small circle to be equal r, then we can put for points on the circle:

$$s = re^{i\theta}, \; ds = ire^{i\theta}\, d\theta, \; s^{\frac{1}{2}} = r^{\frac{1}{2}}\, e^{i\frac{\theta}{2}}$$

$$\therefore \int_{\text{small circle}} \frac{e^{st}\, ds}{\sqrt{s}} = i\int_{r\to 0} e^{rt(\cos\theta + i\sin\theta)}\, r^{\frac{1}{2}} e^{i\frac{\theta}{2}}\, d\theta = 0$$

Next, consider the integral along ED and put for points on ED

$$s = xe^{-i\pi} = -x, \; ds = -dx, \; s^{\frac{1}{2}} = x^{\frac{1}{2}} e^{-i\frac{\pi}{2}} = -ix^{\frac{1}{2}}$$

$$\therefore \int_{ED} \frac{e^{st}\, ds}{\sqrt{s}} = -i\int_{\infty}^{0} x^{-\frac{1}{2}} e^{-xt}\, dx = i\int_{0}^{\infty} x^{-\frac{1}{2}} e^{-xt}\, dx$$

Finally along DC, put

$$s = xe^{i\pi} = -x, \; ds = -dx, \; s^{\frac{1}{2}} = x^{\frac{1}{2}} e^{i\frac{\pi}{2}} = ix^{\frac{1}{2}}$$

$$\therefore \int_{DC} \frac{e^{st}\, ds}{\sqrt{s}} = i\int_{0}^{\infty} x^{-\frac{1}{2}} e^{-xt}\, dx$$

Hence the sum of the three integrals is

$$2i\int_{0}^{\infty} x^{-\frac{1}{2}} e^{-xt}\, dx$$

Put $xt = \lambda^2$ where t is constant. $\therefore t\, dx = 2\lambda\, d\lambda$

$$\therefore 2i\int_{0}^{\infty} x^{-\frac{1}{2}} e^{-xt}\, dx = 2i\int_{0}^{\infty} \frac{\sqrt{t}}{\lambda} e^{-\lambda^2} \frac{2\lambda}{t}\, d\lambda$$

$$= \frac{4i}{\sqrt{t}}\int_{0}^{\infty} e^{-\lambda^2}\, d\lambda = \frac{4i}{\sqrt{t}} \cdot \frac{\sqrt{\pi}}{2} = 2i\sqrt{\frac{\pi}{t}}$$

Hence,

$$L^{-1}\left\{\frac{1}{\sqrt{s}}\right\} = \frac{1}{2\pi i} \times 2i\sqrt{\frac{\pi}{t}} = \sqrt{\frac{1}{\pi t}}$$

11.2.25. Miscellaneous Examples on Laplace Transform

Example 1   If $L^{-1}\left\{f(s)\right\} = F(t)$, determine $L^{-1}\left\{\dfrac{f(s)}{\sinh cs}\right\}$

$$\frac{f(s)}{\sinh cs} = \frac{2f(s)}{e^{cs} - e^{-cs}} = \frac{2f(s)e^{-cs}}{1 - e^{-2cs}}$$

Now $\quad \dfrac{1}{1 - e^{-2cs}} = \sum_{n=0}^{\infty} e^{-2ncs}$

$\therefore \quad \dfrac{f(s)}{\sinh cs} = 2 \sum_{n=0}^{\infty} f(s)e^{(-2nc-c)s}$

$\therefore \quad L^{-1}\left\{\dfrac{f(s)}{\sinh cs}\right\} = 2 \sum_{n=0}^{\infty} F(t-2nc-c)U(t-2nc-c)$

Example 2   Solve the differential equation

$$x''(t) + x(t) = F(t) , \quad x(o) = x'(o) = o \quad \text{in which}$$
$$F(t) = 4 \qquad\qquad o \leqslant t \leqslant 2$$
$$= t + 2 \qquad\qquad t > 2$$

Solution:

$$f(s) = \int_{0}^{2} 4e^{-st} \, dt + \int_{2}^{\infty} (t+2) e^{-st} \, dt$$

$$= 4\left[\frac{e^{-st}}{-s}\right]_{0}^{2} - \frac{1}{s}\int_{2}^{\infty} (t+2) \, de^{-st}$$

$$= 4\left[\frac{e^{-st}}{-s}\right]_{0}^{2} - \frac{1}{s}\left[(t+2)e^{-st} + \frac{e^{-st}}{s}\right]_{2}^{\infty}$$

$$= \frac{4}{s} + \frac{e^{-2s}}{s^2}$$

$\therefore \quad (s^2+1) \, \bar{x} = \dfrac{4}{s} + \dfrac{e^{-2s}}{s^2}$

$\therefore \quad \bar{x} = \dfrac{4}{s(s^2 + 1)} + \dfrac{e^{-2s}}{s^2 (s^2 + 1)}$

$$= 4\left[\frac{1}{s} - \frac{s}{s^2 + 1}\right] + \left[\frac{1}{s^2} - \frac{1}{s^2+1}\right]e^{-2s}$$

$\therefore \quad x(t) = 4 - 4\cos t + [t - 2 - \sin (t-2)] \, U(t-2)$

Example 3   Find $L^{-1}\left\{\dfrac{1}{\sqrt{s}+1}\right\}$

Solution:

$$\frac{1}{\sqrt{s}+1}=\frac{\sqrt{s}-1}{s-1}=-\frac{1}{s-1}+\frac{\sqrt{s}}{s-1}=-\frac{1}{s-1}+\frac{s}{\sqrt{s}}\left(\frac{1}{s-1}\right)$$

$$=-\frac{1}{s-1}+\frac{1}{\sqrt{s}}\left(\frac{s-1+1}{s-1}\right)=-\frac{1}{s-1}+\frac{1}{\sqrt{s}}\left(1+\frac{1}{s-1}\right)$$

$$=-\frac{1}{s-1}+\frac{1}{\sqrt{s}}+\frac{1}{\sqrt{s}\,(s-1)}$$

$$\therefore\ L^{-1}\left\{\frac{1}{\sqrt{s}+1}\right\}=-e^{t}+\frac{1}{\sqrt{\pi t}}+e^{t}\ \mathrm{erf}\,(\sqrt{t})$$

$$=\frac{1}{\sqrt{\pi t}}+e^{t}\,(\mathrm{erf}\,(\sqrt{t})-1)\ =\frac{1}{\sqrt{\pi t}}-e^{t}\,\mathrm{erfc}\,\sqrt{t}$$

Example 4.   **Prove that** $\displaystyle\int_{-\infty}^{\infty}\frac{x\sin tx}{a^{2}+x^{2}}\,dy=\pi e^{-at}\ (a>0,\ t>0).$

By differentiating under the sign of integration w.r.t. a

deduce the value of $\displaystyle\int_{-\infty}^{\infty}\frac{x\sin x}{(a^{2}+x^{2})2}\,dx$

Solution:

$$\int_{-\infty}^{\infty}\frac{x\sin tx}{a^{2}+x^{2}}\,dx=2\int_{0}^{\infty}\frac{x\sin tx}{a^{2}+x^{2}}\,dx,\ \text{ the integrand being}$$

an even function of x.

$$L\left\{\int_{0}^{\infty}\frac{x\sin tx}{a^{2}+x^{2}}\,dx\right\}=\int_{0}^{\infty}e^{-st}\left\{\int_{0}^{\infty}\frac{x\sin tx}{a^{2}+x^{2}}\,dx\right\}dt$$

$$=\int_{0}^{\infty}\frac{x}{a^{2}+x^{3}}\left\{\int_{0}^{\infty}e^{-st}\sin tx\,dt\right\}dx$$

$$=\int_{0}^{\infty}\frac{x}{a^{2}+x^{2}}\cdot\frac{x}{s^{2}+x^{2}}\,dx=\int_{0}^{\infty}\frac{x^{2}\,dx}{(x^{2}+a^{2})\,(x^{2}+s^{2})}$$

$$=\frac{1}{s^{2}-a^{2}}\int_{0}^{\infty}\left[\frac{s^{2}}{x^{2}+s^{2}}-\frac{a^{2}}{x^{2}+a^{2}}\right]dx$$

188

$$= \frac{1}{s^2 - a^2} \left[s \tan^{-1} \frac{x}{s} - a \tan^{-1} \frac{x}{a} \right]_0^\infty$$

$$= \frac{1}{s^2 - a^2} \cdot \frac{\pi}{2} (s-a) = \frac{\pi}{2(s+a)}$$

$$\therefore \int_0^\infty \frac{x \sin tx}{a^3 + x^3} dx = \frac{\pi}{2} e^{-at}, \quad \therefore \int_{-\infty}^\infty \frac{x \sin tx}{a^2 + x^2} dx = \pi e^{-at}$$

Putting t =1 and differentiating under the sign of integration w.r,t a we get

$$- 2a \int_{-\infty}^\infty \frac{x \sin x}{(a^2+x^2)^2} dx = - \pi e^{-a}$$

$$\therefore \int_{-\infty}^\infty \frac{x \sin x}{(a^2+x^2)^2} dx = \frac{1}{2a} \pi e^{-a}$$

Example 5   Solve, using an inversion integral, the differential equation

$$(D^2 - 2D + 2)(D^2 + 2D - 3) x = 0, \ t > 0$$

given that $x(o) = 1$, $x'(o) = 0$, $x''(o) = 6$, $x'''(o) = -14$

Solution:

The equation may be written
$$(D^4 - 5D^2 + 10D - 6) x = 0$$

$$s^4 \bar{x} - s^3 x(o) - s^2 x'(o) - s x''(o) - x'''(o)$$

$$-5 [s^2 \bar{x} - s x(o) - x'(o)] + 10 [s \bar{x} - x(o)] - 6 \bar{x} = 0$$

$$\therefore [s^4 - 5s^2 + 10s - 6] \bar{x} = s^3 + 6s - 14 - 5s + 10$$

i.e $(s^2 - 2s + 2)(s^2 + 2s - 3) \bar{x} = s^3 + s - 4$

$$\therefore \ \bar{x} = \frac{s^3 + s - 4}{(s^2 - 2s + 2)(s+3)(s-1)}$$

$\therefore \ x = $ sum of residues of $\dfrac{(s^3 + s - 4) e^{st}}{(s^2 - 2s + 2)(s+3)(s-1)}$ at its poles.

Residue at $s = 1 \pm i$ is $\frac{1}{2} (1 \mp i) e^{(1 \pm i)t}$

Residue at $s = -3$ is $\frac{1}{2} e^{-3t}$

Residue at $s = 1$ is $-\frac{1}{2} e^t$

Adding we get

$$x = e^t (\cos t + \sin t) - \tfrac{1}{2} e^t + \tfrac{1}{2} e^{-3t}$$

Example 6   **Prove that** $\displaystyle\int_0^\infty J_0(t)\, dt = 1$

We have shown that

$$L\left\{ J_0(t) \right\} = \frac{1}{\sqrt{s^2+1}} \quad (\,2.10\,)$$

$$\therefore \int_0^\infty e^{-st} J_0(t)\, dt = \frac{1}{\sqrt{s^2+1}}$$

Let $s \to 0 +$ we get

$$\int_0^\infty J_0(t)\, dt = 1$$

Example 7   **Find** $L\left\{ t \sinh 3t \right\}$

$$L\left\{ t \sinh 3t \right\} = -\frac{d}{ds}\left\{ \frac{3}{s^2 - 9} \right\} = \frac{6s}{(s^2-9)^2}$$

Example 8   **Evaluate** $\displaystyle\int_0^\infty t e^{-2t} \sin t\, dt$

Solution:

$$L\left\{ t \sin t \right\} = -\frac{d}{ds}\,\frac{1}{s^2+1} = \frac{2s}{(s^2+1)^2}$$

$$\therefore \int_0^\infty e^{-st}\, t \sin t\, dt = \frac{2s}{(s^2+1)^2}$$

Let $s \to 2$ $\therefore \displaystyle\int_0^\infty t e^{-2t} \sin t\, dt = \frac{4}{25}$

190

Example 9   Find $L\left\{F(t)\right\}$ where

$$F(t) = t \qquad 0 < t < 1$$

$$= 0 \qquad 1 < t < 2$$

$$F(t+2) = F(t)$$

$$L\left\{F(t)\right\} = \frac{\displaystyle\int_0^2 e^{-st} F(t)\, dt}{1 - e^{-2s}} = \frac{\displaystyle\int_0^1 e^{-st} t\, dt}{1 - e^{-2s}}$$

$$\int_0^1 e^{-st} t\, dt = -\frac{1}{s}\int_0^1 t\, de^{-st} = -\frac{1}{s}\left[\, t e^{-st} + \frac{e^{-st}}{2} \,\right]_0^1$$

$$= \frac{1}{s}\left[\frac{1}{s} - e^{-s} - \frac{e^{-s}}{s}\right] = \frac{1}{s^2}\left[\, 1 - (s+1)\, e^{-s} \,\right]$$

$$\therefore\ L\left\{F(t)\right\} = \frac{1}{s^2}\ \frac{1 - (s+1)e^{-s}}{1 - e^{-2s}}$$

Example 10   Express in terms of the unit step function

$$F(t) = \cos 2t \qquad 0 < t < \pi,$$

$$= \cos 4t \qquad \pi < t < 2\pi,$$

$$= \cos 6t \qquad t > 2\pi$$

It follows directly from the definition of the unit step function that

$$F(t) = \cos 2t + (\cos 4t - \cos 2t)\, U(t - \pi)$$

$$+ (\cos 6t - \cos 4t)\, U(t - 2\pi)$$

Example 11   Evaluate $\displaystyle\int_0^\infty \frac{e^{-t}\sin t}{t}\, dt$

$$L\left\{\sin t\right\} = \frac{1}{s^2 + 1}$$

$$L\left\{\frac{\sin t}{t}\right\} = \int_s^\infty \frac{dx}{x^2+1} = \left[\tan^{-1}x\right]_s^\infty = \frac{\pi}{2} - \tan^{-1}s$$

$$\therefore \int_0^\infty \frac{e^{-st}\sin t}{t}\,dt = \frac{\pi}{2} - \tan^{-1}s$$

Let s $\to$ 1

$$\therefore \int_0^\infty \frac{e^{-t}\sin t}{t}\,dt = \frac{\pi}{2} - \tan^{-1}1 = \frac{\pi}{4}$$

Example 12   Evaluate $\int_0^t J_0(u)\,J_1(t-u)\,du$

Solution:

$$\int_0^t J_0(u)\,J_1(t-u)\,du = J_0(t)^* \, J_1(t)$$

$$L\left\{\int_0^t J_0(u)\,J_1(t-u)\,du\right\} = \frac{1}{\sqrt{s^2+1}}\left(1 - \frac{s}{\sqrt{s^2+1}}\right)$$

$$= \frac{1}{\sqrt{s^2+1}} - \frac{s}{s^2+1}$$

$$\therefore \int_0^t J_0(u)\,J_1(t-u)\,du = J_0(t) - \cos t$$

Example 13   Solve the integral equation $\int_0^t Y(t-\lambda)\,Y(\lambda)\,d\lambda = 8\sin 2t$

$$Y(t)^* \, Y(t) = 8\sin 2t$$

$$\therefore [y(s)]^2 = \frac{16}{s^2+4} \quad \therefore y(s) = \pm \frac{4}{\sqrt{s^2+4}}$$

Now $L\left\{J_0(at)\right\} = \dfrac{1}{\sqrt{s^2+a^2}}$

$$\therefore Y(t) = \pm 4J_0(2t)$$

Example 14

Find L { Log t }

192

we have shown in 1.2 that $L\left\{ t^k \right\} = \dfrac{\Gamma(k+1)}{s^{k+1}}$ where $k > -1$

i.e $\displaystyle\int_0^\infty e^{-st} t^k \, dt = \dfrac{\Gamma(k+1)}{s^{k+1}}$

Differentiate w,r.t.. k

$$\int_0^\infty e^{-st} t^k \log t \, dt = \frac{\Gamma'(k+1) - \Gamma(k+1) \log s}{s^{k+1}}$$

Put $k = 0$

$$\int_0^\infty e^{-st} \log t \, dt = \frac{\Gamma'(1) - \Gamma(1) \log s}{s}$$

$$\therefore \quad L\left\{ \log t \right\} = \frac{\Gamma'(1) - \log s}{s}$$

Example 15   Solve the integral equation

$$Y(t) = a \sin 8t + 6 \int_0^t Y(\lambda) \sin 8 \, (t-\lambda) \, d\lambda$$

$\therefore \quad Y(t) = a \sin 8t + 6Y(t)^* \sin 8t$

$\qquad y(s) = \dfrac{8a}{s^2+64} + 6y(s) \dfrac{8}{s^2+64}$

$\therefore \quad y(s) \left[1 - \dfrac{48}{s^2+64} \right] = \dfrac{8a}{s^2+64}$

i.e $\quad y(s) \cdot \dfrac{s^2 + 16}{s^2 + 64} = \dfrac{8a}{s^2 + 64}$

$\therefore \quad y(s) = \dfrac{8a}{s^2+16} \quad \therefore \quad Y(t) = 2a \sin 4t$

Example 16

Evaluate $\displaystyle\int_0^\infty \int_0^t \frac{e^{-t} \sin u}{u} \, dt \, du$

Integral $= \int\limits_{0}^{\infty} e^{-t} \left\{ \int\limits_{0}^{t} \frac{\sin u}{u} \, du \right\} dt$

$L\left\{ \sin t \right\} = \frac{1}{s^2+1}$

$L\left\{ \frac{\sin t}{t} \right\} = \int\limits_{s}^{\infty} \frac{dx}{x^2+1} = \left[\tan^{-1}x \right]_{s}^{\infty} = \frac{\pi}{2} - \tan^{-1}s = \tan^{-1}\frac{1}{s}$

$L\left\{ \int\limits_{0}^{t} \frac{\sin u}{u} \, du \right\} = \frac{1}{s} \tan^{-1} \frac{1}{s}$

i.e $\int\limits_{0}^{\infty} e^{-st} \left\{ \int\limits_{0}^{t} \frac{\sin u}{u} \, du \right\} dt = \frac{1}{s} \tan^{-1} \frac{1}{s}$

Let s → 1

$\int\limits_{0}^{\infty} e^{-t} \left\{ \int\limits_{0}^{t} \frac{\sin u}{u} \, du \right\} dt = \tan^{-1} 1 = \frac{\pi}{4}$

$\therefore \int\limits_{0}^{\infty} \int\limits_{0}^{t} e^{-t} \frac{\sin u}{u} \, dt \, du = \frac{\pi}{4}$

Example 17

Evaluate $\int\limits_{0}^{\infty} \frac{e^{-2t} - e^{-4t}}{t} \, dt$

$L\left\{ e^{-2t} \right\} = \frac{1}{s+2}, \ L\left\{ e^{-4t} \right\} = \frac{1}{s+4}$

$\therefore L\left\{ e^{-2t} - e^{-4t} \right\} = \left[\frac{1}{s+2} - \frac{1}{s+4} \right]$

$\therefore L\left\{ \frac{e^{-2t} - e^{-4t}}{t} \right\} = \int\limits_{s}^{\infty} \left[\frac{1}{x+2} - \frac{1}{x+4} \right] dx$

$= \left[\log \frac{x+2}{x+4} \right]_{s}^{\infty} = \log \frac{s+4}{s+2}$

i.e $\int\limits_{0}^{\infty} e^{-st} \frac{e^{-2t} - e^{-4t}}{t} \, dt = \log \frac{s+4}{s+2}$

194

Let $s \to 0$ $\therefore \int_0^\infty \dfrac{e^{-2t} - e^{-4t}}{t} \, dt = \log \dfrac{4}{2} = \log 2$

Example 18

Evaluate $\int_0^\infty e^{-t} \, J_0(t) \, dt$

We have seen that $L\left\{ J_0(t) \right\} = \dfrac{1}{\sqrt{s^2 + 1}}$

$\therefore \int_0^\infty e^{-st} \, J_0(t) \, dt = \dfrac{1}{\sqrt{s^2 + 1}}$

Let $s \to 1$ $\therefore \int_0^\infty e^{-t} \, J_0(t) \, dt = \dfrac{1}{\sqrt{2}}$

Example 19

Evaluate $L^{-1}\left\{ \dfrac{1}{s^2(s^2 + 1)} \right\}$ (i) by resolving into partial

fractions, (ii) by convolution, (iii) from $L^{-1}\left\{ \dfrac{1}{s^2 + 1} \right\}$, (iv) by

using an inversion integral.

(i) $L^{-1}\left\{ \dfrac{1}{s^2(s^2 + 1)} \right\} = L^{-1}\left\{ \dfrac{1}{s^2} - \dfrac{1}{s^2 + 1} \right\} = t - \sin t$

(ii) $L^{-1}\left\{ \dfrac{1}{s^2} \cdot \dfrac{1}{s^2 + 1} \right\} = t * \sin t = \int_0^t (t - \lambda) \sin \lambda \, d\lambda$

$= -\int_0^t (t - \lambda) \, d \cos \lambda = \left[-(t - \lambda)\cos\lambda - \sin\lambda \right]_0^t = -\sin t + t$

(iii) $L^{-1}\left\{ \dfrac{1}{s^2 + 1} \right\} = \sin t$ $\therefore L^{-1}\left\{ \dfrac{1}{s(s^2 + 1)} \right\}$

$= \int_0^t \sin \tau \, d\tau = \left[-\cos \tau \right]_0^t = 1 - \cos t$

$L^{-1}\left\{ \dfrac{1}{s^2(s^2 + 1)} \right\} = \int_0^t (1 - \cos\tau) \, d\tau = \left[\tau - \sin\tau \right]_0^t = t - \sin t$

(iv) $L^{-1}\left\{\dfrac{1}{s^2(s^2+1)}\right\}=$ sum of residues of $\left\{\dfrac{e^{st}}{s^2(s-i)(s+i)}\right\}$

at its poles.

$s=0$ residue $=\dfrac{1}{1!}\left[\dfrac{d}{ds}\left(\dfrac{e^{st}}{s^2+1}\right)\right]_{s=0}$

$$=\left[\dfrac{te^{st}(s^2+1)-2se^{st}}{(s^2+1)^2}\right]_{s=0}=t$$

Sum of residues at $s=\pm i$ is $\left[\dfrac{e^{st}}{s^2(s+i)}\right]_{s=i}+\left[\dfrac{e^{st}}{s^2(s-i)}\right]_{s=-i}$

$$=\dfrac{e^{it}}{-2i}+\dfrac{e^{-it}}{2i}=-\sin t$$

$\therefore\ L^{-1}\left\{\dfrac{1}{s^2(s^2+1)}\right\}=t-\sin t$

Example 20

Solve the equation

$tY''(t)+(t-1)Y'(t)-Y(t)=0$, given that $Y(o)=5$, $Y(\infty)=0$

$$-\dfrac{d}{ds}[s^2y(s)-sY(o)-Y'(o)]-\dfrac{d}{ds}[sy(s)-Y(o)]$$

$$-[sy(s)-Y(o)]-y(s)=0$$

$\therefore\ -\dfrac{d}{ds}[s^2y(s)-5s-A]-\dfrac{d}{ds}[sy(s)-5]-[sy(s)-5]-y(s)=0$

$\therefore\ -s^2\dfrac{dy}{ds}-2sy+5-s\dfrac{dy}{ds}-y-sy+5-y=0$

$$-(s^2+s)\dfrac{dy}{ds}-3sy-2y+10=0$$

$\therefore\ \dfrac{dy}{ds}+\dfrac{3s+2}{s^2+s}\,y=\dfrac{10}{s^2+s}$

which is a linear differential equation of the first order.

Now $\dfrac{3s+2}{s^2+s}=\dfrac{2}{s}+\dfrac{1}{s+1}$

Integrating factor $=e^{\int\left(\frac{2}{s}+\frac{1}{s+1}\right)ds}=e^{2\log s+\log(s+1)}$

$$=e^{\log s^2(s+1)}=s^2(s+1)$$

196

$$\therefore \quad y \, s^2(s+1) = \int \frac{10}{s(s+1)} \, s^2(s+1) \, ds + C$$

$$= 10 \int s \, ds + C = 5s^2 + C$$

$$\therefore \quad y(s) = \frac{5}{s+1} + \frac{C}{s^2(s+1)}$$

$$= \frac{5}{s+1} + C\left(\frac{1}{s^2} - \frac{1}{s} + \frac{1}{s+1}\right)$$

$$\therefore \quad Y(t) = 5e^{-t} + C(t - 1 + e^{-t})$$

and since $Y(t) \to o$ when $t \to \infty$, then $C = o$

and hence $Y(t) = 5e^{-t}$

Example 21   Solve the differential equation

$$\frac{d^2y}{dx^2} - a^2y = \delta(x-b), \; o < x < l, \; o < b < l,$$

with $y = 0$ when $x = 0$ and $x = l$

$$s^2\bar{y} - sy(o) - y'(o) - a^2\bar{y} = e^{-bs}$$

Put $y'(o) = A$ $\therefore (s^3 - a^2)\bar{y} = A + e^{-bs}$

$$\bar{y} = \frac{A}{s^2 - a^2} + \frac{e^{-bs}}{s^3 - a^2}$$

$$y = \frac{A}{a} \sinh ax + \frac{1}{a} \sinh a(x-b) \, U(x-b)$$

Since $y = o$ when $x = l$ we get

$$o = \frac{A}{a} \sinh al + \frac{1}{a} \sinh a(l-b)$$

$$\therefore \quad A = -\frac{\sinh a(l-b)}{\sinh al}$$

$$y = -\frac{\sinh a(l-b) \sinh ax}{a \sinh al} + \frac{1}{a} \sinh a(x-b) \, U(x-b)$$

i.e $y = -\dfrac{\sinh a(l-b) \sinh ax}{a \sinh al} \qquad o < x < b$

$$= -\frac{\sinh a(l-b) \sinh ax}{a \sinh al} + \frac{l}{a} \sinh a(x-b)$$

$$= -\frac{\sinh ab \sinh a(l-x)}{a \sinh al} \qquad b < x < l$$

(1) Find (i) $L^{-1}\left\{\dfrac{5e^{-3s}}{s} - \dfrac{e^{-s}}{s}\right\}$, (ii) $L^{-1}\left\{\dfrac{e^{-4s}}{(s+2)^3}\right\}$

[Ans. (i) $5U(t-3) - U(t-1)$, (ii) $\tfrac{1}{2}(t-4)^2 e^{-2(t-4)} U(t-4)$]

(2) If $F(t)$ is to be continuous for $t \geqslant o$ and

$$F(t) = L^{-1}\left\{\frac{(1-e^{-2s})(1-3e^{-2s})}{s^2}\right\}$$

evaluate $F(1)$, $F(3)$, $F(5)$.

[Ans. $1, -1, -4$]

(3) Find the Laplace transform of

$$F(t) = E \sin \omega t \qquad o < t < \frac{\pi}{\omega}$$

$$= o \qquad\qquad t > \frac{\pi}{\omega}$$

[Ans. $f(s) = \dfrac{E\omega}{s^2 + \omega^2}\left(1 + e^{-\frac{s\pi}{\omega}}\right)$]

(4) Find $L\left\{t^2 \sin kt\right\}$, $L\left\{t^2 \cos kt\right\}$

[Ans. $\dfrac{2k(3s^2 - k^2)}{(s^2+k^2)^3}$ $s>0$; $\dfrac{2s(s^2-3k^2)}{(s^2+k^2)^3}$, $s>o$]

(5) Find the Laplace transform of the periodic function $\psi(t,c)$ where

$$\psi(t,c) = 1 \qquad o < t < c \quad,$$
$$= o \qquad c < t < 2c \quad,$$
$$\psi(t+2c) = \psi(t,c)$$

[Ans. $\dfrac{1}{s(1+e^{-cs})}$]

(6) Evaluate $L\left\{\dfrac{\sin kt}{t}\right\}$ [Ans. $\tan^{-1}\dfrac{k}{s}$, $s>o$]

(7) Evaluate $L\left\{\dfrac{\sinh kt}{t}\right\}$ [Ans. $\tfrac{1}{2}\log\dfrac{s+k}{s-k}$, $s>k>o$]

(8) Evaluate $L\left\{\dfrac{1-\cosh kt}{t}\right\}$ [Ans. $\tfrac{1}{2}\log\left(1-\dfrac{k^2}{s^2}\right)$, $s>k>o$]

(9) Evaluate $F(t) = L^{-1} \left\{ \dfrac{1}{s^3 (1-e^{-2s})} \right\}$ and compute $F(5)$.

$$\left[\text{Ans. } F(t) = \tfrac{1}{2} \sum_{n=0}^{\infty} (t-2n)^2\ U(t-2n), \quad F(5) = 17.5 \right]$$

(10) If $\Phi(t) = L^{-1} \left\{ \dfrac{3}{s^4 \sinh 3s} \right\}$, compute $\Phi(10)$ [Ans 344]

(11) Prove that $L^{-1} \left\{ f(s) \tanh cs \right\} = F(t)$

$$+ 2 \sum_{n=1}^{\infty} (-1)^n\ F(t-2nc)\ U(t-2nc) \quad c>0, s>0.$$

(12) Prove that

$$L^{-1} \left\{ \dfrac{f(s)}{\cosh cs} \right\} = 2 \sum_{n=0}^{\infty} (-1)^n$$

$$F(t-2nc-c)\ U(t-2nc-c), \quad c>0, \quad s>0.$$

(13) Compute $x(1)$ and $x(4)$ for the function $x(t)$ which satisfies the boundary-value problem

$$x''(t) + 2x'(t) + x(t) = 2 + (t-3)\ U(t-3)$$

given that $x(o) = 2$ and $x'(o) = 1$

$$\left[\text{Ans. } x(1) = 2 + \dfrac{1}{e}, \quad x(4) = 1 + \dfrac{3}{e} + \dfrac{4}{e^4} \right]$$

(14) Solve the differential equation

$$x''(t) + 2x'(t) + x(t) = F(t); \quad x(o) = x'(o) = o$$

$$\left[\text{Ans. } x(t) = \int_0^t \lambda\, e^{-\lambda}\ F(t-\lambda)\ d\lambda \right]$$

(15) Solve the differential equation

$$y''(t) + 4y'(t) + 13\ y(t) = F(t); \quad y(o) = o, y'(o) = o.$$

$$\left[\text{Ans. } y(t) = \tfrac{1}{3} \int_0^t e^{-2\lambda} \sin 3\lambda\ F(t-\lambda)\ d\lambda \right]$$

Solve the differential equations

(16)
$$x''(t) + 6x'(t) + 9x(t) = F(t); \quad x(o) = A, x'(o) = B.$$

$$\left[\text{Ans. } x(t) = e^{-3t}\ [A + (B + 3A)\ t] \right.$$

$$\left. + \int_0^t \lambda e^{-3\lambda}\ F(t-\lambda)\ d\lambda \right]$$

(17) $F(t) = 1 + 2 \int_0^t F(t-\lambda) \, e^{-2\lambda} \, d\lambda$

[Ans. $F(t) = 1 + 2t$]

(18) $F(t) = 4t^2 - \int_0^t F(t-\lambda) \, e^{-\lambda} \, d\lambda$

[Ans. $F(t) = -1 + 2t + 2t^2 + e^{-2t}$]

(19) $F(t) = 8t^3 - 3 \int_0^t F(\lambda) \, \sin(t-\lambda) \, d\lambda$

[Ans. $F(t) = 2t^3 + 3 - 3 \cos 2t$]

Solve the following differential equations and check by using an inversion integral:

(20) $(D^3+1) x = 1$, $x(o) = x'(o) = x''(o) = o$

[Ans. $x = 1 - \dfrac{1}{3} e^{-t} - \dfrac{2}{3} e^{\frac{1}{2}t} \cos \frac{1}{2} t \sqrt{3}$]

(21) $(D+1)(D+2)(D+3) x = 1 + t + t^2$

given that $x(o) = x'(o) = x''(o) = o$

[Ans. $\dfrac{35}{54} - \dfrac{4}{9} t + \dfrac{1}{6} t^2 - e^{-t} + \frac{1}{2} e^{-2t} - \dfrac{4}{27} e^{-3t}$]

(22) $D(D-1)x = t^2$ where $x(o) = x_0$ and $x'(o) = x_1$

[Ans. $(x_0 - x_1) + x_1 e^t - 2 (1 + t + \dfrac{t^2}{2!} + \dfrac{t^3}{3!} - e^t)$]

(23) $(D^2 - 3D + 2) x = e^t$ where $x(o) = x_0$ and $x'(o) = x_1$

[Ans. $(x_1 - x_0 + 1) e^{2t} + (2x_0 - x_1 - 1 - t) e^t$]

(24) Indicate which of the following are null functions

(i) $F(t) = 2, \ t = 7$,(ii) $F(t) = 4 \quad 1 \leqslant t \leqslant 3$

$= 0$ otherwise

[Ans. (i) is a null function but (ii) is not.]

(25) Given that $L\left\{ \sin \sqrt{t} \right\} = \dfrac{\sqrt{\pi}}{2s^{3/2}} e^{-\frac{1}{4s}}$, show that

$$L\left\{\frac{\cos \sqrt{t}}{\sqrt{t}}\right\} = \frac{\sqrt{\pi}}{s^{1/2}}\; e^{-\frac{1}{4s}}$$

(26) If $L\left\{F(t)\right\} = \dfrac{e^{-1/s}}{s}$ Find $L\left\{e^{-t} F(4t)\right\}$

$$\left[\;\text{Ans.}\quad \frac{e^{-\frac{4}{s+1}}}{s+1}\;\right]$$

(27) Show that $L\left\{\displaystyle\int_0^t \frac{1-e^{-\tau}}{\tau}\; d\tau\right\} = \dfrac{1}{s}\log\left(1+\dfrac{1}{s}\right)$

(28) Find $L\{F(t)\}$ where

$$F(t) = t^2 \qquad o<t<2,$$

$$F(t+2) = F(t)$$

$$\left[\;\text{Ans.}\quad \frac{2-2e^{-2s}-4se^{-2s}-4s^2e^{-2s}}{s^3(t-e^{-2s})}\;\right]$$

(29) Express in terms of the unit step function

$$F(t)=t^3 \qquad o<t<3$$
$$= 5t^2 \qquad t>3$$

$$[\;\text{Ans.}\quad F(t) = t^3 + (5t^2 - t^3)\; U(t-3)\;]$$

(30) Prove that $\displaystyle\int_0^\infty t^3 e^{-t}\sin t\; dt = o$

(31) Find $L\left\{F(t)\right\}$ where

$$F(t) = \cos t \qquad o<t<\pi$$
$$= \sin t \qquad t>\pi$$

$$\left[\;\text{Ans.}\quad \frac{s+(s-1)e^{-\pi s}}{s^2+1}\;\right]$$

(32) Prove that $L\left\{\sin^5 t\right\} = \dfrac{120}{(s^2+1)(s^2+9)(s^2+25)}$

(33) Prove that $\displaystyle\int_0^\infty \cos x^2\; dx = \tfrac{1}{2}\sqrt{\dfrac{\pi}{2}}$

(37) Find by convolution

i) $L^{-1} \left\{ \dfrac{1}{(s+2)^2 (s-2)} \right\}$

$$\left[\text{Ans. } \frac{1}{16} (e^{2t} - e^{-2t} - 4t\, e^{-2t}) \right]$$

ii) $L^{-1} \left\{ \dfrac{s^2}{(s^2+4)^2} \right\}$ $\quad \left[\text{Ans. } \frac{1}{2} t \cos 2t + \frac{1}{4} \sin 2t \right]$

iii) $L^{-1} \left\{ \dfrac{1}{(s+1) (s^2+1)} \right\}$

$$\left[\text{Ans. } \frac{1}{2} (\sin t - \cos t + e^{-t}) \right]$$

(35) Using Heaviside's expansion formula, find

$$L^{-1} \left\{ \frac{2s + 5}{(s+2) (s+3)} \right\}$$

$$[\text{Ans. } e^{-2t} + e^{-3t}]$$

(36) Prove that $L^{-1} \left\{ \dfrac{1}{s} \cos \dfrac{1}{s} \right\} = 1 - \dfrac{t^2}{(2!)^2}$

$$+ \frac{t^4}{(4!)^2} - \frac{t^6}{(6!)^2} + \cdots$$

(37) Solve the equation

$$tY''(t) + 2Y'(t) + tY(t) = o$$

given that $Y(o +) = 1$ and $y(\pi) = o$ $\quad \left[\text{Ans. } Y(t) = \dfrac{\sin t}{t} \right]$

(38) Solve the equation $Y''(t) + tY'(t) - Y(t) = o,$
given that $Y(o) = o,$ $Y'(o) = 1$ $\quad [\text{Ans. } Y(t) = t]$

(39) Solve the equation $tY''(t) + (1-2t) Y'(t) - 2Y(t) = o,$
given that $Y(o) = 1,$ $Y'(o) = 2.$ $[\text{Ans. } Y(t) = e^{2t}]$

(40) Solve the integral equation

$$Y(t) = t^2 + \int_o^t Y(\lambda) \sin (t-\lambda) \, d\lambda$$

$$\left[\text{Ans } Y(t) = t^2 + \frac{1}{12} t^4 \right]$$

(41) Solve for $Y(t)$, the second order difference equation

$Y(t) - (a+b) Y(t \cdot h) + ab\, Y(t-2h) = F(t),$ $\quad h > o$

where $Y(t) = o$ and $F(t) = o$ when $t < o$ in the case when
(i) $a \neq b$, (ii) $a = b$

$\left[\text{Ans. } \text{i)} \right.$

$$Y(t) = \frac{1}{a-b} \sum_{n=0}^{\infty} (a^{n+1} - b^{n+1})\, F(t-nh)\, U(t-nh)$$

$$\text{ii) } Y(t) = \sum_{n=0}^{\infty} (n+1)\, a^n\, F(t-nh)\, U(t-nh) \Big]$$

(42) Solve the integral equation

$$Y(t) = 6t + \int_0^t Y(t-\lambda) \sin \lambda \; d\lambda$$

[Ans, $Y(t) = t^3 + 6t$]

Solve the equations

(43)

$$Y''(t) + 2tY'(t) - 4Y(t) = 1, \; Y(o) = Y'(o) = o$$

$$\left[\text{Ans.} \; Y(t) = \frac{t^2}{2} \right]$$

(44)

$$tY''(t) - (1+t) Y'(t) + 2Y(t) = t-1$$

given that $Y(o) = o$ [Ans. $Y(t) = t + At^2$]

(45)

$$Y'(t) + k^2 \int_0^t Y(\lambda) \cosh k \; (t-\lambda \; d\lambda = o$$

$$\left[\text{Ans.} \; Y(t) = C \left(1 - \frac{k^2 t^2}{2} \right) \right]$$

(46)

Prove that $L \left\{ \dfrac{e^t - \cos t}{t} \right\} = \frac{1}{2} \log \dfrac{s^2 + 1}{(s-1)^2}$

(47) Prove that $\int_0^\infty \dfrac{\cos t \, x}{x^2 + a^2} \, dx = \dfrac{\pi}{2a} e^{-at}$, $a > o$, $t \geq o$

(48) Prove that $\int_0^\infty \dfrac{1 - \cos 2tx}{x^2} \, dx = \pi t$

(49) Find using an inversion integral $L^{-1} \left\{ \dfrac{2s + 1}{s \; (s^2 + 1)} \right\}$

[Ans. $1 - \cos t + 2 \sin t$]

(50) Prove that $L \left\{ t \; J_o(at) \right\} = \dfrac{s}{(s^2 + a^2)^{3/2}}$

(51) Find $L \left\{ t \; J_1(t) \right\}$ $\left[\text{Ans.} \; \dfrac{1}{(s^2 + 1)^{5/2}} \right]$

(52) Solve the integral equation

$$Y(t) = \cos t - 3 \int_0^t Y(\lambda) \sin (t-\lambda) \; d\lambda$$

[Ans. $Y(t) = \cos 2t$]

(53) Solve the difference equations

(i) $Y(t) - Y(t-2) = t$ where $Y(t) = 0$ when $t < 0$

(ii) $Y'(t) - Y(t-1) = t$ where $Y(t) = 0$ when $t \leq 0$

[Ans. (i) $Y(t) = t + (t-2)U(t-2) + (t-4)U(t-4) + \cdots$

(ii) $Y(t) = \dfrac{t^2}{2!} + \dfrac{(t-1)^3}{3!}U(t-1) + \dfrac{(t-2)^4}{4!}U(t-2) + \cdots$]

(54) Find the Laplace transform of the wave $F(c,t)$ defined by

$F(c,t) = 0 \qquad 0 < t < c$

$\qquad = t - c \qquad c < t < 2c$

$F(t+2c) = F(t)$
$\qquad\qquad\qquad$ [Ans. $\dfrac{e^{-cs} - (cs+1)\, e^{-2cs}}{s^2(1-e^{-2cs})}$]

(55) Find $L^{-1}\left\{ \dfrac{1}{s^2(s-1)} \right\}$ (i) by partial fractions

(ii) by convolution $\quad$ (iii) from $L^{-1}\left\{ \dfrac{1}{s-1} \right\}$

(iv) by an inversion integral [Ans. $e^t - (t+1)$]

(56) Solve $Y(t) + 2 \displaystyle\int_0^t Y(\lambda) \cosh(t-\lambda)d\lambda = \sinh t$

[Ans. $Y(t) = \dfrac{1}{\sqrt{2}}\, e^{-t} \sinh \sqrt{2}\, t$]

(57) Find $L^{-1}\left\{ \dfrac{1}{s(s^2-1)} \right\}$ using an inversion integral.

[Ans. $-1 + \tfrac{1}{2} e^t + \tfrac{1}{2} e^{-t}$]

(58) Solve $Y'(t) = \sin t + \displaystyle\int_0^t Y(t-\lambda) \cos \lambda\, d\lambda$, given that

$Y(0) = 0$
$\qquad\qquad\qquad$ [Ans. $Y(t) = \dfrac{t^2}{2}$]

(59) Find the Laplace transform of $F(c,t)$ defined by

$F(c,t) = 1 \qquad 0 < t < c$

$\qquad = 0 \qquad c < t < 2c$

$\qquad = 1 \qquad 2c < t < 3c$

$F(c,t+3c) = F(c,t)$

[Ans. $L\left\{ F(c,t) \right\} = \dfrac{1-e^{-cs} + e^{-2cs} - e^{-3cs}}{s(1-e^{-3cs})}$]

(60) Solve $Y'(t) = \sinh t - \displaystyle\int_0^t Y(t-\lambda) \cosh \lambda\, d\lambda,\ Y(0) = 0$

[Ans $Y(t) = \tfrac{1}{2} t^2$]

(61) Solve the difference equation

$Y(t) - 4Y(t-h) + 4Y(t-2h) = t^2$

[Ans. $Y(t) = t^2 + 4(t-h)^2 U(t-h) + 12(t-2h)^2 U(t-2h) + \cdots$]

204

(62) Find the Laplace transform of the wave $F(c,t)$ defined by

$$F(c,t) = \quad 2 \qquad o < t < c$$
$$= -2 \qquad c < t < 2c$$
$$= \quad 2 \qquad 2c < t < 3c$$

$$F(c, t+3c) = F(c,t)$$

$$\left[\text{Ans. } \frac{2[1-2e^{-cs} + 2e^{-2cs} - e^{-3cs}]}{s(1-e^{-3cs})} \right]$$

(63) Solve $Y'(t) + \int_0^t Y(\lambda) \cosh (t-\lambda)\, d\lambda = o$

$$[\text{Ans. } Y(t) = c(1-t^2/2)]$$

(64) Find the Laplace transform of

$$F(t) = o \qquad o < t < \frac{a}{3}$$
$$= c \qquad \frac{a}{3} < t < \frac{2a}{3}$$
$$= o \qquad \frac{2a}{3} < t < a$$

$$F(t+a) = F(t)$$

$$\left[\text{Ans. } \frac{c(e^{-\frac{a}{3}s} - e^{-\frac{2}{3}as})}{s(1-e^{-as})} \right]$$

(65) Evaluate $\int_0^\infty \frac{\cos 2t - \cos 4t}{t}\, dt$ [Ans. log 2]

(66) Evaluate $\int_{-\infty}^\infty e^{-t} U(t-3) dt$ [Ans. e^{-3}]

(67) Find $L\left\{ te^{-3t} J_0(t\sqrt{2}) \right\}$ $\left[\text{Ans. } \frac{s+3}{(s^2+6s+11)^{3/2}} \right]$

(68) Find using an inversion integral $L^{-1}\left\{ \frac{5s-2}{s^2(s+2)(s-1)} \right\}$

$$[\text{Ans. } t - 2 + e^t + e^{-2t}]$$

(69) Solve the difference equation

$$Y'(t) + Y(t-1) = t^2$$

given that $Y(t) = o$ for $t \le o$.

$$\left[\text{Ans: } Y(t) = \frac{2t^3}{3!} - \frac{2(t-1)^4}{4!} U(t-1) + \frac{2(t-2)^5}{5!} U(t-2) - ... \right]$$

(70) Find using an inversion integral $L^{-1}\left\{ \frac{1}{s^2(s^2+16)} \right\}$

$$\left[\text{Ans. } \frac{1}{64} [4t - \sin 4t] \right]$$

11.3. Electrical Applications of the Laplace Transformation

Applications 1.

Find the current I at time t in a circuit consisting of a resistance R and inductance L in series with a condenser of capacity C when a constant E.M.F is applied at time t = 0, the initial values of charge and current being zero.

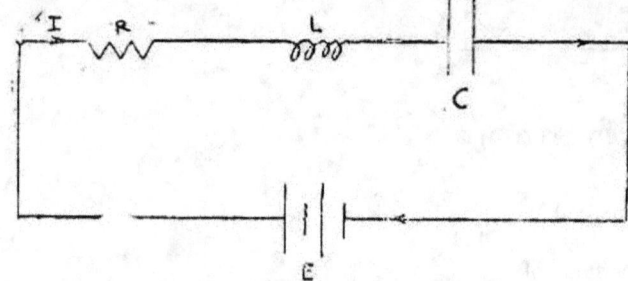

The differential equation satisfied by the current I is

$$IR + L \frac{dI}{dt} + \frac{1}{C} \int_0^t I(\tau) \, d\tau = E$$

with the initial conditions $I(0) = 0$, $Q(0) = 0$. The subsidiary equation is given by:

$$Ri(s) + Lsi(s) + \frac{i(s)}{Cs} = \frac{E}{s}.$$

$$\therefore \left(R + Ls + \frac{1}{Cs} \right) i(s) = \frac{E}{s}$$

$$i(s) = \frac{E}{s\left(R + Ls + \frac{1}{Cs} \right)} = \frac{E}{Ls^2 + Rs + \frac{1}{C}} = \frac{E}{L} \cdot \frac{1}{s^2 + \frac{R}{L}s + \frac{1}{CL}}$$

$$= \frac{E}{L} \cdot \frac{1}{\left(s + \frac{R}{2L} \right)^2 + \frac{1}{CL} - \frac{R^2}{4L^2}} = \frac{E}{L} \cdot \frac{1}{\left(s + \frac{R}{2L} \right)^2 + n^2}$$

where $n^2 = \frac{1}{CL} - \frac{R^2}{4L^2}$

case (i) $n^2 + $ ve $I = \frac{E}{nL} e^{-\frac{R}{2L}t} \sin nt$

case (ii) $n^2 = o$ $I = \dfrac{E}{nL}\, t\, e^{-\frac{R}{2L}t}$

case (iii) $n^2 -$ ve and is equal to $- \mu^2$

$$i(s) = \frac{E}{L} \cdot \frac{1}{\left(s + \frac{R}{2L}\right)^2 - \mu^2}$$

$\therefore$ $I = \dfrac{E}{\mu L}\, e^{-\frac{R}{2L}t}\, \sinh \mu t$

Application 2.

Find the current in a series circuit of resistance R, inductance L and capacity C in which an E.M.F. V is applied at t = 0, the initial values of the current I and charge Q on the condenser being respectively I_o and Q_o.

The differential equation of the current is

$$L \frac{dI}{dt} + RI + \frac{Q}{C} = V$$

I and Q are connected by I = dQ/dt

The subsidiary equations are

$$L\,[\,s\,i(s) - I_o\,] + R\,i(s) + \frac{q(s)}{C} = v(s)$$

$$i(s) = sq(s) - Q_o$$

Eliminating q(s) we get

$$\left(Ls + R + \frac{1}{Cs}\right) i(s) = v(s) + LI_o - \frac{Q_o}{Cs}$$

which on inversion gives the current I,

Application 3.

A battery of E M.F. E is connected at time t =0 to a series circuit of resistance R., inductance L and capacity C. The initial values of the current and charge are zero. If the battery is short-circuited at t = T find the current I at any instant t.

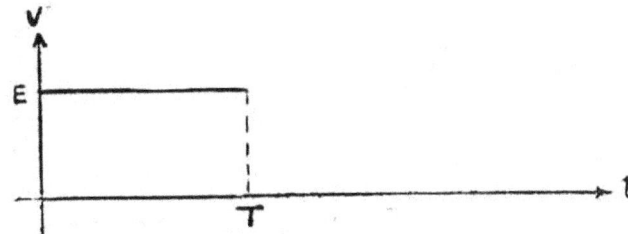

$$IR + L \frac{dI}{dt} + \frac{1}{C} \int_{o}^{t} I(\tau)\, d\tau = V$$

where $\quad V = E \qquad o < t < T,$

$\qquad\qquad = o \qquad\quad t > T$

$\therefore \quad v(s) = \dfrac{E}{s}\left(1 - e^{-Ts}\right)$

$$= \dfrac{E}{Ls^2 + Rs + \dfrac{1}{C}}\left(1 - e^{-Ts}\right)$$

$$Ri(s) + Lsi(s) + \dfrac{1}{Cs}\, i(s) = \dfrac{E}{s}\left(1 - e^{-Ts}\right)$$

$$\therefore \quad i(s) = \dfrac{E}{s\left(R + Ls + \dfrac{1}{Cs}\right)}\left(1 - e^{-Ts}\right)$$

$$= \dfrac{E}{Ls^2 + Rs + \dfrac{1}{C}} - \dfrac{E}{Ls^2 + Rs + \dfrac{1}{C}}\, e^{-Ts}$$

$$= \dfrac{E}{L}\cdot\dfrac{1}{\left(s + \dfrac{R}{2L}\right)^2 + \dfrac{1}{CL} - \dfrac{R^2}{4L^2}} - \dfrac{E}{L}\cdot\dfrac{1}{\left(s + \dfrac{R}{2L}\right)^2 + \dfrac{1}{CL} - \dfrac{R^2}{4L^2}}\, e^{-Ts}$$

$$\therefore \quad I = \dfrac{E}{nL}\, e^{-\frac{R}{2L}}\sin nt \quad o < t < T$$

$$= \dfrac{E}{nL}\, e^{-\frac{R}{2L}t}\sin nt - \dfrac{E}{nL}\, e^{-\frac{R}{2L}(t-T)},\ \sin n(t-T),\ t > T$$

$$\text{where } n^2 = \dfrac{1}{CL} - \dfrac{R^2}{4L^2} > o.$$

Application 4.

A periodic E. M. F. of resonance frequency E sin(nt) is applied at time $t = 0$ to a series circuit consisting of inductance L and capacity C where $n^2 = \dfrac{1}{CL}$.

Find the current the circuit at time t assuming initial zero current and charge.

$$L \frac{dI}{dt} + \frac{1}{C} \int_0^t I(\tau)d\tau = E \sin nt$$

with the conditions $I(0) = 0$, $Q(0) = 0$

$$Ls\, i(s) + \frac{1}{Cs}\, i(s) = \frac{En}{s^2+n^2}$$

$$\therefore \quad i(s) = \frac{En}{\left(Ls + \frac{1}{Cs}\right)(s^2+n^2)} = \frac{Ens}{\left(Ls^2 + \frac{1}{C}\right)(s^2+n^2)}$$

$$= \frac{En}{L} \cdot \frac{s}{\left(s^2 + \frac{1}{CL}\right)(s^2 + n^2)} = \frac{En}{L} \frac{s}{(s^2 + n^2)^2}$$

and since $L^{-1}\left\{\dfrac{s}{(s^2 + k^2)^2}\right\} = \dfrac{1}{2k} t \sin kt$ \qquad 1.13

then $i(s) = \dfrac{En}{L} \cdot \dfrac{1}{2n} t \sin nt$

$$= \frac{E}{2L} t \sin nt$$

Application 5.

Find the current at time t when a periodic E.M.F. E sin *wt* is applied at time t = 0 to an inductive resistance L, R the initial current being zero;

$$L \frac{dI}{dt} + RI = E \sin \omega t$$

with $I(0) = 0$.

$$(Ls + R)\, i(s) = \frac{E\omega}{s^2 + \omega^2}$$

$$\therefore \quad i(s) = \frac{E\omega}{(Ls + R)(s^2 + \omega^2)} = \frac{E\omega}{L} \cdot \frac{1}{\left(s + \frac{R}{L}\right)(s^2+\omega^2)}$$

Put $\dfrac{R}{L} = \mu$

$$i(s) = \frac{E\omega}{L} \cdot \frac{1}{(s + \mu)(s^2 + \omega^2)}$$

$$= \frac{E\omega}{L(\mu^2 + \omega^2)}\left[\frac{1}{s + \mu} - \frac{s - \mu}{s^2 + \omega^2}\right]$$

$$\therefore \quad I(t) = \frac{E\omega}{L(\mu^2 + \omega^2)}\left[e^{-\mu t} - \cos \omega t + \frac{\mu}{\omega} \sin \omega t\right]$$

Application 6.

An E.M.F. E_1 for $o < t < T$ and E_2 for $t > T$ where E_1 and E_2 are constants is applied to a series circuit L, R, C. Find the current at any time t assuming zero initial current and charge.

$$L \frac{dI}{dt} + RI + \frac{1}{C} \int_0^t I(\tau) \, d\tau = V(t)$$

where $\quad V(t) = E_1 \qquad o < t < T$

$$= E_2 \qquad t > T$$

$$L \left\{ V(t) \right\} = \frac{E_1}{s} - \frac{E_1 - E_2}{s} e^{-Ts}$$

$$\therefore \quad \left(Ls + R + \frac{1}{Cs} \right) i(s) = \frac{E_1}{s} \frac{E_1 - E_2}{s} e^{-Ts}$$

$$i(s) = \frac{E_1}{Ls^2 + Rs + \frac{1}{C}} - \frac{E_1 - E_2}{Ls^2 + Rs + \frac{1}{C}} e^{-Ts}$$

$$= \frac{E_1}{L} \cdot \frac{1}{s^2 + \frac{R}{L}s + \frac{1}{CL}} - \frac{E_1 - E_2}{L} \cdot \frac{1}{s^2 + \frac{R}{L}s + \frac{1}{CL}} e^{-Ts}$$

Now $\quad s^2 + \frac{R}{L}s + \frac{1}{CL} = \left(s + \frac{R}{2L} \right)^2 + \frac{1}{CL} - \frac{R^2}{4L^2}$

$$= (s + \mu)^2 + n^2 \quad \text{where} \quad \mu = \frac{R}{2L}, \; n^2 = \frac{1}{CL} - \frac{R^2}{4L^2} > 0$$

$$\therefore \quad i(s) = \frac{E_1}{L} \cdot \frac{1}{(s+\mu)^2 + n^2} - \frac{E_1 - E_2}{L} \cdot \frac{1}{(s+\mu)^2 + n^2} e^{-Ts}$$

$$\therefore \quad I(t) = \frac{E_1}{nL} e^{-\mu t} \sin nt \qquad o < t < T \; ,$$

$$= \frac{E_1}{nL} e^{-\mu t} \sin nt - \frac{E_1 - E_2}{nL} e^{-\mu(t-T)} \sin n(t-T), t > T.$$

Application 7.

The two circuits shown in the diagram are coupled by mutual inductance M.A constant E.M.F. E is applied to the primary at time $t = 0$ with zero initial conditions, the primary resistance being neglected. Find secondary current at any time t.

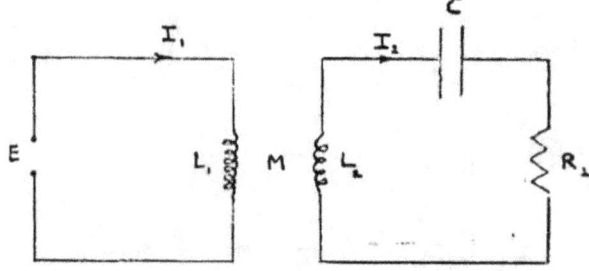

For the primary we have

$$L_1 \frac{dI_1}{dt} + M \frac{dI_2}{dt} = E \quad \dots \quad (1)$$

with the initial condition $I_1(0) = 0$.

For the secondary we have

$$I_2R_2 + L_2 \frac{dI_2}{dt} + \frac{1}{C} \int_0^t I_2(\tau)\, d\tau + M \frac{dI_1}{dt} = 0 \quad \dots \quad (2)$$

with the initial conditions $I_2(0) = 0$, $Q(0) = 0$.
The subsidiary equations are

$$L_1 s i_1 + M s i_2 = \frac{E}{s}$$

$$i_2 R_2 + L_2 s i_1 + \frac{1}{Cs}\, i_2 + M s i_1 = 0$$

i.e

$$L_1 s i_1 + M s i_2 = \frac{E}{s} \quad \dots\dots (3)$$

$$M s i_1 + \left(R_2 + L_2 s + \frac{1}{Cs}\right) i_2 = 0 \quad \dots \quad (4)$$

Eliminating i_1 we get

$$i_2 = \frac{-ME}{(L_1 L_2 - M^2)s^2 + L_1 R_2 s + \frac{L_1}{C}}$$

$$= -\frac{ME}{L_1 L_2 - M^2} \cdot \frac{1}{s^2 + \dfrac{L_1 R_2 s}{L_1 L_2 - M^2} + \dfrac{L_1}{C(L_1 L_2 - M^2)}}$$

$$= -\frac{ME}{L_1 L_2 - M^2} \cdot \frac{1}{(s+a)^2 + \beta^2}$$

Where

$$2a = \frac{L_1 R_2}{L_1 L_2 - M^2} \; , \; \beta^2 = \frac{L_1}{C(L_1 L_2 - M^2)} - a^2$$

$$\therefore I_2 = -\frac{ME}{L_1 L_2 - M^2}\, e^{-at}\, \frac{1}{\beta}\, \sin \beta t$$

Application 8.

In the following network find the total current I at any instant t, the initial currents and charges are zero.

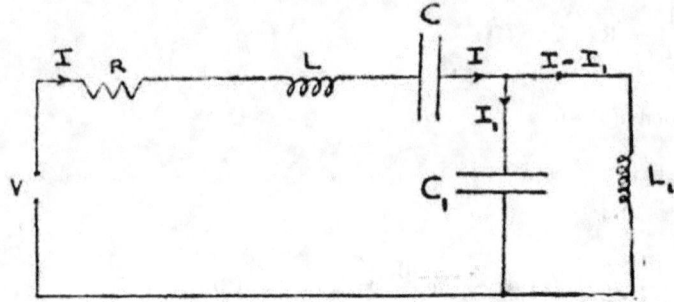

Applying Kirchoff's voltage law to the circuit on the left which states that the impressed voltage equate the sum of the voltage drops across the elements in that circuit we get

$$IR + L\frac{dI}{dt} + \frac{1}{C}\int_0^t I(\tau)\, d\tau + \frac{1}{C_1}\int_0^t I_1(\tau)\, d\tau = V \quad \ldots \quad (i)$$

Again, applying the voltage law to the circuit on the right we get

$$L_1 \frac{d}{dt}(I - I_1) - \frac{1}{C_1}\int_0^t I_1(\tau)\, d\tau = 0 \quad \ldots \quad (ii)$$

The subsidiary equations are :

$$Ri + Lsi + \frac{1}{Cs} i + \frac{1}{C_1 s} i_1 = v(s)$$

$$L_1(si - si_1) - \frac{1}{C_1 s} i_1 = 0$$

i.e $$\left(R + Ls + \frac{1}{Cs}\right) i + \frac{1}{C_1 s} i_1 = v(s) \quad \ldots \quad (iii)$$

$$L_1 si - \left(L_1 s + \frac{1}{C_1 s}\right) i_1 = 0 \quad \ldots \quad (iv)$$

Eliminating i_1 between (iii) and (iv) we get

$$i(s) = \frac{v(s) . s(L_1 s^2 + C_1^{-1})}{p(s)}$$

where $p(s) = (L_1 s^2 + C_1^{-1})(Ls^2 + Rs + C^{-1} + C_1^{-1}) - C_1^{-2}$

Inverting we get the required current I.

Application 9.

In the following network, the switch S is closed at time t = o when both condensers are charged to voltage E. Find the current I.

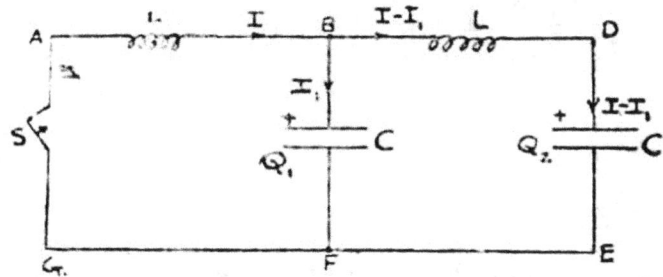

Applying the voltage law to the circuit ABFGA we get

$$L \frac{dI}{dt} + \frac{Q_1}{C} = 0 \quad \text{wher } I_1 = \frac{dQ_1}{dt}$$

Applying the voltage law to the circuit ABDEFGA we get

$$L \frac{dI}{dt} + L \frac{d}{dt} (I - I_1) + \frac{Q_2}{C} = 0$$

where $\quad I - I_1 = \dfrac{dQ_2}{dt}$

$I(o) = I_1(o) = o. \quad Q_1(o) = Q_2(o) = EC$

$Lsi + \dfrac{q_1}{C} = 0 \quad \text{where } i_1 = sq_1 - Q_1(o) = sq_1 - EC.$

$Lsi = L[si - si_1] + \dfrac{q}{C} = 0$

where $i - i_1 = sq_2 - EC$

$\therefore \quad Lsi + \dfrac{1}{C} \left[\dfrac{i_1}{s} + \dfrac{EC}{s} \right] = 0$

$Lsi + Lsi - Lsi_1 + \dfrac{1}{C} \left[\dfrac{i}{s} - \dfrac{i_1}{s} + \dfrac{EC}{s} \right] = 0$

i.e $\quad Lsi + \dfrac{i_1}{Cs} = - \dfrac{E}{s} \qquad \dots \text{(i)}$

$$\left(2Ls + \frac{1}{Cs} \right) i = \left(Ls + \frac{1}{Cs} \right) i_1 = - \frac{E}{s} \quad \dots \text{(ii)}$$

Eliminating i_1 between (i) and (ii) we get

$$i = - \frac{EC (CLs^2 + 2)}{C^2L^2s^4 + 3CLs^2 + 1}$$

Putting $n^2 = \dfrac{1}{CL}$ we get.

$$i = - \frac{E}{L} \frac{s^2 + 2n^2}{s^4 + 3n^2s^2 + n^4}$$

$$= - \frac{E}{2L\sqrt{5}} \left[\frac{1 + \sqrt{5}}{s^2 + \frac{1}{2}(3 - \sqrt{5}) n^2} - \frac{1 - \sqrt{5}}{s^2 + \frac{1}{2}(3 + \sqrt{5}) n^2} \right]$$

213

$$\therefore I = -E\sqrt{\frac{C}{10L}}\left\{\frac{1+\sqrt{5}}{\sqrt{3-\sqrt{5}}}\sin nt\left[\tfrac{1}{2}(3-\sqrt{5})\right]^{\frac{1}{2}}\right.$$

$$\left.-\frac{1-\sqrt{5}}{\sqrt{3+\sqrt{5}}}\sin nt\left[\tfrac{1}{2}(3+\sqrt{5})\right]^{\frac{1}{2}}\right\}$$

Application 10.

In the network shown below, determine the character of the current $I_1(t)$ assuming that each current is zero when the switch is closed.

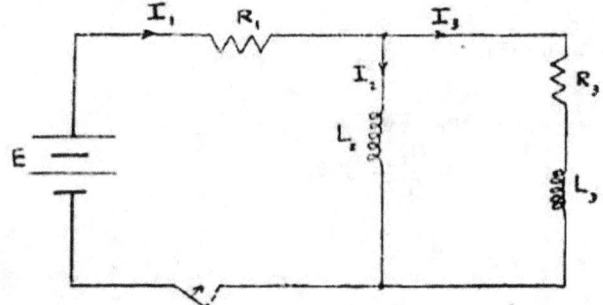

Since the algebraic sum of the currents at any junction is zero, then

$$I_1 - I_2 - I_3 = 0 \quad \ldots \quad (1)$$

Applying the voltage law to the circuit on the left we get

$$R_1 I_1 + L_2 \frac{dI_2}{dt} = E \quad \ldots \quad (2)$$

Applying again the voltage low to the outside circuit we get

$$R_1 I_1 + R_3 I_3 + L_3 \frac{dI_3}{dt} = E \quad \ldots \quad (3)$$

Transforming we get

$$i_1 - i_2 - i_3 = 0 \quad \ldots \ldots \quad (4)$$

$$R_1 i_1 + sL_2 i_2 = \frac{E}{s} \quad \ldots \quad (5)$$

$$R_1 i_1 + (R_3 + sL_3) i_3 = \frac{E}{s} . \quad (6)$$

$$\therefore i_1(s) = \frac{\begin{vmatrix} 0 & -1 & -1 \\ \dfrac{E}{s} & sL_2 & 0 \\ \dfrac{E}{s} & 0 & R_3 + sL_3 \end{vmatrix}}{\triangle} = \frac{E}{s}\frac{R_3 + s(L_2 + L_3)}{\triangle}$$

where

$$\triangle = \begin{vmatrix} 1 & -1 & -1 \\ R_1 & sL_2 & 0 \\ R_1 & 0 & R_3 + sL_3 \end{vmatrix} = \begin{vmatrix} 1 & 0 & 0 \\ R_1 & sL_2 + R_1 & R_1 \\ R_1 & R_1 & R_1 + R_3 + sL_3 \end{vmatrix}$$

i. o $\Delta = L_2L_3s^2 + (R_1L_3 + R_3L_2 + R_1L_3) s + R_1R_3$

Since we are interested in the factors of Δ, we consider the equation $\Delta = 0$. Since all coefficients of this equation are positive, hence it cannot have any positive roots . Its discriminant is

$$(R_1L_2 + R_3L_2 + R_1L_3)^2 - 4L_2L_3R_1R_3$$

which can be written

$$R_1^2L_2^2 + 2R_1L_2 (R_3L_2 + R_1L_3) + (R_3L_2 - R_1L_3)^2$$

which is positive. Hence the equation $\Delta = 0$ has two negative distinct roots $-a_1, -a_2$ (say),

$$\therefore \Delta \equiv L_2L_3(s+a_1) (s+a_2)$$

$$\therefore i_1(s) = \frac{E}{s} \quad \frac{R_3 + s (L_2+L_3)}{L_2L_3 (s+a_1) (s+a_2)}$$

$$= \frac{A_0}{s} + \frac{A_1}{s+a_1} + \frac{A_2}{s+a_2}$$

$$\therefore I_1(t) = A_0 + A_1e^{-a_1t} + A_2e^{-a_2t}$$

Application 11.

In the network shown below, derive the equations satisfied by the currents I1, I2, I3 and the chugs Q3 assuming that all initial currents and charges are zero and obtain the transformed equations.

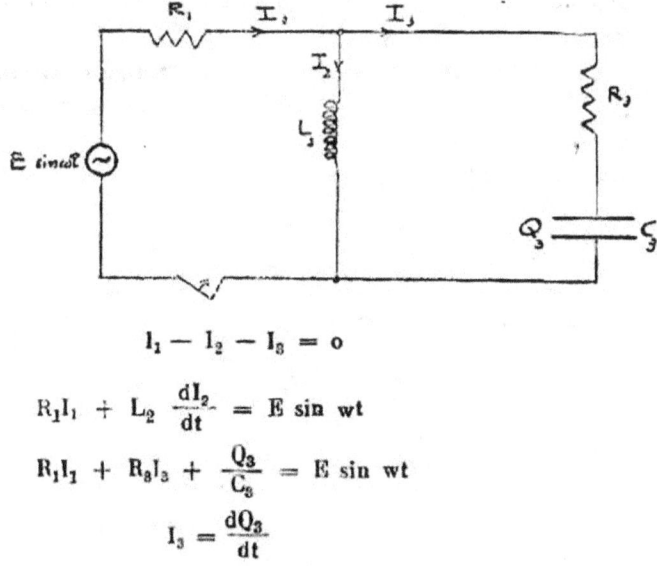

$$I_1 - I_2 - I_3 = 0$$

$$R_1I_1 + L_2 \frac{dI_2}{dt} = E \sin wt$$

$$R_1I_1 + R_3I_3 + \frac{Q_3}{C_3} = E \sin wt$$

$$I_3 = \frac{dQ_3}{dt}$$

Initial conditions are

$$I_1(0) = I_2(0) = I_3(0) = 0, Q_3(0) = 0$$

Transforming we get

$$i_1 - i_2 - i_3 = 0$$

$$R_1 i_1 + sL_2 i_2 = \frac{E\omega}{s^2 + \omega^2}$$

$$R_1 \, i_1 + R_3 \, i_3 + \frac{q_3}{C_3} = \frac{E\omega}{s^2 + \omega^2}$$

$$i_3 = s \, q_3 \quad \text{or} \quad q_3 = \frac{i_3}{s}$$

The solution can then be completed as in the previous example,

Application 12.
_____

An impulsive E.M.F. $E_0\delta(t)$ is applied at $t = 0$ to an L, C, R, circuit in series with zero initial currents and charges. Find the current at *any* instant t,

$$L \frac{dI}{dt} + IR + \frac{1}{C} \int_0^t I(\tau) \, d\tau = E_0\delta(t)$$

$$\therefore \left(Ls + R + \frac{1}{Cs} \right) i(s) = E_0$$

$$i(s) = \frac{E_0 s}{Ls^2 + Rs + \frac{1}{C}}$$

$$= \frac{E_0}{L} \cdot \frac{s}{s^2 + \frac{R}{L} s + \frac{1}{CL}}$$

$$= \frac{E_0}{L} \cdot \frac{s}{\left(s + \frac{R}{2L}\right)^2 + \frac{1}{CL} - \frac{R^2}{4L^2}}$$

$$= \frac{E_0}{L} \frac{s + \frac{R}{2L}}{\left(s + \frac{R}{2L}\right)^2 + \frac{1}{CL} - \frac{R^2}{4L^2}}$$

$$- \frac{E_0}{L} \frac{\frac{R}{2L}}{\left(s + \frac{R}{2L}\right)^2 + \frac{1}{CL} - \frac{R^2}{4L^2}}$$

$$\therefore \quad I(t) = \frac{E_0}{L} e^{-\frac{R}{2L}t} \cos nt - \frac{E_0 R}{2L^2 n} e^{-\frac{R}{2L}t} \sin nt$$

where

$$n^2 = \frac{1}{CL} - \frac{R^2}{4L^2} > 0.$$

11.4. Dynamical Applications of Laplace Transforms

Application 1

A particle is projected vertically upwards at time $t = 0$ with velocity v_0 from the origin under the action of gravity and a resistance equal to 2 km times the velocity. Find its displacement at any instant t.

216

The equation of motion of the mass is

$$m\ddot{x} = -2km\dot{x} - mg$$

with the initial conditions

$$x(o) = 0. \quad \dot{x}(o) = v_o$$

$$\therefore \quad \ddot{x} + 2k\dot{x} = -g$$

$$s^2\bar{x} - sx(o) - \dot{x}(o) + 2k[s\bar{x} - x(o)] = -\frac{g}{s}$$

$$(s^2 + 2ks)\bar{x} = v_o - \frac{g}{s}$$

$$\therefore \quad \bar{x} = \frac{v_o}{s(s+2k)} - \frac{g}{s^2(s+2k)}$$

$$= \frac{v_o}{2k}\left[\frac{1}{s} - \frac{1}{s+2k}\right] - g\left[\frac{1}{2ks^2} - \frac{1}{4k^2s} + \frac{1}{4k^2(s+2k)}\right]$$

$$\therefore \quad x = \frac{v_o}{2k}\left(1 - e^{-2kt}\right) - g\left[\frac{t}{2k} - \frac{1}{4k^2} + \frac{1}{4k^2}e^{-2kt}\right]$$

i.e

$$x(t) = -\frac{gt}{2k} + \frac{g + 2kv_o}{4k^2}\left(1 - e^{-2kt}\right)$$

Application 2.

A particle of mass m moves in a vertical plane under the action of gravity and a force directed towards the origin and equal to μr where r is the distance of the particle from the origin. If the particle is projected from the point (a, 0) vertically upwards with velocity v_o, find the coordinates of the particle at any instant t.

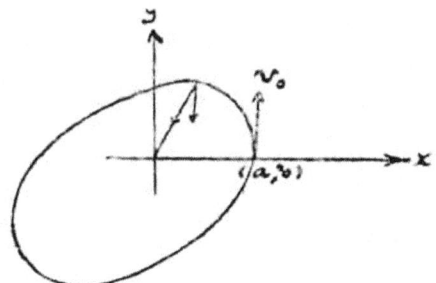

$$m\ddot{x} = -\mu x \, , \quad m\ddot{y} = -\mu y - mg$$

Put $\dfrac{\mu}{m} = n^2$

$$\therefore \quad \ddot{x} + n^2x = 0, \quad \ddot{y} + n^2y = -g$$

The initial conditions are

217

$$x(o) = a \,, \quad y(o) = o \,, \quad \dot{x}(o) = o \,, \quad \dot{y}(o) = v_o$$

$$s^2 \bar{x}(s) - s x(o) - \dot{x}(o) + n^2 \bar{x}(s) = o$$

$$(s^2 + n^2) \bar{x} = as$$

$$\therefore \quad \bar{x} = \frac{as}{s^2 + n^2} \qquad \therefore \quad x(t) = a \cos n t$$

$$s^2 \bar{y}(s) - s y(o) - \dot{y}(o) + n^2 \bar{y}(s) = -\frac{g}{s}$$

$$s^2 \bar{y} - v_o + n^2 \bar{y} = -\frac{g}{s}$$

$$(s^2 + n^2) \bar{y} = v_o - \frac{g}{s}$$

$$\bar{y} = \frac{v_o}{s^2 + n^2} - \frac{g}{s(s^2 + n^2)}$$

Now $L^{-1} \left\{ \dfrac{1}{s(s^2 + n^2)} \right\} = \dfrac{1}{n^2} (1 - \cos n t)$

$$\therefore \quad y(t) = \frac{v_o}{n} \sin n t - \frac{g}{n^2} (1 - \cos n t)$$

Application 3.

A particle of mass m moves in a straight line under a resistance $2\mu m$ times the velocity and a restoring force $m\lambda$ times the displacement where $\lambda > \mu^2$. If it is projected at time t = 0o with velocity v_o at distance x_o; from its equilibrium position, find its displacement at any subsequent instant t.

Equation of motion of the mass is

$$m\ddot{x} = -2\mu m \dot{x} - m\lambda x$$

i.e $\quad \ddot{x} + 2p\dot{x} + \lambda x = o$

with the initial conditions

$$x(o) = x_o \,, \quad \dot{x}(o) = v_o$$

$$[\, s^2 \bar{x} - s x(o) - \dot{x}(o) \,] + 2\mu \,[\, s \bar{x} - x(o) \,] \lambda \bar{x} = o$$

$$\therefore \quad (s^2 + 2\mu s + \lambda) \bar{x} = s x_o + v_o + 2\mu x_o$$

$$\bar{x} = \frac{s x_o + v_o + 2\mu x_o}{(s+\mu)^2 + \lambda - \mu^2}$$

Put $\lambda - \mu^2 = n^2$

$$\therefore \quad \bar{x} = \frac{x_o(s+\mu) + v_o + \mu x_o}{(s+\mu)^2 + n^2}$$

$$= \frac{x_o(s+\mu)}{(s+\mu)^2 + n^2} + \frac{v_o + \mu x_o}{(s+\mu)^2 + n^2}$$

$$\therefore \quad x(t) = x_o e^{-\mu t} \cos n t + \frac{v_o + \mu x_o}{n} e^{-\mu t} \sin n t$$

218

$$= \frac{1}{n} \, e^{-\mu t} \, [\, nx_o \cos nt + (x_o + \mu x_o) \sin nt \,]$$

Application 4

A spring of stiffness k is placed on a smooth horizontal table. One end of the spring is fixed at a point O on the table and to the other end is attached a mass m. The system is
initially at rest with the spring unstretched. A constant force F is applied to the mass for a time t_o and then removed. Find the mass at any time t.

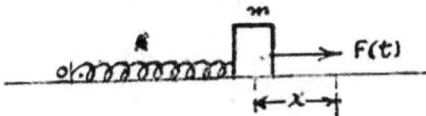

Equation of motion of the mass is

$$m \ddot{x} = -kx + F(t)$$

where $F(t) = F_o \qquad o < t < t_o$

$$= o \qquad t > t_o$$

The initial conditions are

$$x(o) = \dot{x}(o) = o$$

$$\ddot{x} = -\frac{k}{m} x + \frac{1}{m} F(t).$$

Put $\dfrac{k}{m} = n^2$

$$\therefore \ddot{x} + n^2 x = \frac{1}{m} F(t)$$

Transforming and noticing that

$$L \left\{ F(t) \right\} = \frac{F_o}{s} \, (1 - e^{-t_o s})$$

We get

$$(s^2 + n^2) \, x = \frac{F_o}{ms} \, (1 - e^{-t_o s})$$

$$\bar{x} = \frac{F_o}{ms(s^2 + n^2)} \, (1 - e^{-t_o s})$$

Now $L^{-1} \left\{ \dfrac{1}{s(s^2 + n^2)} \right\} = \dfrac{1}{n^2} \, (1 - \cos nt) = \dfrac{2}{n^2} \sin^2 \dfrac{nt}{2}$

$$\therefore \quad x(t) = \frac{2F_o}{mn^2} \sin^2 \frac{nt}{2} \qquad\qquad o < t < t_o$$

$$= \frac{2F_o}{mn^2} \sin^2 \frac{nt}{2} - \frac{2F_o}{mn^2} \sin^2 \frac{n}{2} \, (t - t_o) \qquad t > t_o$$

i.e $x(t) = \dfrac{2F_\circ}{k} \sin^2 \dfrac{nt}{2}$ $\qquad$ $o < t < t_\circ$

$\qquad\quad = \dfrac{2F_\circ}{k} \left[\sin^2 \dfrac{nt}{2} - \sin^2 \dfrac{n}{2} (t-t_\circ) \right]$ $\qquad$ $t > t_\circ$

Application 5

Two particles each of mass m are connected by a spring of stiffness k and they are free to move in a straight line on a smooth horizontal table. At time t = o when both particles are at rest and the spring is unstrained a constant force P is applied to one of them in the direction towards the other particle. Find the displacement of the other particle from its initial position at any subsequent instant t.

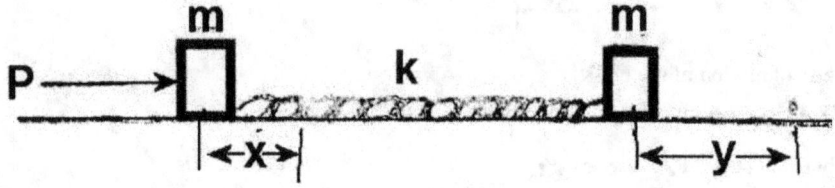

The equations of motion of the two masses are

$$m\ddot{x} = k(y-x) + P$$

$$m\ddot{y} = -k(y-x)$$

Putting $k/m = n^2$ we get

$$\ddot{x} = n^2(y-x) + \dfrac{P}{m}$$

$$\ddot{y} = -n^2(y-x)$$

The initial conditions are

$$x(o) = y(o) = \dot{x}(o) = \dot{y}(o) = o$$

$\therefore \qquad (s^2 + n^2)\,\bar{x} - n^2\bar{y} = \dfrac{P}{ms}$

$\qquad\quad - n^2\bar{x} + (s^2 + n^2)\,\bar{y} = o$

Eliminate $\bar{x}$

$\therefore \quad [\,(s^2+n^2)^2 - n^4\,]\,\bar{y} = \dfrac{n^2 P}{ms}$

$\therefore \quad s^2(s^2+2n^2)\,\bar{y} = \dfrac{n^2 P}{ms}$

$\qquad \bar{y} = \dfrac{n^2 P}{m} \cdot \dfrac{1}{s^3(s^2+2n^2)}$

$\qquad\quad = \dfrac{P}{4m} \left[\dfrac{2}{s^3} - \dfrac{1}{n^2 s} + \dfrac{s}{n^2(s^2+2n^2)} \right]$

$\therefore \quad y(t) = \dfrac{P}{4m} \left[t^2 - \dfrac{m}{k} + \dfrac{m}{k} \cos t \sqrt{\dfrac{2k}{m}} \right]$

220

Application 6

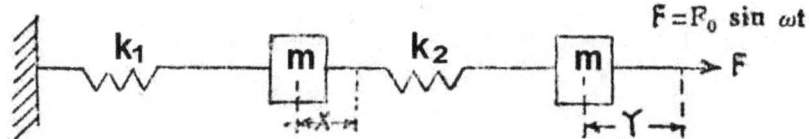

Two masses are equal each being equal to m. The stiffnesses k_1, k_2 of the two springs are connected by the relation $k_1 / k_2 = 3 / 2$. The system is initially at rest with the two springs unstrained. A periodic force $F = F_0 \sin \omega t$ where $\omega^2 = k_2 / m$ is applied to the right mass. Find the displacement X(t) of the left mass at any instant t.

Solution:

$$m\ddot{X} = k_2(Y-X) - k_1 X$$

$$m\ddot{Y} = -k_2(Y-X) + F_0 \sin \omega t$$

$$\therefore \ddot{X} = \omega^2(Y-X) - \frac{3}{2}\omega^2 X$$

$$\ddot{Y} = -\omega^2(Y-X) + \frac{F_0}{m}\sin \omega t$$

Initial conditions are

$$X(o) = Y(o) = \dot{X}(o) = \dot{Y}(o) = 0$$

$$\ddot{X} + \frac{5}{2}\omega^2 X - \omega^2 Y = 0$$

$$-\omega^2 X + \ddot{Y} + \omega^2 Y = \frac{F_0}{m}\sin \omega t$$

$$\therefore \left(D^2 + \frac{5}{2}\omega^2\right) X - \omega^2 Y = 0$$

$$-\omega^2 X + \left(D^2 + \omega^2\right) Y = \frac{F_0}{m}\sin \omega t$$

$$\therefore \left(s^2 + \frac{5}{2}\omega^2\right) x - \omega^2 y = 0$$

$$-\omega^2 x + (s^2 + \omega^2) y = \frac{F_0 \omega}{m} \cdot \frac{1}{s^2 + \omega^2}$$

Eliminating y we get

$$\left[\left(s^2 + \frac{5}{2}\omega^2\right)(s^2 + \omega^2) - \omega^4\right] x = \frac{F_0 \omega^3}{m} \cdot \frac{1}{s^2 + \omega^2}$$

$$\left[s^4 + \frac{7}{2}s^2\omega^2 + \frac{3}{2}\omega^4\right] x = \frac{F_0 \omega^5}{k_2} \cdot \frac{1}{s^2 + \omega^2}$$

$$\therefore \left(s^2 + \tfrac{1}{2}\omega^2\right)(s^2 + 3\omega^2) x = \frac{F_0 \omega^5}{k_2} \cdot \frac{1}{s^2 + \omega^2}$$

$$x(s) = \frac{F_0 \omega^5}{k_2} \cdot \frac{1}{(s^2 + \tfrac{1}{2}\omega^2)(s^2 + 3\omega^2)(s^2 + \omega^2)}$$

$$= \frac{F_0 \omega^5}{k_2}\left[\frac{4}{5\omega^4} \cdot \frac{1}{s^2 + \tfrac{1}{2}\omega^2} + \frac{1}{5\omega^4(s^2 + 3\omega^2)} - \frac{1}{\omega^4(s^2 + \omega^2)}\right]$$

$$= \frac{F_0 \omega}{5k_2}\left[\frac{4}{s^2 + \tfrac{1}{2}\omega^2} + \frac{1}{s^2 + 3\omega^2} - \frac{5}{s^2 + \omega^2}\right]$$

$$\therefore X(t) = \frac{F_0 \omega}{5k_2} \left[\frac{4\sqrt{2}}{\omega} \sin \sqrt{\tfrac{1}{2}}\ \omega t + \frac{1}{\sqrt{3}\ \omega} \sin \sqrt{3}\ \omega t - \frac{5}{\omega} \sin \omega t \right]$$

i.e $15\,k_2 X(t) = F_0 [12 \sqrt{2}\ \sin (\sqrt{\tfrac{1}{2}}\ \omega t) + \sqrt{3}\ \sin (\sqrt{3}\ \omega t) - 15 \sin \omega t]$

Application 7

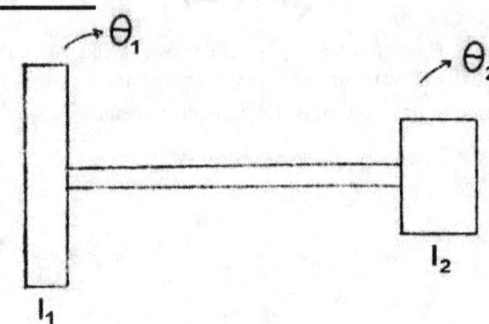

Two flywheels of moments of inertia I1 and I2 are connected by an elastic shaft of torsional stiffness A i.e., the couple per radian relative twist of the flywheels is A. The whole system is rotating with a constant angular velocity ω when at time t = 0, a constant retarding couple P is applied to the wheel I1 , Find the angular velocity of the wheel I2 at any instant t.

Solution:

Let θ_1 be the angular displacement of the flywheel I_1 and θ_2 that of I_2 .

Equations of motion are

$$I_1 \ddot{\theta}_1 = \lambda(\theta_2 - \theta_1) - P$$

$$I_2 \ddot{\theta}_2 = - \lambda(\theta_2 - \theta_1)$$

Initial conditions can be taken as

$$\theta_1(o) = \theta_2(o) = o \ , \ \dot{\theta}_1(o) = \dot{\theta}_2(o) = \omega$$

$$\therefore \ I_1 [s^2 \bar{\theta}_1 - s\theta_1(o) - \dot{\theta}_1(o)] = \lambda (\bar{\theta}_2 - \bar{\theta}_1) - \frac{P}{s}$$

$$I_2 [s^2 \bar{\theta}_2 - s\theta_2(o) - \dot{\theta}_2(o)] = - \lambda (\bar{\theta}_2 - \bar{\theta}_1)$$

$$(I_1 s^2 + \lambda)\ \bar{\theta}_1 - \lambda \bar{\theta}_2 = I_1 \omega - \frac{P}{s}$$

$$- \lambda \bar{\theta}_1 + (I_2 s^2 + \lambda)\bar{\theta}_2 = I_2 \omega$$

Eliminate $\bar{\theta}_1$

$$[(I_1 s^2 + \lambda)(I_2 s^2 + \lambda) - \lambda^2]\bar{\theta}_2 = \lambda\ (I_1 \omega - \frac{P}{s}) + I_2 \omega (I_1 s^2 + \lambda)$$

$$\therefore \ s^2 [I_1 I_2 s^2 + \lambda (I_1 + I_2)]\bar{\theta}_2 = \omega [I_1 I_2 s^2 + \lambda(I_1 + I_2)] - \frac{\lambda P}{s}$$

$$\bar{\theta}_2 = \frac{\omega}{s^2} - \frac{\lambda P}{s^2 [I_1 I_2 s^2 + \lambda(I_1 + I_2)]}$$

Let $\phi = \dot{\theta}_2$

222

$$\bar{\phi} = s\bar{\theta}_2 - \theta_2(o) = \frac{\omega}{s} - \frac{\lambda P}{s^2[I_1 I_2 s^2 + \lambda(I_1 + I_2)]}$$

$$\frac{\omega}{s} - \frac{\lambda P}{I_1 I_2 s^2(s^2 + n^2)}$$

where $n^2 = \dfrac{\lambda(I_1 + I_2)}{I_1 I_2}$

$$\therefore \bar{\phi} = \frac{\omega}{s} - \frac{\lambda P}{I_1 I_2 n^2}\left[\frac{1}{s^2} - \frac{1}{s^2 + n^2}\right]$$

$$= \frac{\omega}{s} - \frac{P}{I_1 + I_2}\left[\frac{1}{s^2} - \frac{1}{s^2 + n^2}\right]$$

$$\therefore \phi = \omega - \frac{Pt}{I_1 + I_2} + \frac{P}{n(I_1 + I_2)}\sin nt$$

Application 8

In previous application, let the retarding couple P be applied for time T only.

Solution:

Here the transform of P is
$$\frac{P}{s}(1 - e^{-Ts})$$
instead of P / s in the last problem.

$$\text{Thus } \bar{\phi} = \frac{\omega}{s} - \frac{P(1 - e^{-Ts})}{I_1 + I_2}\left[\frac{1}{s^2} - \frac{1}{s^2 + n^2}\right]$$

$$\therefore \phi = \omega - \frac{Pt}{I_1 + I_2} + \frac{P}{n(I_1 + I_2)}\sin nt \qquad o < t < < T$$

$$= \omega - \frac{Pt}{I_1 + I_2} + \frac{P}{n(I_1 + I_2)}\sin nt + \frac{P(t - T)}{I_1 + I_2}$$

$$- \frac{P}{n(I_1 + I_2)}\sin n(t - T) \, , \, t > T.$$

Application 9

The tautochrone is the shape taken by a smooth wire through the origin in the vertical x-y plane, such that when a bead is constrained to slide on it, time of descent of the bead from any point on the wire at which it is at rest to the origin which is its lowest point is constant. i.e., independent of the starting point.

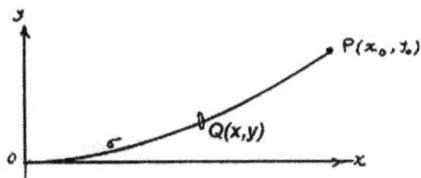

Solution:

223

Let P(x0,yo) be the starting print i.e., the bead is at rest at P and starts sliding and let Q(x,y) be any point on the wire between O and P where O is the origin taken at the lowest point of the wire. Let σ = arc OQ measured from O and m = mass of bead.

The loss of potential energy when the bead falls through the vertical distance yo - y is equal to the gain of the kinetic energy.

$$\therefore \tfrac{1}{2} m \left(\frac{d\sigma}{dt}\right)^2 = mg \ (y_0 - y)$$

$$\therefore \frac{d\sigma}{dt} = -\sqrt{2g(y_0 - y)}$$

The negative sign is taken since σ decreases as t increases. Let T be the time of descent from P to O, then

$$T = -\int_{y_0}^{o} \frac{d\sigma}{\sqrt{2g(y_0 - y)}}$$

$$\text{i.e } T = \int_{0}^{y_0} \frac{d\sigma}{\sqrt{2g(y_0 - y)}}$$

Now when the equation of the curve is known, we can express the length of the arc σ in terms of y and consequently the differential of the arc d σ can be expressed in terms of y and dy — Hence we can put d σ = F(y) dy

$$\therefore T\sqrt{2g} = \int_{0}^{y_0} \frac{F(y)dy}{\sqrt{(y_0 - y)}}$$

This is an integral equation of the convolution type in which the unknown function F(y) has to be determined such that T is constant that is independent of yo. Now the integral equation can be written

$$T\sqrt{2g} = F(y)*y^{-\frac{1}{2}}$$

Transforming, we get

$$\frac{T\sqrt{2g}}{s} = f(s) \sqrt{\frac{\pi}{s}}$$

$$\therefore f(s) = T \sqrt{\frac{2g}{\pi}} \cdot \frac{1}{\sqrt{s}} \ .$$

Inverting we get

$$F(y) = T \sqrt{\frac{2g}{\pi}} \cdot \frac{1}{\sqrt{\pi y}}$$

$$= \frac{T}{\pi} \sqrt{2g} \cdot \frac{1}{\sqrt{y}}$$

Now $\dfrac{d\sigma}{dy} = \dfrac{\sqrt{dx^2 + dy^2}}{dy} = \sqrt{1 + \left(\dfrac{dx}{dy}\right)^2} = F(y)$

$$= \frac{T}{\pi} \sqrt{2g} \cdot \frac{1}{\sqrt{y}}$$

$$\text{i.e } \sqrt{1 + \left(\frac{dx}{dy}\right)^2} = \frac{T}{\pi} \sqrt{2g} \cdot \frac{1}{\sqrt{y}}$$

$$1 + \left(\frac{dx}{dy}\right)^2 = \frac{2gT^2}{\pi^2} \cdot \frac{1}{y}$$

Put $c = \frac{2gT^2}{\pi^2}$

$$\therefore \quad 1 + \left(\frac{dx}{dy}\right)^2 = \frac{c}{y}$$

i.e $\left(\frac{dx}{dy}\right)^2 = \frac{c-y}{y}$ $\quad \therefore \frac{dx}{dy} = \sqrt{\frac{c-y}{y}}$

i.e $x = \int \sqrt{\frac{c-y}{y}} \, dy + d$

Put $y = c \sin^2 \theta$

$$x = \int \sqrt{\frac{c \cos^2\theta}{c \sin^2\theta}} \cdot 2c \sin\theta \cos\theta \, d\theta + d$$

$$= 2c \int \cos^2\theta \, d\theta + d$$

$$= c \int (1 + \cos 2\theta) \, d\theta + d$$

$$= c \left[\theta + \frac{\sin 2\theta}{2} \right] + d$$

$$= \frac{c}{2} [2\theta + \sin 2\theta] + d$$

and $y = \frac{c}{2} [1 - \cos 2\theta]$

Now since the curve passes through origin $d = 0$ and the parametric equations of the curve can be put in the form

$$x = a [\phi + \sin \phi]$$
$$y = a [1 - \cos \phi]$$

where $a = \frac{c}{2} = \frac{gT^2}{\pi^2}$ and $\phi = 2\theta$

These are the parametric equations of a cycloid. Hence the wire assumes the shape of a cycloid in which the radius of the generating circle is $\frac{gT^2}{\pi^2}$.

Application 10

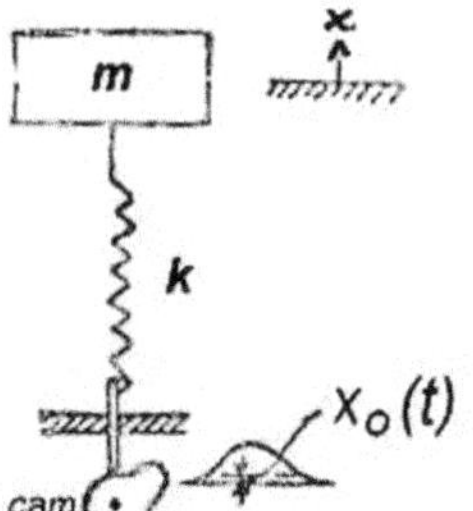

The figure shows a spring - mass system, the lower end of which undergoes a motion xo(t) prescribed by a cam.

The equation of motion of the mass m is

$$m\ddot{x} = k [x_o(t) - x]$$

$$\therefore \quad \ddot{x} = \frac{k}{m} [x_o(t) - x]$$

225

Put $w^2 = k/m$

$$\therefore \quad \ddot{x} + w^2 x = w^2 x_o(t)$$

If the system is started from rest, the subsidiary equation for the displacement of m is

$$(s^2 + w^2) \, x(s) = w^2 x_o(s)$$

i.e $\quad \overline{x}(s) = \dfrac{w^2 \overline{x}_o(s)}{s^2 + w^2}$

When $x_o(t)$ is specified. $\overline{x}_o(s)$ can be determined and consequently $\overline{x}(s)$ and x(t).

To obtain the force exerted on in by the spring in which case we are concerned with the relative motion y = xo(t) — x of the ends of the spring, we have

$$F = ky = m\ddot{x}$$

The subsidiary equation of which is

$$\overline{F}(s) = ms^2 \, \overline{x}(s) = \dfrac{m\omega^2 s^2 \overline{x}_o(s)}{s^2 + w^2}$$

Application 11

A uniform rod of length 2a and mass m is at rest on a smooth horizontal table, At time t = 0, it is set in motion by a blow of impulse P at one end of the rod perpendicular to the rod. Find and solve its equation of motion.

Solution:

Let x be the linear displacement of the middle point of the rod and θ the angular displacement of the rod about its centre.

$$m\ddot{x} = P \, \delta(t)$$

$$\tfrac{1}{3} ma^2 \ddot{\theta} = P a \, \delta(t)$$

Transforming we get

$$ms^2 \overline{x} = P, \quad \tfrac{1}{3} mas^2 \overline{\theta} = P$$

i.e $\quad \overline{x} = \dfrac{P}{ms^2}, \quad \overline{\theta} = \dfrac{3P}{mas^2}$

$$\therefore \quad x(t) = \dfrac{P}{m} t, \quad \theta(t) = \dfrac{3 \, Pt}{ma}$$

11.5. Structural Applications

11.5.1. Deflection of beams

The differential equation for the deflection y of a team is

$$EI \, \dfrac{d^2 y}{dx^2} = -M$$

where M is the bending moment at a point x of the beam, F is its Young's modulus and I the moment of inertia of the cross section of the beam about its neutral axis.

For transverse loading w per unit length of the beam, the above differential equation becomes:

$$EI \, \dfrac{d^4 y}{dx^4} = w$$

provided E and I are constants.

A concentrated load W at $x = a$ may be considered as a distributed load w per unit length of the beam such that

$$w = W\delta(x-a) = WU'(x-a)$$

where δ is the Dirac delta function. Its transform is

$$\bar{w} = We^{-as}$$

A constant load wo per unit length in $o < x < a$ and zero load for $x > a$ may be written

$$w = w_o [1 - U(x-a)]$$

$$\text{and } \bar{w} = \frac{w_o}{s}\left(1 - e^{-as}\right)$$

The transform of a couple

$$\mu\,\delta'(x-a) = \mu\,U''(x-a)$$

of moment μ applied at $x = a$ is μse^{-as}

Now consider the general beam equation

$$EI\,\frac{d^4y}{dx^4} = f(x)$$

$$\text{or} \quad \frac{d^4y}{dx^4} = \frac{1}{EI}\,f(x)$$

where $f(x)$ represents the load per unit length at a point x of the beam.

The subsidiary equation is given by

$$s^4\,\bar{y}(s) - s^3 y(o) - s^2 y'(o) - s y''(o) - y'''(o) = \frac{1}{EI}\,\bar{f}(s)$$

$$\text{i.e} \quad \bar{y}(s) = \frac{y(o)}{s} + \frac{y'(o)}{s^2} + \frac{y''(o)}{s^3} + \frac{y'''(o)}{s^4} + \frac{1}{EI}\,\frac{\bar{f}(s)}{s^4}$$

Inverting we get

$$y(x) = y(o) + y'(o)\,x + y''(o)\,\frac{x^2}{2!} + y'''(o)\,\frac{x^3}{3!} + \frac{1}{EI}\,L^{-1}\left\{\frac{\bar{f}(s)}{s^4}\right\}$$

In practical problems some of the quantities $y(0)$, $y'(0)$, $y''(0)$. $y'''(0)$ are known if we are given at $x = 0$ the deflection or slope or bending moment or shearing force. The remaining quantities are determined from conditions at other points of the beam.

The following are applications of the above principles.

Application 1
_____

A beam is hinged at its ends $x = o$ and $x = l$. It carries a uniformly distributed lead w perr unit length. Find the static deflection at any point on the beam.

$$EI\,\frac{d^4y}{dx^4} = w_o \quad \therefore \quad \frac{d^4y}{dx^4} = \frac{w_o}{EI}$$

$$s^4\,\bar{y} - s^3 y(o) - s^2 y'(o) - s y''(o) - y'''(o) = \frac{w_o}{EIs}$$

227

$y(o) = y''(o) = o$ and put $y'(o) = A$ and $y'''(o) = B$

$$s^4 \bar{y} = As^2 + B + \frac{w_o}{EIs}$$

$$\therefore \quad \bar{y} = \frac{A}{s^2} + \frac{B}{s^4} + \frac{w_o}{EIs^5}$$

$$y = Ax + B\frac{x^3}{3!} + \frac{w_o}{EI}\frac{x^4}{4!}$$

we have now to determine A and B from the other two conditions namely $y(l) = y''(l) = 0$,

$$o = Al + B\frac{l^3}{6} + \frac{w_o}{EI}\frac{l^2}{24}$$

$$o = Bl + \frac{w_o}{EI}\frac{l^2}{2}$$

$$\therefore B = -\frac{w_o}{EI}\frac{l}{2} , A = \frac{w_o l^3}{24EI}$$

$$\therefore y = \frac{w_o l^3}{24EI}x - \frac{w_o l}{12EI}x^3 + \frac{w_o}{24EI}x^4$$

$$= \frac{w_o}{24EI}(l^3 x - 2lx^3 + x^4)$$

$$= \frac{w_o}{24EI}x(l-x)(l^2 + lx - x^2)$$

Application 2

A beam of length $2\,l$ has both its ends built in. It carries a load $w(x)$ per wilt length such that

$$W(x) = w_o \qquad o < x < l$$

$$= o \qquad l < x < 2l$$

Find the static deflection y at any point x.

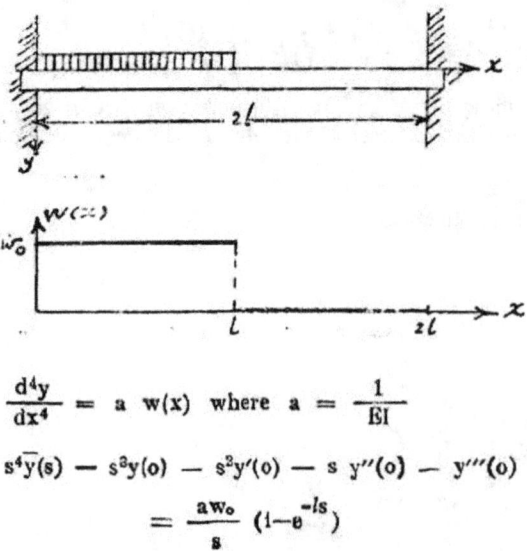

$$\frac{d^4y}{dx^4} = a\,w(x) \quad \text{where} \quad a = \frac{1}{EI}$$

$$s^4\bar{y}(s) - s^3y(o) - s^2y'(o) - s\,y''(o) - y'''(o)$$

$$= \frac{aw_o}{s}(1 - e^{-ls})$$

228

Now $y(0) = y'(0) = 0$ and let $y''(0) = A$, $y'''(0) = B$

$$\therefore \quad s^4\bar{y} = As + B + \frac{aw_0}{s}(1-e^{-ls})$$

$$\bar{y} = \frac{A}{s^3} + \frac{B}{s^4} + \frac{aw_0}{s^5}(1-e^{-ls})$$

$$y = \frac{Ax^2}{2!} + \frac{Bx^3}{3!} + aw_0\frac{x^4}{4!} \qquad 0 < x < l,$$

$$= \frac{Ax^2}{2!} + \frac{Bx^3}{3!} + aw_0\frac{x^4}{4!} - \frac{aw_0}{4!}(x-l)^4 \qquad x > l$$

Now when $x = 2l$, $y = 0$, $y' = 0$.

$$o = 2Al^2 + \frac{4}{3}Bl^3 + \frac{2}{3}aw_0\,l^4 - \frac{aw_0}{24}\,l^4$$

$$o = 2Al + 2Bl^2 + \frac{4}{3}aw_0\,l^3 - \frac{aw_0}{6}\,l^3$$

from which $\quad A = \frac{11}{48}\,l^2aw_0$, $B = -\frac{13}{16}\,law_0$

$$\therefore \quad \frac{y}{aw_0} = \frac{11}{96}\,l^2x^2 - \frac{13}{96}\,lx^3 + \frac{1}{24}\,x^4 - \frac{1}{24}(x-l)^4\,U(x-l).$$

Application 3

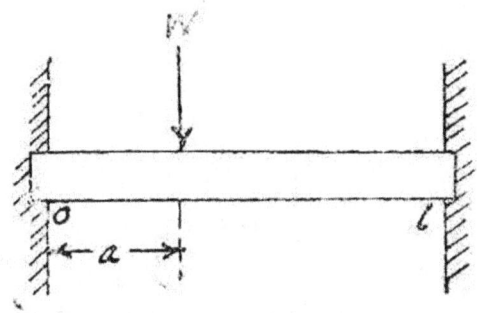

A beam of length l is clamped horizontally at its ends $x = o$ and $x = l$ and carries a concentrated load W at $x = a$. Find the static deflection y at any point x.

$$EI\,\frac{d^4y}{dx^4} = W\,\delta(x-a)$$

$$\therefore \quad \frac{d^4y}{dx^4} = \frac{W}{EI}\,\delta(x-a)$$

$$s^4\bar{y} - s^3y(o) - s^2y'(o) - sy''(o) - y'''(o) = \frac{W}{EI}\,e^{-as}$$

Now $y(o) = y'(o) = o$ and put $y''(o) = A$ and $y'''(o) = B$

$$\therefore \quad s^4\bar{y} = As + B + \frac{W}{EI}\,e^{-as}$$

$$\bar{y} = \frac{A}{s^3} + \frac{B}{s^4} + \frac{W}{EI} \frac{e^{-as}}{s^4}$$

$$y = A \frac{x^2}{2} + B\frac{x^3}{6} + \frac{W}{6EI} (x-a)^3 U(x-a)$$

Now at $x = l$ we have $y = o$, $y' = o$

$$o = Al^2 + \frac{B}{3} l^3 + \frac{W}{3EI} (l-a)^3$$

$$o = Al + \tfrac{1}{2} Bl^2 + \frac{W}{2EI} (l-a)^2$$

$$\therefore A = \frac{wa(l-a)^2}{EI\,l^2}, \quad B = - \frac{w(l-a)^2 (l+2a)}{EI\,l^3}$$

$$\therefore EIy = \frac{wa(l-a)^2 x^2}{2l^2} - \frac{w(l-a)^2 (l+2a)x^3}{6l^3} + \frac{w(x-a)^3}{6}U(x-a).$$

Application 4

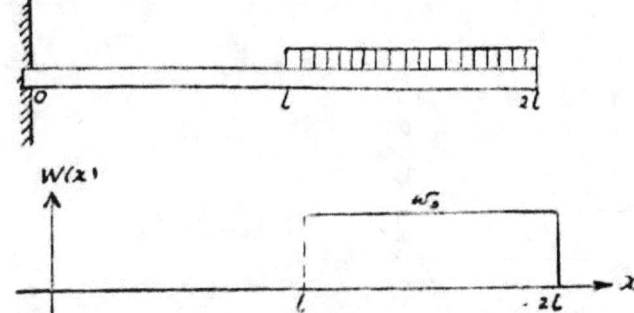

A cantilever of length $2l$ has its end $x=0$ built in while the end $x=2l$ is free. It carries a load $w(x)$ per unit length which is zero over $o < x < l$ and equal to a constant W_0 over $l < x < 2l$. *Find* the static deflection at any point x of the cantilever.

$$EI \frac{d^4y}{dx^4} = w(x)$$

where
$$w(x) = o \quad o < x < l$$
$$= w_0 \quad l < x < 2l.$$

Extend the definition of w(x) so as to be equal to w o for all x>o

$$\frac{d^4y}{dx^4} = aw(x) \quad \text{where} \quad a = \frac{1}{EI}$$

$$s^4\bar{y} - s^3y(o) - s^2y'(o) - sy''(o) - y'''(o) = \frac{aw_0}{s} e^{-ls}$$

230

Now $y(o) = 0$, $y'(o) = 0$ and put $y''(o) = A$, $y'''(o) = B$

$$\therefore \ s^4\bar{y} = As + B + \frac{aw_o}{s}e^{-ls}$$

$$\therefore \ \bar{y} = \frac{A}{s^3} + \frac{B}{s^4} + \frac{aw_o}{s^5}e^{-ls}$$

$$y = A\frac{x^2}{2} + B\frac{x^3}{6} + \frac{aw_o}{24}(x-l)^4 U(x-l)$$

The constants A and B are determined *from* the conditions:
$Y'''(2l) = Y''(2l) = 0$ since there is no bending shearing force at $x = 2l$

For $s > l$ we have

$$Y''(x) = A + Bx + \tfrac{1}{2}aw_o(x-l)^2.$$

$$Y'''(x) = B + aw_o(x-l)$$

$$\therefore \ 0 = A + 2Bl + \tfrac{1}{2}aw_o l^2$$

$$0 = B + aw_o l$$

$$\therefore \ A = \frac{3}{2}aw_o l^2 \ , \ B = - aw_o l$$

$$\therefore \ y = aw_o\left[\frac{3}{4}l^2 x^2 - \frac{1}{6}lx^3 + \frac{1}{24}(x-l)^4 U(x-l)\right], \ 0 \leqslant x \leqslant 2l$$

Application 5

A beam of length l is clamped horizontally at $x=0$ and freely hinged at $x= l$ It carries a load wx per unit length in $0<x<l/2$ and a load $w(l-x)$ in $l/2<x<l$. Find the static deflection at any point x.

$$\frac{d^4y}{dx^4} = \frac{1}{EI}\left[wx - 2w\left(x-\frac{l}{2}\right)U\left(x-\frac{l}{2}\right)\right], \ 0<x<l$$

$$s^4\bar{y} - s^3 y(o) - s^2 y'(o) - sy''(o) - y'''(o)$$

$$= \frac{1}{EI}\left[\frac{w}{s^2} - \frac{2w}{s^2}e^{-\frac{l}{2}s}\right]$$

$y(o) = y'(o) = 0$ and put $y''(o) = A$ and $y'''(o) = B$

$$s^4\bar{y} - As - B = \frac{1}{EI}\left[\frac{w}{s^2} - \frac{2w}{s^2}e^{-\frac{l}{2}s}\right]$$

$$\bar{y} = \frac{B}{s^4} + \frac{A}{s^3} + \frac{w}{EIs^6} - \frac{2w}{EIs^6}e^{-\frac{l}{2}s}$$

$$\therefore \ y = \frac{1}{6}Bx^3 + \frac{1}{2}Ax^2 + \frac{wx^5}{5!EI} - \frac{2w\left(x-\frac{l}{2}\right)^5}{5!\ EI}U\left(x-\frac{l}{2}\right)$$

Now we must have $y = y'' = 0$ when $x = l$,

$$\therefore \ \frac{1}{6}Bl^3 + \frac{1}{2}Al^2 + \frac{15}{16}\frac{wl^5}{5!\ EI} = 0$$

$$Bl + A + \frac{3}{4}\frac{wl^3}{3!\ EI} = 0$$

Substituting the values of A and B from these two equations in the expression for y we get

231

$$\text{EIy} = \frac{5}{256} \text{ w } x^2 l^3 - \frac{7}{256} \text{ w } x^{3/2} + \frac{1}{120} \text{ w} x^6$$

$$- \frac{1}{60} \text{ w } (x - \tfrac{1}{2} l)^5 \text{ U}\left(x - \frac{l}{2}\right).$$

Application 6

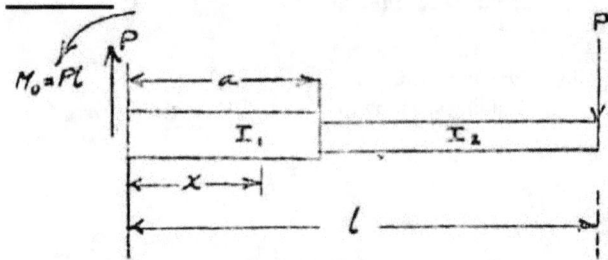

Determine the static deflection at any point of the cantilever beam shown above.

Solution:

We shall use the moment equation instead of the loading equation and the change in the stiffness is accounted for by a function of the form.

$$\frac{1}{EI} [1 + k \text{ U}(x-a)]$$

The moment equation is

$$\frac{d^2y}{dx^2} = \frac{M}{EI_1} [1 - k \text{ U}(x-a)]$$

where $\frac{1}{EI}$ over the section $0 \leqslant x \leqslant a$ is $\frac{1}{EI_1}$ and that over the section $x \geqslant a$ is

$$\frac{1}{EI_1} (1-k) = \frac{1}{EI_2}$$

from which $\quad k = 1 - \dfrac{I_1}{I_2}$

$$\therefore \quad \frac{d^2y}{dx^2} = \frac{P(l-x)}{EI_1} [1 - k \text{U}(x-a)]$$

Since $y(o) = y'(o) = o$, the subsidiary equation becomes

$$EI_1 \text{ } s^2\bar{y}(s) = \frac{Pl}{s} - \frac{P}{s^2} - \frac{Plke^{-as}}{s} + Pke^{-as}\left(\frac{a}{s} + \frac{1}{s^2}\right)$$

where $\quad L \left\{ x \text{ U}(x-a) \right\} = e^{-as}\left(\frac{a}{s} + \frac{1}{s^2}\right)$

Inverting we get

$$EI_1 \text{ } y(x) = Pl\frac{x^2}{2} - P\frac{x^3}{6}$$

$$- Pk \left[\frac{l(x-a)^2}{2} - \frac{(x-a)^3}{6} - \frac{a(x-a)^2}{2} \right] \text{U}(x-a)$$

Application 7

_____

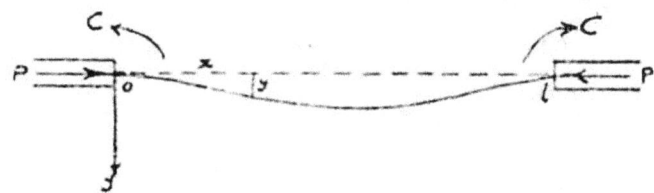

Find the critical loads for a strut clamped at both ends,

Solution:
Let the fixing couple at each end be C.

$$EI \frac{d^2y}{dx^2} = C - Py$$

$$EI [s^2\bar{y}(s) - sy(o) - y'(o)] = \frac{C}{s} - P\bar{y}(s)$$

$$\therefore \quad (EIs^2 + P)\bar{y} = \frac{C}{s}$$

$$\bar{y} = \frac{C}{s(EIs^2+P)} = \frac{C}{EIs\left(s^2+\frac{P}{EI}\right)}$$

$$= \frac{C}{EIs(s^2+n^2)} \quad \text{where } n^2 = \frac{P}{EI}$$

$$\therefore \quad y = \frac{C}{EI} \cdot \frac{1}{n^2} (1 - \cos nx)$$

i.e. $\quad y = \frac{C}{P} (1 - \cos nx)$

at $\quad x = l$, $\quad y = o \quad \therefore 1 - \cos nl = o \quad \cos nl = 1$

$\therefore \quad nl = 2\pi, 4\pi, \ldots$

$$n = \frac{2\pi}{l}, \frac{4\pi}{l}, \ldots$$

First critical load when $n = \frac{2\pi}{l}$ is given by

$$P = n^2EI = \frac{4\pi^2 EI}{l^2}$$

Application 8

_____

Critical loads for non-uniform columns.

233

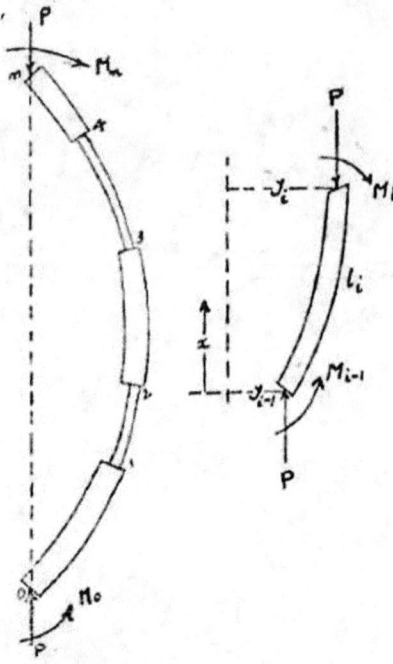

The operational method for treating columns of several sections is by shifting the origin to the end of each section.

Thus considering the i th part of the column shown in the above diagram. We have

$$E_i I_i \frac{d^2y}{dx^2} = -P(y-y_{i-1}) + M_{i-1}$$

Dividing by $E_i I_i$ and putting $\frac{P}{E_i I_i} = a_i^2$ we get

$$\frac{d^2y}{dx^2} = -a_i^2 (y - y_{i-1}) + \frac{M_{i-1}}{E_i I_i}$$

$$\therefore s^2 \overline{y}(s) - sy_{i-1} - y'_{i-1} = -a_i^2 \overline{y}(s) + \frac{a_i^2 y_{i-1}}{s} + \frac{M_{i-1}}{E_i I_i s}$$

$$\therefore (s^2 + a_i^2)\, \overline{y}(s) = sy_{i-1} + y'_{i-1} + \frac{a_i^2 y_{i-1}}{s} + \frac{M_{i-1}}{E_i I_i s}$$

$$\therefore \overline{y}(s) = \frac{y_{i-1}}{s} + \frac{y'_{i-1}}{s^2 + a_i^2} + \frac{M_{i-1}}{E_i I_i s(s^2 + a_i^2)}$$

Inverting we get

$$y(x) = y_{i-1} + \frac{y'_{i-1}}{a_i} \sin a_i x + \frac{M_{i-1}}{E_i I_i a_i^2} (1 - \cos a_i x)$$

234

Now in this equation and in the two equations obtained by differentiating it twice w.r.t x put $x = l_i$. so as to obtain the deflection, slope and bending moment at i in terms of the corresponding quantities at i —1.

$$y_i = y_{i-1} + \frac{y'_{i-1}}{a_i} \sin a_i l_i + \frac{M_{i-1}}{E_i I_i a_i^2}(1 - \cos a_i l_i)$$

$$y'_i = y'_{i-1} \cos a_i l_i + \frac{M_{i-1}}{E_i I_i a_i} \sin a_i l_i$$

$$M_i = -y'_{i-1} E_i I_i a_i \sin a_i l_i + M_{i-1} \cos a_i l_i$$

which may *be* expressed in matrix form as follows:

$$
\begin{bmatrix} y_i \\ y'_i \\ M_i \end{bmatrix}
=
\begin{bmatrix}
1 & \frac{1}{a_i}\sin a_i l_i & \frac{1}{E_i I_i a_i^2}(1-\cos a_i l_i) \\
0 & \cos a_i l & \frac{1}{E_i I_i a_i}\sin a_i l_i \\
0 & -E_i I_i a_i \sin a_i l_i & \cos a_i l_i
\end{bmatrix}
\begin{bmatrix} y_{i-1} \\ y'_{i-1} \\ M_{i-1} \end{bmatrix}
$$

In the same way, by repeated application, we can

$$y_{i-1} , y'_{i-1} , M_{i-1}$$

in terms of y_{i-2} , y'_{i-2} , M_{i-2} and so on.

Thus by multiplying the successive matrices we can obtain y_n , y'_n , M_n in terms of y_o , y'_o and M_o in the form

$$
\begin{bmatrix} y_n \\ y' \\ M_n \end{bmatrix}
=
\begin{bmatrix}
P_{11} & P_{12} & P_{13} \\
0 & P_{22} & P_{23} \\
0 & P_{32} & P_{33}
\end{bmatrix}
\begin{bmatrix} y_o \\ y'_o \\ M_o \end{bmatrix}
$$

and the critical loads can be obtained by substituting the boundary conditions in this equation. Thus for example in the case of a column whose ends are both built in we have

$$y_n = y_o = 0 \quad \text{and} \quad y'_n = y'_o = 0$$

and since

$$y_n = P_{11}y_o + P_{12}y'_o + P_{13}M_o$$

And

$$y'_n = P_{22}y'_o + P_{23}M_o$$

$$\therefore P_{13} = 0, P_{23} = 0$$

If the column is hinged at both ends, then

$$y_n = y_c = 0 \quad \text{and} \quad M_n = M_o = 0$$

Now

$$y_n = P_{11}y_o + P_{12}y'_o + P_{13}M_o .$$

$$M_n = P_{32} y'_o + P_{33} M_o$$

$$\therefore \quad P_{12} = 0, \quad P_{32} = 0$$

Application 9

A uniform beam with its ends x= 0 and x =./ built in carries a uniform load w per unit length. At x=a, there is an elastic support which provides a reaction equal to λ times the deflection. Determine the deflection at any point x of the beam.

Solution:

If R is the reaction at the support then

$$R = \lambda y(a)$$

$$EI \frac{d^4y}{dx^4} = w - R\delta(x-a)$$

$$\therefore EI \left[s^4\bar{y} - s^3 y(o) - s^2 y'(o) - sy''(o) - y'''(o)\right] = \frac{w}{s} - Re^{-as}$$

$$y(o) = y'(o) = 0 \text{ and put } y''(o) = A, \ y'''(o) = B, \text{ then}$$

$$\bar{y} = \frac{A}{s^3} + \frac{B}{s^4} + \frac{w}{EIs^5} - \frac{R}{EIs^4} e^{-as}$$

$$y = \tfrac{1}{2} Ax^2 + \frac{1}{6} Bx^3 + \frac{wx^4}{24EI} - \frac{R(x-a)^3}{6 \ EI} U(x \cdot a)$$

The unknowns A, B, R can be determined as follows :

Since y(l) = y'(1)= 0, then

$$o = \tfrac{1}{2} Al^2 + \frac{1}{6} Bl^3 + \frac{wl^4}{24 \ EI} - \frac{R(l-a)^3}{6 \ EI} \quad \ldots \quad (i)$$

$$o = Al + \tfrac{1}{2} Bl^2 + \frac{wl^3}{6EI} - \frac{R(l-a)^2}{2 \ EI} \quad \ldots \quad (ii)$$

Since at x = a, y = R/ λ then

$$\frac{R}{\lambda} = \tfrac{1}{2} Aa^2 + \frac{1}{6} Ba^3 + \frac{wa^4}{24EI} \quad \ldots \ldots \quad (iii)$$

Equations (i), (ii), (iii) determine A,B,R.

11.5.2. Exercises on Laplace Transform in practical applications

(1) A periodic e.m.f. E sin cut is applied at time t=o to an RC circuit in series. If the initial current and charge are zero, find the current I at any instant t.

(2) In problem (1) replace E sin *wt* by

$$E \left[U(t-t_o) - U(t-t_1)\right], \quad t_1 > t_o > o.$$

(3) An e.m.f. E(t) is applied at time t o to an RC circuit in series. The initial current and charge are zero. Find the charge and current at any time t if
(i) E Eo a constant.
(ii)

$$E = E_o e^{-\alpha t}, \quad \alpha > o.$$

(4) Same as in problem (3) but with $E(t) = E_o \delta(t)$ where $\delta(t)$ is the Dirac delta function.

(5) An electric circuit of an inductance L in series with a condenser of capacity C. At t=o an e.m.f. given by

$$E(t) = \frac{E_0 t}{T_0} \qquad o < t < T_0$$

$$= o \qquad t > T_0$$

is applied. Assuming zero initial current and charge, find the charge at any instant t.

(6) An e. .f. $E \cos(\omega t + a)$ is applied at time $t = o$ to a series circuit of capacity C and inductance L. Find the current at any time t, assuming zero initial current and charge.

(7) In the circuit shown, calculate I_1, I_2, I_3 if E is constant and the initial currents and charges are zero.

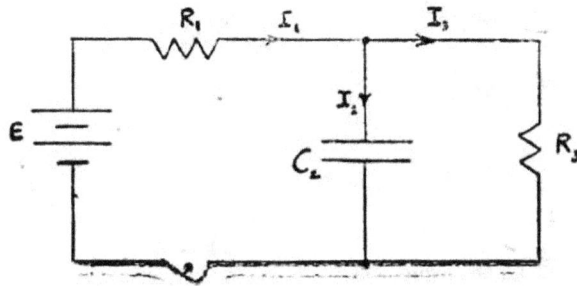

(8) In the circuit shown, the initial currents and charge are zero,

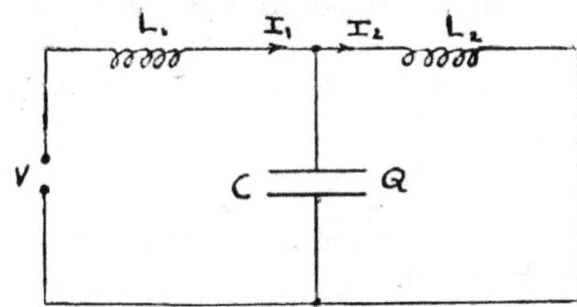

Find I_1 and I_2 when $V = E_o \sin \omega t$.

(9) In the circuit shown, set up the equations for the determination of the currents I_1, I_2, I_3 and the charge Q_3. Transform the problem into algebraic form, assuming E constant and initial currents and charges to be zero.

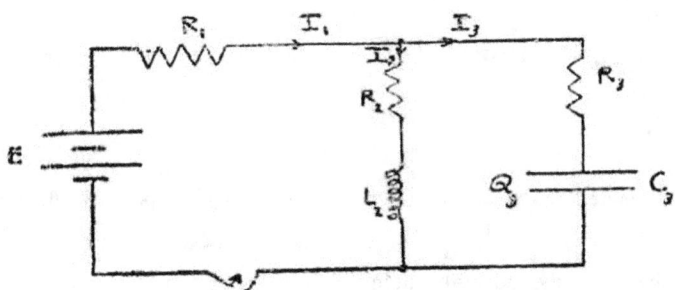

(10) A spring of stiffness k is placed on a smooth horizontal table. One end is fixed to a point 0 on the table and to the other end is attached a mass m. A force F(t), t >o acts on the mass. The differential equation of motion of the mass is

$$m\ddot{X} + kX = F(t)$$

Assuming $X(o) = a$, $\dot{X}(o) = o$ find $X(t)$ under the following conditions :

(i) $F(t) = F_o$ for $t > o$ where F_o is constant.

(ii) $F(t) = F_o e^{at}$ where $a > o$

(iii) $F(t) = F_o \sin \omega t$ where $\omega \neq \sqrt{\dfrac{k}{m}}$

(iv) $F(t) = F_o \sin \omega t$ where $\omega = \sqrt{\dfrac{k}{m}}$

(v) $F(t) = F_o U(t-T)$

(vi) $F(t) = F_o \delta(t-T)$.

(11) Same as in problem (10) but with initial conditions
$X(o) = \dot{X}(o) = o$. Find $X(t)$ when $F(t) = F_o\delta(t)$.

(12) A particle of mass m, at rest at the origin, is set in motion at t o by a blow of impulse P. Find its displacement at time t.

(13) A particle moves along a straight line such that its displacement x from a fixed point at time t is given by

$$\ddot{x} + 2\dot{x} + 4x = 20 \sin 4t$$

with initial conditions

$$x(o) = \dot{x}(o) = o.$$

Solve the equation and state which term in the result is the transient term and which is the steady state term. Find the amplitude and period of the steady state.

(14) Solve the equation of motion $\quad m\ddot{x} = F(t)$

of a particle with initial conditions
$x(o) = \dot{x}(o) = o$
where F(t) is given by

$$F(t) = \frac{2F_o}{T}\, t , \qquad o < t < \frac{T}{2}$$

$$= -\frac{2F_o}{T}\, (t-T), \frac{T}{2} < t < T$$

$$= o \qquad t > T$$

(15) A beam with its ends built in at x =o and x = l carries a uniform load W_o per unit length. Find the deflection at any point x.

(16) Work the same problem with the end x =0 built in and the end x = l hinged.

(17) A beam with its ends x = o and x = l hinged carries a load w(x) per unit length given by :

238

$$w(x) = 0 \qquad 0 < x < \frac{l}{4}$$

$$= w_0 \qquad \frac{l}{4} < x < l,$$

find the static deflection at any point x.

(18) A cantilever beam with its end x=o built in and the end x =1 free carries a concentrated load P_0 at x = 1/3, find the static deflection.

(19) A beam with its ends x=o and x=1 hinged carries a concentrated load Po at x = 1/4 . Find the deflection.

(20) A cantilever beam clamped at x = o and free at x= l carries a uniform load w_0 per unit length. Find the deflection.

11.6. Using Laplace Transformation in solving Linear Partial Differential Equations.

The Laplace transformation can be used with advantage in solving linear partial differential equations. It is found that on applying this transformation the partial differential equation transforms to an ordinary differential equation, We illustrate by the following examples.

Example 1   Solve the partial differential equation $2x \dfrac{\partial Y}{\partial t} + \dfrac{\partial Y}{\partial x} = 2x$ ... **(1)**

given that $\qquad\qquad\qquad Y(x,o) = 1 , Y(0,t) = 1 \qquad\qquad ..,(2)$

Writing equation (1) in the form

$$2xY_t \ (x,t) + Y_x \ (x,t) = 2x$$

and noticing that

$$L\left\{ Y_x \ (x,t) \right\} = \int_0^\infty e^{-st} \frac{\partial}{\partial x} Y(x,t)dt = \frac{\partial}{\partial x} \int_0^\infty e^{-st} Y(x,t) \, dt$$

$$= y_x \ (x,s)$$

we get on transforming (1) w.r.t. t

$$2x \ [sy(x,s) - Y(x,o)] + y_x \ (x,s) = \frac{2x}{s}$$

$$\therefore \ 2x \ [sy(x,s) - 1] + y_x \ (x,s) = \frac{2x}{s}$$

$$\frac{dy}{dx} + 2xsy = 2x + \frac{2x}{s}$$

$$= 2x \ (1 + \frac{1}{s})$$

This is a linear differential equation of the first order. The integrating factor is

$$e^{\int 2xsdx} = e^{sx^2}$$

$$\therefore \ ye^{sx^2} = \int 2x \ (1 + \frac{1}{s}) \ e^{sx^2} dx + C$$

$$= \frac{1}{s} (1 + \frac{1}{s}) \ e^{sx^2} + C$$

239

i.e $y(x,s) = \frac{1}{s}(1 + \frac{1}{s}) + Ce^{-sx^2}$ .... (3)

Transforming (2) we get $y(0,s) = 1/s$

$$\frac{1}{s} = \frac{1}{s}(1 + \frac{1}{s}) + C \quad \therefore C = -\frac{1}{s^2}$$

$$y(x,s) = \frac{1}{s}(1 + \frac{1}{s}) - \frac{1}{s^2}e^{-sx^2}$$

$$= \frac{1}{s} + \frac{1}{s^2} - \frac{1}{s^3}e^{-x^2s}$$

$$\therefore Y(x,t) = 1 + t \qquad\qquad o \leqslant t \leqslant x^2$$

$$= 1 + t - (t - x^2) \qquad t \geqslant x^2$$

i.e $Y(x,t) = 1 + t \qquad\qquad o \leqslant t \leqslant x^2$

$$= 1 + x^2 \qquad\qquad t \geqslant x^2$$

Example 2

Find the solution of $\quad \frac{\partial U}{\partial x} = 2\frac{\partial U}{\partial t} + U$, $U(x,o) = 4e^{-2x}$

which is bounded for $x > o$, $t > o$.

$$U_x(x,t) = 2U_t(x,t) + U(x,t)$$

$$\therefore u_x(x,s) = 2[su(x,s) - U(x,o)] + u(x,s)$$

$$= 2[su(x,s) - 4e^{-2x}] + u(x,s)$$

$$\frac{du}{dx} - (2s+1)u = -8e^{-2x}$$

which is a linear differential equation of the first order. Integrating factor is

$$e^{\int - (2s+1)dx} = e^{-(2s+1)x}$$

$$\therefore ue^{-(2s+1)x} = -8\int e^{-(2s+1)x}\, e^{-2x}\, dx + C$$

$$= -8\int e^{-(2s+3)x}\, dx + C$$

$$= \frac{8}{2s+3}e^{-(2s+3)x} + C$$

$$\therefore u(x,s) = \frac{8e^{-2x}}{2s+3} + Ce^{(2s+1)x}$$

Since $U(x,t)$ must be bounded as $x \rightarrow \infty$, we must have $u(x,s)$ also bounded as $x \rightarrow \infty$. $\qquad \therefore C = 0$

240

$$u(x,s) = \frac{8}{2s+3} e^{-2x} = \frac{4}{s+\frac{3}{2}} e^{-2x}$$

$$\therefore U(x,t) = 4e^{-\frac{3}{2}t} e^{-2x} = 4e^{-\left(\frac{3}{2}t+2x\right)}$$

Example 3

Solve the partial differential equation

$$xY_x(x,t) + Y_t(x,t) + Y(x,t) = x F(t)$$

with the boundary conditions $Y(x,0) = Y(0,t) = 0$

Solution:

$$xy_x(x,s) + sy(x,s) - Y(x,0) + y(x,s) = xf(s)$$

$$\therefore \quad x\frac{dy}{dx} + sy + y = xf(s)$$

i.e $\quad \dfrac{dy}{dx} + \dfrac{s+1}{x} y = f(s)$

Integrating factor is

$$e^{\int \frac{s+1}{x} dx} = e^{(s+1)\log x} = x^{s+1}$$

$$\therefore \quad yx^{s+1} = \int f(s) x^{s+1} dx + C = f(s) \frac{x^{s+2}}{s+2} + C$$

$$\therefore \quad y(x,s) = f(s) \frac{x}{s+2} + \frac{C}{x^{s+1}}$$

and since $Y(0,t) = 0$, $\therefore y(0,s) = 0$, $\therefore C = 0$

$$y(x,s) = x. \frac{1}{s+2} f(s)$$

$$\therefore \quad Y(x,t) = x\left(e^{-2t} * F(t)\right) = x \int_0^t e^{-2(t-\lambda)} F(\lambda)d\lambda$$

241

$$= xe^{-2t} \int_0^t e^{2\lambda} F'(\lambda)d\lambda$$

Example 4

Solve the partial differential equation

$$U_{xx}(x,t) - 2U_{tx}(x,t) + U_{tt}(x,t) = 0 \quad (0<x<1 \ , \ t>0)$$

with the boundary conditions

$$U(x,0) = U_t(x,0) = U(0,t) = 0 \ , \ U(1,t) = F(t) \ t>0.$$

Solution:

$$L\left\{U_{xx}(x,t)\right\} = \int_0^\infty e^{-st} \frac{\partial^2}{\partial x^2} U(x,t)dt$$

$$= \frac{\partial^2}{\partial x^2} \int_0^\infty e^{-st} U(x,t)dt = \frac{\partial^2}{\partial x^2} u(x,s) = u_{xx}(x,s)$$

$$\therefore \ u_{xx}(x,s) - 2\frac{\partial}{\partial x}[su(x,s) - U(x,0)] + s^2 u(x,s) - sU(x,0) - U_t(x,0) = 0$$

$$\therefore \ \frac{d^2u}{dx^2} - 2s\frac{du}{dx} + s^2 u = 0$$

$$(D^2 - 2sD + s^2)\,u = 0 \quad \text{i.e} \quad (D-s)^2 u = 0$$

$$\therefore \ u(x,s) = e^{sx}(A + Bx)$$

and since $U(0,t) = 0$ $\therefore u(0,s) = 0$ and hence $A = 0$

$$\therefore \ u(x,s) = Bxe^{xs}$$

Since $U(1,t) = F(t)$ $\therefore u(1,s) = f(s)$

$$\therefore \ f(s) = Be^s \qquad \text{i.e } B = e^{-s} f(s)$$

$$\therefore \ u(x,s) = e^{-s} f(s) \ x e^{xs} \qquad \text{i.e } u(x,s) = xf(s) e^{-(1-x)s}$$

$$\therefore \ U(x,t) = 0 \qquad\qquad 0<t< 1-x$$

$$= x\,F(t-1+x) \qquad\qquad t> 1-x$$

Example 5

Find a bounded solution of

$$\frac{\partial V}{\partial t} = k\frac{\partial^2 V}{\partial x^2} \qquad x>0 \ , \ t>0$$

subject to the boundary conditions

242

$$V(o,t) = F(t) \ , \ V(x,o) = o$$

$$sv(x,s) - V(x,o) = kv_{xx}(x,s)$$

$$\frac{d^2v}{dx^2} - \frac{s}{k} v = o$$

$$\therefore \quad v(x,s) = A e^{\sqrt{s/k}\,x} + B e^{-\sqrt{s/k}\,x}$$

Since the solution is bounded, then $A = 0$

$$v(x,s) = B e^{-\sqrt{s/k}\,x}$$

$$v(x,s) = f(s)\, e^{-\sqrt{s/k}\,x}$$

Now $\quad L^{-1}\left\{ e^{-a\sqrt{s}} \right\} = \dfrac{a}{2\sqrt{\pi t^3}}\, e^{-\frac{a^2}{4t}}$

$$\therefore \quad V(x,t) = F(t) * \frac{x}{2\sqrt{\pi k t^3}}\, e^{-\frac{x^2}{4kt}}$$

$$= \frac{x}{2\sqrt{\pi k}} \int_o^t \frac{F(\lambda)}{(t-\lambda)^{3/2}}\, e^{-\frac{x^2}{4k(t-\lambda)}}\, d\lambda$$

11.6.1. Transverse vibrations of a stretched string under gravity.

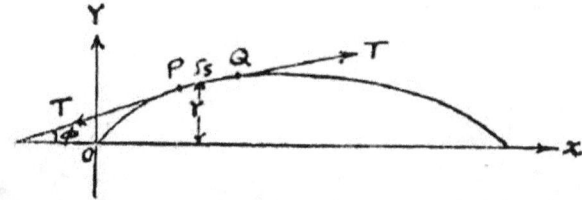

Let T be the tension in the string and m mass per unit length.
Consider an element PQ of the string of length δs.
Component of the tension at P resolved in the direction OY is

$$- T \sin \phi = - T \frac{\partial Y}{\partial s}$$

Component of the tension at Q resolved in the direction OY is

$$T \frac{\partial Y}{\partial s} + \frac{\partial}{\partial s}\left(T \frac{\partial Y}{\partial s} \right) \delta s$$

Resultant force due to the two tensions in the direction OY is

$$\frac{\partial}{\partial s}\left(T \frac{\partial Y}{\partial s} \right) \delta s = T \frac{\partial^2 Y}{\partial s^2} \delta s$$

$$\therefore \quad m\delta s \frac{\partial^2 Y}{\partial t^2} = T \frac{\partial^2 Y}{\partial s^2} \delta s - m \delta s\, g$$

$$\frac{\partial^2 Y}{\partial t^2} = \frac{T}{m} \frac{\partial^2 Y}{\partial s^2} - g. \qquad \text{Put } \frac{T}{m} = a^2$$

$$\therefore \frac{\partial^2 Y}{\partial t^2} = a^2 \frac{\partial^2 Y}{\partial s^2} - g$$

Assuming small lateral displacements and gradients we can replace

$$\frac{\partial^2 Y}{\partial s^2} \text{ by } \frac{\partial^2 Y}{\partial x^2}$$

$$\therefore \frac{\partial^2 Y}{\partial t^2} = a^2 \frac{\partial^2 Y}{\partial x^2} - g$$

i.e $Y_{tt}(x,t) = a^2 Y_{xx}(x,t) - g$

Example 1.

A semi - infinite stretched string of negligible weight has its distant end fixed while the end x =o is initially at the origin and moves along the Y — axis such that Y(t) =
$C \sin \omega t$, t > o. The string is initially along the x-axis with no initial velocity. When the end x = o starts to move find the shape of the string at any subsequent instant.

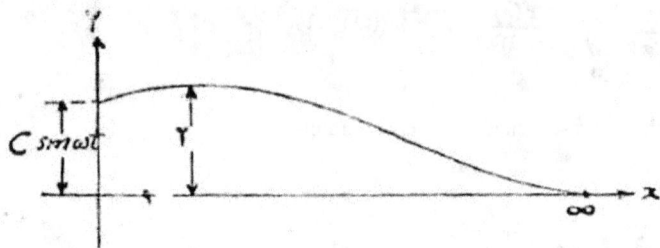

Differential equation of motion of the siring is

$$Y_{tt}(x,t) = a^2 Y_{xx}(x,t) \quad \cdots \quad (1)$$

At t=o displacements and velocities are zero.

$$Y(x,o) = o, \quad Y_t(x,o) = o \quad \cdots \quad (2)$$

At x=o, $Y(o,t) = C \sin \omega t \quad \cdots \quad (3)$

At x= ∞ , $\lim_{x \to \infty} Y(x,t) = o \quad \cdots \quad (4)$

$$\therefore y(x,s) = A e^{\frac{s}{a}x} + B e^{-\frac{s}{a}x}$$

Transforming (1) w.r.t. t we get

$$s^2 y(x,s) - sY(s,o) - Y_t(x,o) = a^2 y_{xx}(x,s)$$

$$\therefore s^2 y = a^2 \frac{d^2 y}{dx^2} \quad \text{i.e} \quad \frac{d^2 y}{dx^2} - \frac{s^2}{a^2} y = o$$

$$\therefore y(x,s) = A e^{\frac{s}{a}x} + B e^{-\frac{s}{a}x}$$

Equation (4) transforms into

244

$$\lim_{x \to \infty} y(x,s) = 0 \quad \therefore \ A = 0$$

$$y(x,s) = Be^{-\frac{s}{a}x}$$

Equation (3) transforms into $y(o,s) = \dfrac{C\omega}{s^2+\omega^2}$

$$\therefore \quad B = \frac{C\omega}{s^2+\omega^2}$$

$$y(x,s) = \frac{C\omega}{s^2+\omega^2} \cdot e^{-\frac{s}{a}x}$$

$$Y(x,t) = 0 \qquad\qquad 0 < t < \frac{x}{a}$$

$$= C \sin \omega\left(t - \frac{x}{a}\right) \quad t > \frac{x}{a}$$

The motion may be interpreted as follows:

A point at distance x from the origin remains at rest until a time t = x/a elapses when the motion at the origin is transferred to the point. The time t = x/a is the time necessary for a wave traveling with velocity a to describe a distance x.

Example 2

A semi - infinite string has its end x = o fixed while the distant and is looped around a vertical support that cannot exert any vertical force upon the string i.e.,

$$T\frac{\partial Y}{\partial x} = 0$$

at that end. The string is initially supported along the x – axis and at time t.= o the support is removed and the string moves downwards under the action of gravity. Find the shape of the string at any instant.

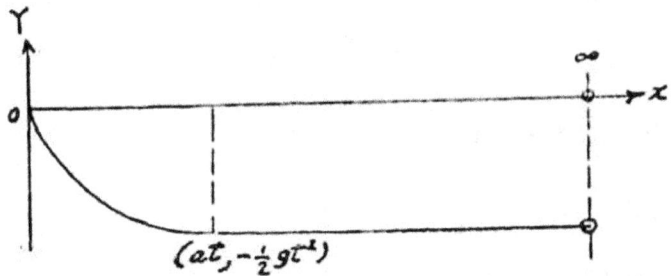

$$\left(at, -\tfrac{1}{2}gt^2\right)$$

Solution:

$$Y_{tt}(x,t) = a^2 Y_{xx}(x,t) - g \quad \cdots \quad (1)$$

At time t = 0, displacements and velocities at all points of the string are zero.

$$\therefore \quad Y(x,0) = 0 \ , \ Y_t(x,o) = 0 \qquad \cdots \quad (2)$$

At the end x= 0 , Y = 0 for all t, i.e.,

245

$Y(0,t) = 0$ .... (3)

At the distant end $x = \infty$, the vertical component of the tension is zero.

$$\therefore \quad \lim_{x \to \infty} Y_x(x,t) = 0 \qquad \qquad \cdot \; \cdot \; \cdot \; (4)$$

Transforming (1) w.r.t. t

$$s^2 y(x,s) = u^2 y_{xx}(x,s) - \frac{g}{s}$$

$$\therefore \quad \frac{d^2 y}{dx^2} - \frac{s^2}{a^2} y = \frac{g}{a^2 s}$$

$$y(x,s) = Ae^{\frac{s}{a}x} + Be^{-\frac{s}{a}x} - \frac{g}{s^3}$$

Equation (4) transforms into

$$\lim_{x \to \infty} y_x(x,s) = 0 \qquad \therefore \quad A = 0$$

$$\therefore \quad y(x,s) = Be^{-\frac{s}{a}x} - \frac{g}{s^3}$$

Again equation (g) transforms into

$$y(0,s) = 0 \quad \therefore \quad B = \frac{g}{s^3}$$

$$y(x,s) = \frac{g}{s^3}\left(e^{-\frac{s}{a}x} - 1\right)$$

$$= -\frac{g}{2}\left(\frac{2}{s^3} - \frac{2}{s^3} e^{-\frac{x}{a}s}\right)$$

$$\therefore \quad Y(x,t) = -\tfrac{1}{2} gt^2 \qquad \qquad 0 < t < \frac{x}{a}$$

$$= -\tfrac{1}{2} gt^2 + \tfrac{1}{2} g\left(t - \frac{x}{a}\right)^2 \qquad t > \frac{x}{a}$$

i.e $\quad Y(x,t) = -\tfrac{1}{2} gt^2 \qquad \qquad 0 < t < \frac{x}{a}$

$$= -\frac{g}{2a^2}(2axt - x^2) \qquad t > \frac{x}{a}$$

$$\therefore \quad Y(x,t) = -\frac{g}{2a^2}(2axt - x^2) \qquad x < at$$

$$= -\tfrac{1}{2} gt^2 \qquad \qquad x > at$$

Considering the $f$irst equation

$$x^2 - 2axt = \frac{2a^2 Y}{g}$$

246

$$\therefore \quad (x - at)^2 = \frac{2a^2Y}{g} + a^2t^3$$

$$\text{i.e} \quad (x - at)^3 = \frac{2a^2}{g} (Y + \tfrac{1}{2} gt^2)$$

which is a parabola vertex

$$(at , - \tfrac{1}{2} gt^2)$$

and whose axis is parallel to ()Y.

In the instantaneous position of the string, we notice that, at any time t, all elements of the string to the right of the point x at move like freely falling bodies.

Example 3

A semi infinite stretched string has its distant end fixed on the x — axis while the end x = 0 is looped around the Y - axis which exerts, no vertical force on the loop. The string is initially at rest in the position $Y = e^{-x}$, released from this position with no external forces acting on it. Find the displacement of the string at any point x and at any instant t.

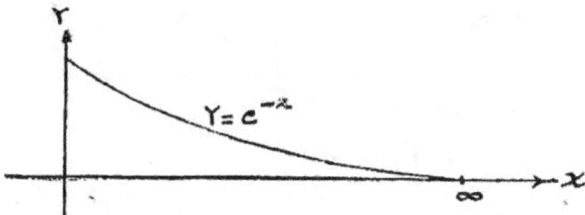

Solution:

The equation of motion of the string is

$$Y_{tt}(x,t) = a^2 Y_{xx}(x,t)$$

Boundary conditions are

$$Y(x,0) = e^{-x} , \quad Y_t(x,0) = 0 , \quad Y_x(0,t) = 0 , \quad \lim_{x \to \infty} Y(x,t) = 0.$$

$$s^2y(x,s) - s Y(x,0) - Y_t(x,0) = a^2 y_{xx}(x,s)$$

$$\therefore \quad s^2y(x,s) - se^{-x} = a^2 y_{xx}(x,s)$$

$$\frac{d^2y}{dx^2} - \frac{s^2}{a^2} y = - \frac{s}{a^2} e^{-x}$$

$$\therefore \quad y(x,s) = Ae^{\frac{s}{a}x} + Be^{-\frac{s}{a}x} + \frac{1}{D^2 - \frac{s^2}{a^2}} \left(- \frac{s}{a^2} e^{-x} \right)$$

247

$$y(x,s) = Ae^{\frac{s}{a}x} + Be^{-\frac{s}{a}x} - \frac{s}{a^2 - s^2}e^{-x}$$

$$\lim_{x \to \infty} y(x,s) = 0 \qquad \therefore \ A = 0$$

$$y(x,s) = Be^{-\frac{s}{a}x} - \frac{s}{a^2 - s^2}e^{-x}$$

$$y_x(0,s) = 0 \ \text{and} \ y_x(x,s) = -\frac{s}{a}Be^{-\frac{s}{a}x} + \frac{s}{a^2-s^2}e^{-x}$$

$$\therefore \ y_x(0,s) = -\frac{s}{a}B + \frac{s}{a^2-s^2} = 0 \ \therefore \ B = \frac{a}{a^2-s^2}$$

$$\therefore \ y(x,s) = -\frac{a}{s^2-a^2}e^{-\frac{s}{a}x} + \frac{s}{s^2-a^2}e^{-x}$$

$$\therefore \ Y(x,t) = e^{-x}\cosh at \qquad\qquad 0 < t < \frac{x}{a}$$

$$= e^{-x}\cosh at - \sinh a(t - \frac{x}{a}) \qquad t > \frac{x}{a}$$

$$\text{i.e} \ Y(x,t) = e^{-x}\cosh at + \sinh(x - at) \qquad x < at$$

$$= e^{-x}\cosh at \qquad\qquad x > at.$$

Example 4

A string is stretched between the two fixed points x = 0 and x =1. The string has initially the shape
$$Y = b \sin \frac{\pi x}{l}$$
and is released from rest at t = 0 in that position. Find the shape of the string at any subsequent instant.

Solution:

The differential equation of motion of the string is

$$Y_{tt}(x,t) = a^2 Y_{xx}(x,t) \qquad\qquad 0 < x < l, \ t > 0.$$

The boundary conditions are

$$Y(x,0) = b \sin\frac{\pi x}{l}, \ Y_t(x,0) = 0, \ Y(0,t) = Y(l,t) = 0$$

$$s^2 y(x,s) - sY(x,0) - Y_t(x,0) = a^2 y_{xx}(x,s)$$

$$\therefore \ s^2 y - bs \sin\frac{\pi x}{l} = a^2 \frac{d^2 y}{dx^2}$$

248

$$\text{i e } \quad \frac{d^2y}{dx^2} - \frac{s^2}{a^2}\, y = - \frac{bs}{a^2}\sin\frac{\pi x}{l}$$

$$y(x,s) = A\cosh\frac{s}{a}\,x + B\sinh\frac{s}{a}\,x + \frac{1}{\dfrac{D^2-s^2}{a^2}}\left(-\frac{bs}{a^2}\sin\frac{\pi x}{l}\right)$$

$$= A\cosh\frac{s}{a}\,x + B\sinh\frac{s}{a}\,x + \frac{1}{-\dfrac{\pi^2}{l^2}-\dfrac{s^2}{a^2}}\left(-\frac{bs}{a^2}\sin\frac{\pi x}{l}\right)$$

$$= A\cosh\frac{s}{a}\,x + B\sinh\frac{s}{a}\,x + \frac{bl^2s}{\pi^2a^3 + l^2s^2}\sin\frac{\pi x}{l}$$

Since $Y(o,t) = o \quad \therefore y(o,s) = o \quad \therefore A = o$

$$y(x,s) = B\sinh\frac{s}{a}\,x + \frac{bl^2s}{\pi^2a^2+l^2s^2}\sin\frac{\pi x}{l}$$

Since $Y(l,t) = o \quad \therefore y(l,s) = o$

$$o = B\sinh\frac{s}{a}\,l \qquad \therefore B = o$$

Hence $\quad y(x,s) = \dfrac{bl^2s}{\pi^2a^2 + l^2s^2}\sin\frac{\pi x}{l}$

$$= bl^2\sin\frac{\pi x}{l}\cdot\frac{1}{l^2}\frac{s}{s^2 + \dfrac{\pi^2a^2}{l^2}}$$

$$= b\sin\frac{\pi x}{l}\cdot\frac{s}{s^2 + \dfrac{\pi^2a^2}{l^2}}$$

$$\therefore \quad Y(x,t) = b\sin\frac{\pi x}{l}\cos\frac{\pi a}{l}\,t$$

11.6.2. Longitudinal vibrations of bars

Let one end O of an elastic bar be fixed and let A be the cross sectional area of the bar, p its density and E its Young's modulus.

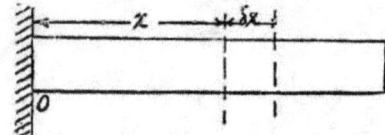

Consider an element of length δx of the bar, Its mass is $\rho A\ \delta x$. Suppose that the length x extends a distance Y then a length δx extends a distance

$$\frac{\partial Y}{\partial x}\ \delta x$$

The strain in the element is

$$\left(\frac{\partial Y}{\partial x}\ \delta x\right)/\delta x \quad \text{i.e} \quad \frac{\partial Y}{\partial x}$$

$$\text{Stress} = E\ \frac{\partial Y}{\partial x} \quad . \quad \text{Tension} = EA\ \frac{\partial Y}{\partial x}$$

This is at a distance x. Tension at

$$x+\delta x \text{ is } EA\frac{\partial Y}{\partial x} + \frac{\partial}{\partial x}\left(EA\frac{\partial Y}{\partial x}\right)\delta x.$$

$$\therefore \quad \text{Resultant tension in element} = EA\ \frac{\partial^2 Y}{\partial x^2}\ \delta x. \text{ This should}$$

be equal to $\rho A\ \delta x\ \dfrac{\partial^2 Y}{\partial t^2}$.

$$\therefore \quad \rho A\ \delta x\ \frac{\partial^2 Y}{\partial t^2} = EA\ \frac{\partial^2 Y}{\partial x^2}\ \delta x$$

$$\text{i.e} \quad \frac{\partial^2 Y}{\partial t^2} = \frac{E}{\rho}\ \frac{\partial^2 Y}{\partial x^2} \quad . \quad \text{Put} \quad \frac{E}{\rho} = a^2$$

$$\therefore \quad Y_{tt}\ (x,t) = a^2\ Y_{xx}\ (x,t).$$

Example 1

An elastic bar of length /, has its end x=o fixed while the other end x=1 is acted upon by a constant force Fo per unit area parallel to the bar. The bar is initially unstrained and at rest. Find the longitudinal displacement of the end x = / at any instant.

$$Y_{tt}\ (x,t) = a^2 Y_{xx}\ (x,t) \quad . \quad . \quad . \quad . \quad (1)$$

$$\text{At } t=o \quad Y\ (x,o) = o,\ Y_t\ (x,o) = o \quad . \quad . \quad (2)$$

250

At $x=0$ $Y(0,t) = 0$ . . . . (3)

At $x=l$, stress $= Fo$

$$\therefore \quad EY_x(l,t) = Fo \quad . \; . \; . \; . \; (4)$$

Transforming (1) we get

$$s^2y(x,s) - sY(x,0) - Y_t(x,0) = a^2y_{xx}(x,s)$$

$$\therefore \quad s^2y(x,s) = a^2y_{xx}(x,s)$$

$$\frac{d^2y}{dx^2} - \frac{s^2}{a^2}y = 0$$

$$y(x,s) = A\cosh\frac{s}{a}x + B\sinh\frac{s}{a}x$$

Transforming (3) we get

$$y(0,s) = 0 \quad \therefore \quad A = 0$$

$$\therefore \quad y(x,s) = B\sinh\frac{s}{a}x$$

Transforming (4) we get

$$Ey_x(l,s) = \frac{Fo}{s}$$

$$E.B\frac{s}{a}\cosh\frac{s}{a}l = \frac{F}{s}$$

$$\therefore \quad B = \frac{aF_o}{E}\cdot\frac{1}{s^2}\operatorname{sech}\frac{s}{a}l$$

$$y(x,s) = \frac{aF_o}{E}\cdot\frac{1}{s^2}\operatorname{sech}\frac{s}{a}l\sinh\frac{s}{a}x$$

$$\therefore \quad y(l,s) = \frac{aF_o}{E}\frac{1}{s^2}\tanh\frac{s}{a}l$$

Now we already know that the Laplace transform of the triangular wave $H(c,t)$ defined by

$$H(c,t) = t \qquad 0 < t < c$$

$$= 2c - t \qquad c < t < 2c$$

$$H(c,t+2c) = H(c,t)$$

is given by

$$L\left\{H(c,t)\right\} = \frac{1}{s^2}\tanh\frac{cs}{2}$$

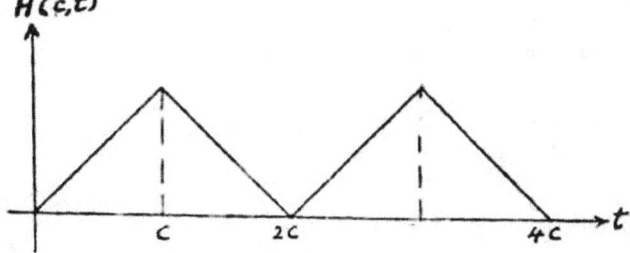

251

Hence $Y(l,t) = \dfrac{aF_o}{E} \; H\left(\dfrac{2l}{a} \, , \, t\right)$

Example 2

A uniform vertical rod of mass m and length l is clamped at the end $x = o$ and is mass loaded by a mass M at its other end $x = l$. Investigate the behavior of the system to an arbitrary excitation $F(t)$ applied to the mass M.

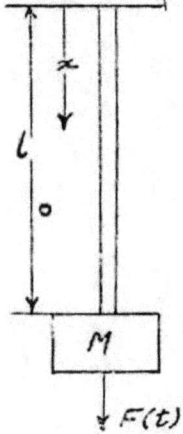

Solution:

The partial differential equation satisfied by the displacement Y of an element at distance x from the top is

$$\dfrac{\partial^2 Y}{\partial t^2} = a^2 \, \dfrac{\partial^2 Y}{\partial x^2} \quad \text{where } a = \sqrt{\dfrac{E}{\rho}}$$

i.e $Y_{tt}(x,t) = a^2 \, Y_{xx}(x,t)$

and since the system is initially at rest, its transform is

$s^2 y(x,s) = a^2 y_{xx}(x,s)$

i.e $\dfrac{d^2 y}{dx^2} - \dfrac{s^2}{a^2} y = o$

$\therefore \; y(x,s) = B \cosh \dfrac{s}{a} x + C \sinh \dfrac{s}{a} x$

and since $Y(o,t) = o$ $\therefore$ $y(o,s) = o$ $\therefore$ $B = o$ and hence

$\qquad y(x,s) = C \sinh \dfrac{s}{a} x$

The constant C can be determined from the boundary condition at $x = l$.

$$M \left[\dfrac{\partial^2 Y}{\partial t^2} \right]_{x=l} = F(t) - AE \left[\dfrac{\partial Y}{\partial x} \right]_{x=l}$$

where A is the cross sectional area of the rod.

$$\therefore \quad Ms^2 y(l,s) = f(s) - AEy_x \ (l,s)$$

Now, since $y(x,s) = C \sinh \frac{s}{a} x$

$$\therefore \quad y(l,s) = C \sinh \frac{s}{a} l$$

and $y_x(l,s) = C \frac{s}{a} \cosh \frac{s}{a} l$

$$\therefore \quad Ms^2 \ C \sinh \frac{s}{a} l = f(s) - AEC \frac{s}{a} \cosh \frac{s}{a} l$$

$$C \left(Ms^2 \sinh \frac{s}{a} l + A E \frac{s}{a} \cosh \frac{s}{a} l \right) = f(s)$$

$$\therefore \quad C = \frac{f(s)}{Ms^2 \sinh \frac{s}{a} l + AE \frac{s}{a} \cosh \frac{s}{a} l}$$

$$\therefore \quad y(x,s) = \frac{f(s) \sinh \frac{s}{a} x}{Ms^2 \sinh \frac{sl}{a} + AE \frac{s}{a} \cosh \frac{sl}{a}}$$

$$\therefore \quad Y(x,t) = \frac{1}{2\pi i} \int_{\gamma - i\infty}^{\gamma + i\infty} \frac{e^{st} \ f(s) \sinh \frac{s}{a} x \ ds}{Ms^2 \sinh \frac{sl}{a} + AE \frac{s}{a} \cosh \frac{sl}{a}}$$

11.6.3. Partial differential equations of transmission lines

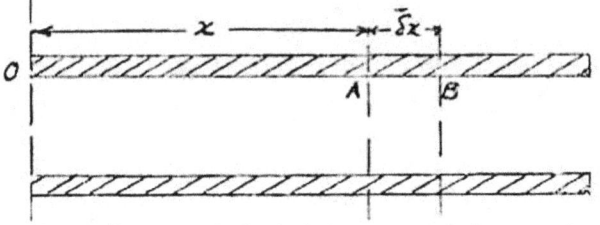

Let L,C,R,G be respectively the inductance, capacitance, resistance find dielectric conductance or leakance, all per unit length of a transmission line. Consider an element of length §x of this line. Let the potential at A be V, then the potential at B is

$$V + \frac{\partial V}{\partial x} \delta x$$

The drop of potential is

$$- \frac{\partial V}{\partial x} \delta x.$$

This is partly due to resistance $R\delta x$ and partly due to inductance $L\delta x$

$$\therefore \quad IR\delta x + L\delta x \frac{\partial I}{\partial t} = - \frac{\partial V}{\partial x} \delta x.$$

i.e $\left(L\dfrac{\partial}{\partial t} + R \right) I = -\dfrac{\partial V}{\partial x}$ . . . (1)

Next consider an interval of time δt Charge of electricity entering the element through section A in this interval is $I\delta t$. Charge leaving the element out of the section B in the same interval is

$\left(I + \dfrac{\partial I}{\partial x}\,\delta x \right) \delta t.$

Hence net outflow of charge from these two sections is

$\dfrac{\partial I}{\partial x}\,\delta x\delta t.$

To this we should add the charge leaking from core to sheath namely

$G\delta x V\delta t.$

Hence the total outflow of charge from element is

$\dfrac{\partial I}{\partial x}\,\delta x\delta t + G\delta x V\delta t$

This will cause a drop of potential

$-\dfrac{\partial V}{\partial t}\,\delta t$

$\therefore\quad \dfrac{\dfrac{\partial I}{\partial x}\,\delta x\delta t + G\delta x V\delta t}{C\delta x} = -\dfrac{\partial V}{\partial t}\,\delta t$

from which we get

$\left(C\dfrac{\partial}{\partial t} + G \right) V = -\dfrac{\partial I}{\partial x}$ . . . (2)

Now operating on both sides of this equation by

$L\dfrac{\partial}{\partial t} + R$

we get

$\left(L\dfrac{\partial}{\partial t} + R \right)\left(C\dfrac{\partial}{\partial t} + G \right) V = -\left(L\dfrac{\partial}{\partial t} + R \right)\dfrac{\partial I}{\partial x}$

$= -\dfrac{\partial}{\partial x}\left(L\dfrac{\partial}{\partial t} + R \right) I$

$= -\dfrac{\partial}{\partial x}\left(-\dfrac{\partial V}{\partial x} \right) = \dfrac{\partial^2 V}{\partial x^2}$

$\therefore\quad CL\dfrac{\partial^2 V}{\partial t^2} + (CR + LG)\dfrac{\partial V}{\partial t} + RGV = \dfrac{\partial^2 V}{\partial x^2}$

This is the partial differential equation satisfied by V and we have a similar equation satisfied by I namely

$CL\dfrac{\partial^2 I}{\partial t^2} + (CR + LG)\dfrac{\partial I}{\partial t} + RGI = \dfrac{\partial^2 I}{\partial x^2}.$

Now let us apply the Laplace transformation to the above equations.

Write equations (I) and (2) in the form:

$$LI_t(x,t) + RI(x,t) = -V_x(x,t)$$

$$CV_t(x,t) + GV(x,t) = -I_x(x,t)$$

Transforming w.r.t. t we get:

$$L\left\{ si(x,s) - I(x,o) \right\} + Ri(x,s) = -v_x(x,s)$$

$$C\left\{ sv(x,s) - V(x,o) \right\} + Gv(x,s) = -i_x(x,s)$$

$$\therefore \ (Ls+R)\, i(x,s) = LI(x,o) - v_x(x,s) \ . \ . \ . \ . \ (3)$$

$$(Cs+G)\, v(x,s) = CV(x,o) - i_x(x,s) \ . \ . \ . \ . \ (4)$$

Eliminating i(x,s) between (3) and (4) by differentiating (3) partially w.r.t. x and substituting for i(x,s) from (4) we get

$$\frac{d^2v}{dx^2} - (Ls+R)\,(Cs+G)\, v = LI_x(x,o) - C(Ls+R)V(x,o)$$

Putting $q^2 = (Ls+R)(Cs+G)$ we get

$$\frac{d^2v}{dx^2} - q^2v = LI_x(x,o) - C(Ls+R)V(x,o)$$

Now suppose we have a semi— infinite line $x \geq o$ with the following conditions:

(i) Initial currents and charges are zero.

(ii) $\lim\limits_{x \to \infty} V(x,t) = o$

(iii) $V(o,t) = F(t)$

Hence the differential equation in v(x,s) reduces to

$$\frac{d^2v}{dx^2} - q^2v = o \quad \text{where } q^2 = (Ls+R)(Cs+G)$$

and we have to solve his equation subject to the two conditions

$$\lim\limits_{x \to \infty} v(x,s) = o \quad \text{and} \quad v(o,s) = f(s)$$

$$v(x,s) = Ae^{qx} + Be^{-qx}$$

and since v(o,s) = f(s) then f(s) =B

Hence

$$v(x,s) = f(s)e^{-qx}$$

i.e $v(x,s) = f(s)e^{-\sqrt{(Ls+R)(Cs+G)}\,x}$

The inverse transform in the general case is a complicated one. The following special but important cases will now be investigated.

Case I.

The lossless line $R = 0$, $G = 0$

Here $v(x,s) = f(s)e^{-\sqrt{CL}\,sx}$. Put $u = \dfrac{1}{\sqrt{CL}}$

$\therefore$ $v(x,s) = f(s)\, e^{-\frac{x}{u}s}$

$\therefore$ $V(x,t) = 0 \qquad 0 < t < \dfrac{x}{u}$

$\qquad\qquad = F\left(t - \dfrac{x}{u}\right) \qquad t > \dfrac{x}{u}$

This means that a point at distance x from the origin remains at zero potential until a time $t = x/u$ elapses when the voltage F(t) at x = 0 is transferred to the point. The time x/u is the time necessary for a wave traveling with velocity u to describe the distance x.

Case II.

The Heaviside's distortionless line $CR = LG$.

Put $\dfrac{R}{L} = \dfrac{G}{C} = \lambda \qquad \therefore$ $R = \lambda L$, $G = \lambda C$

$\therefore$ $q^2 = (Ls+R)(Cs+G) = (Ls+\lambda L)(Cs+\lambda C) = CL(s+\lambda)^2$

$\therefore$ $q = \sqrt{CL}\,(s+\lambda) = \dfrac{s+\lambda}{u} \qquad$ where $u = \dfrac{1}{\sqrt{CL}}$

$\therefore$ $v(x,s) = f(s)\, e^{-\frac{s+\lambda}{u}x}$

i.e $v(x,s) = f(s)\, e^{-\frac{\lambda x}{u}}\, e^{-\frac{x}{u}s}$

$\therefore$ $V(x,t) = 0 \qquad\qquad\qquad 0 < t < \dfrac{x}{u}$

$\qquad\qquad = e^{-\frac{\lambda x}{u}}\, F\left(t - \dfrac{x}{u}\right) \qquad t > \dfrac{x}{u}$

Hence we have a solution similar to the previous case but with attenuation as we move along the line due to the existence of the factor
$e^{-\frac{\lambda x}{u}}$

Case III.

The ideal submarine cable L=0, G=0.

Since $q^2 = (Ls+R)(Cs+G)$ then $q^2 = RCs$

$\therefore$ $q = \sqrt{RCs} = \sqrt{\dfrac{s}{k}} \qquad$ where $k = \dfrac{1}{RC}$

$$\therefore \quad v(x,s) = f(s) \; e^{-\sqrt{s/k} \; x}$$

Suppose that F(t) E where E is constant, then f(s) = E / s

$$\therefore \quad v(x,s) = \frac{E}{s} \; e^{-\sqrt{s/k} \; x}$$

$$\text{Now} \quad L^{-1} \left\{ \frac{1}{s} \; e^{-\sqrt{s/k} \; x} \right\} = \text{erfc} \; \frac{x}{2\sqrt{kt}}$$

$$\therefore \quad V(x,t) = E \; \text{erfc} \left(\frac{x}{2\sqrt{kt}} \right)$$

This problem, as we shall see, has an analogous problem in linear heat flow.

11.6.4. Conduction of heat

Let p = density of conducting medium, c its specific heat and K its thermal conductivity.
Consider an infinitesimal rectangular parallelepiped whose centre P is the point (x,y,z) and whose faces are parallel to the coordinate planes and let the lengths of its sides be $\delta x, \; \delta y, \; \delta z.$

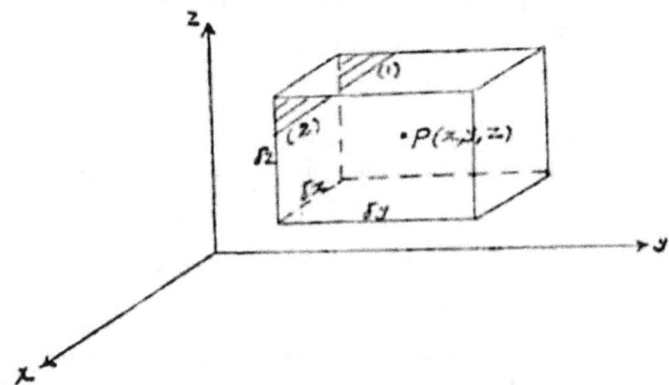

The quantity of heat entering the element in time δt through the face (1) is

$$- K \; \delta y \delta z \left[\frac{\partial U}{\partial x} - \frac{\partial^2 U}{\partial x^2} \; \frac{\delta x}{2} \right] \delta t$$

The quantity of heal leaving the element in time δt from face (2) is

$$- K \; \delta y \delta z \left[\frac{\partial U}{\partial x} + \frac{\partial^2 U}{\partial x^2} \; \frac{\delta x}{2} \right] \delta t$$

Hence net gain of heat from the two faces is

$$K \; \delta x \delta y \delta z \delta t \; \frac{\partial^2 U}{\partial x^2}$$

We have two similar contributions from the remaining two pairs of parallel faces. Hence total gain of heat in the element is

$$K \; \delta x \delta y \delta z \delta t \left[\frac{\partial^2 U}{\partial x^2} + \frac{\partial^2 U}{\partial y^2} + \frac{\partial^2 U}{\partial z^2} \right]$$

This will increase the temperature U by

$$\frac{\partial U}{\partial t} \, \delta t$$

$$K \; \delta x \delta y \delta z \delta t \left[\frac{\partial^2 U}{\partial x^2} + \frac{\partial^2 U}{\partial y^2} + \frac{\partial^2 U}{\partial z^2} \right] = \rho \delta x \delta y \delta z \, c. \, \frac{\partial U}{\partial t} \, \delta t$$

$$\frac{\partial^2 U}{\partial x^2} + \frac{\partial^2 U}{\partial y^2} + \frac{\partial^2 U}{\partial z^2} = \frac{\rho c}{K} \frac{\partial U}{\partial t}$$

$$\text{i.e} \quad \nabla^2 U = \frac{\rho c}{K} \frac{\partial U}{\partial t}$$

Put

$$\frac{K}{\rho c} = k$$

where k is known as the thermal diffusivity of the material .

$$\therefore \; \frac{\partial U}{\partial t} = k \; \nabla^2 U$$

This is known as the heat equation or equation of diffusion. If the flow of heat is uniflow i.e., in one direction say the x - axis then the last equation becomes

$$\frac{\partial U}{\partial t} = k \; \frac{\partial^2 U}{\partial x^2}$$

When the flow of heat is steady i.e., independent of the time, the heat equation reduces to
$$\nabla^2 U = o$$

i.e., the temperature U satisfies Laplace's equation.

Example 1

A semi-infinite solid $x \geq o$ is initially at zero temperature. A constant temperature Uo >0 is applied at time $t = 0$ to the face $x = 0$ and is maintained. Find the temperature distribution at any point x of the solid and at any instant.

Solution:

Here we have to solve the beat equation

$$\frac{\partial U}{\partial t} = k \frac{\partial^2 U}{\partial x^2} \quad x > o, \, t > o$$

subject to the boundary conditions

$$U(x,o) = o, \; U(o,t) = U_o$$

Transforming the differential equation w.r.t. t we get

$$su(x,s) - U(x,o) = ku_{xx}(x,s)$$

$$\frac{d^2 u}{dx^2} - \frac{s}{k} u = o$$

258

$$\therefore \quad u(x,s) = Ae^{\sqrt{s/k}\,x} + Be^{-\sqrt{s/k}\,x}$$

Since U(x,t) is bounded and consequently u(x,s) is also bounded, then A =0 and

$$u(x,s) = Be^{-\sqrt{s/k}\,x}$$

Since $U(o,t) = U_o$ then transforming we get $u(o,s) = \dfrac{U_o}{s}$

$$\therefore \quad \frac{U_o}{s} = B \text{ and hence } u(x,s) = \frac{U_o}{s}e^{-\sqrt{s/k}\,x}$$

$$\therefore \quad U(x,t) = U_o \; \text{erfc}\left(\frac{x}{2\sqrt{kt}}\right)$$

Example 2

Same as in problem I but with U(o,t) =V(t).

$$u(o,s) = v(s) \qquad \therefore \; B = v(s)$$

$$\therefore \; u(x,s) = v(s)\, e^{-\sqrt{s/k}\,x}$$

Now $L^{-1}\left\{ e^{-\sqrt{s/k}\,x} \right\} = \dfrac{x}{2\sqrt{\pi k}}\, t^{-3/2}\, e^{-\frac{x^2}{4kt}}$

Hence according to a theorem on convolution we have

$$U(x,t) = \frac{x}{2\sqrt{\pi k}} \int_0^t V(t-\lambda)\, \lambda^{-3/2}\, e^{-\frac{x^2}{4k\lambda}}\; d\lambda \; .$$

Example 3

A solid is bounded by the infinite planes x=o and x=1.

These planes are kept at zero temperature. If at time the distribution of temperature is

$$U(x,o) = 4 \sin 2\pi x,$$

Find the temperature at any point of the solid and at any instant.

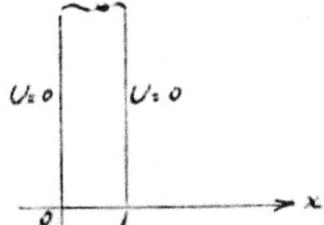

Solution:

$$\frac{\partial U}{\partial t} = k \frac{\partial^2 U}{\partial x^2}$$

$$U(0,t) = 0, \quad U(1,t) = 0, \quad U(x,0) = 4 \sin 2\pi x$$

$$s\, u(x,s) - U(x,0) = k\, u_{xx}(x,s)$$

$$\frac{d^2 u}{dx^2} - \frac{s}{k} u = -\frac{4}{k} \sin 2\pi x$$

$$\therefore u(x,s) = A e^{\sqrt{s/k}\, x} + B e^{-\sqrt{s/k}\, x} + \frac{4}{s + 4\pi^2 k} \sin 2\pi x$$

Now $u(0,s) = 0 \quad \therefore 0 = A + B$

$$u(1,s) = 0 \quad \therefore 0 = A e^{\sqrt{s/k}} + B e^{-\sqrt{s/k}}$$

$$\therefore A = B = 0$$

$$u(x,s) = \frac{4}{s + 4\pi^2 k} \sin 2\pi x$$

$$\therefore U(x,t) = 4 e^{-4\pi^2 k t} \sin 2\pi x$$

Example 4

A semi-infinite insulated bar is initially at zero temperature. The bar is placed along the positive x -.axis. At t = o a quantity of heat is instantaneously generated at the point x = a where a is positive. Find the temperature distribution along the bar at any instant.

Solution:

The partial differential equation satisfied by the temperature U *is*

$$\frac{\partial U}{\partial t} = k \frac{\partial^2 U}{\partial x^2} \quad (x > 0, t > 0)$$

The boundary conditions are

(i) $U(x,0) = 0$, (ii) $U(x,t)$ is bounded, (iii) $U(a,t) = Q\delta(t)$
where Q is a constant and $\delta(t)$ is the Dirac delta function.

$$s u(x,s) - U(x,0) = k\, u_{xx}(x,s)$$

$$\therefore \frac{d^2 u}{dx^2} - \frac{s}{k} u = 0 \quad \therefore u(x,s) = A e^{\sqrt{s/k}\, x} + B e^{-\sqrt{s/k}\, x}$$

From the boundary condition $A = 0$

$$\therefore u(x,s) = B e^{-\sqrt{s/k}\, x}$$

Transforming $U(a,t) = Q\delta(t)$ we get

$$u(a,s) = Q \qquad \therefore \ Q = Be^{-\sqrt{s/k}\, a} \qquad \text{i.e } B = Q\, e^{\sqrt{s/k}\, a}$$

$$u(x,s) = Q\, e^{\sqrt{s/k}\, a} \ . \ e^{-\sqrt{s/k}\, x}$$

$$\text{i e } \ u(x,s) = Q\, e^{-\sqrt{s/k}\,(x-a)}$$

$$\text{Now } L^{-1}\left\{ e^{-a\sqrt{s}} \right\} = \frac{a}{2\sqrt{\pi t^3}}\, e^{-\frac{a^2}{4t}} \quad , \quad a > 0$$

$$\therefore U(x,t) = Q\, \frac{x-a}{2\sqrt{\pi k t^3}}\, e^{-\frac{(x-a)^2}{4kt}}$$

Example 5

A semi-infinite solid $x \geq 0$ with zero initial temperature has a constant heat flux C applied to the face x=0 so that $-KU_x(0,t) = C$.

Find the temperature at any instant and at any point x of the solid.

Solution:

$$\frac{\partial U}{\partial t} = k\, \frac{\partial^2 U}{\partial x^2} \quad , \quad U(x,0) = 0 \ , \quad -K U_x(0,t) = C$$

$$su(x,s) - U(x,0) = k\, u_{xx}(x,s)$$

$$\therefore \ \frac{d^2 u}{dx^2} - \frac{s}{k}\, u = o \quad \therefore \ u(x,s) = Ae^{\sqrt{s/k}\, x} + Be^{-\sqrt{s/k}\, x}$$

From the boundedness condition $A = o$ $\therefore$ $u(x,s) = Be^{-\sqrt{s/k}\, x}$

Transforming the condition $-KU_x(o,t) = C$ w.r.t. t we get

$$-Ku_x(o,s) = \frac{C}{s} \quad \therefore \ u_x(o,s) = -\frac{C}{Ks}$$

$$\text{Now } u_x(x,s) = -B\sqrt{s/k}\ e^{-\sqrt{s/k}\, x}$$

$$\therefore \quad u_x(o,s) = -B\sqrt{s/k}$$

$$\therefore \ -\frac{C}{Ks} = -B\sqrt{s/k} \quad \text{i.e } B = \frac{C\sqrt{k}}{Ks^{3/2}}$$

$$\therefore \ u(x,s) = \frac{C\sqrt{k}}{Ks^{3/2}}\ e^{-\sqrt{s/k}\, x}$$

$$\text{Now } \ L^{-1}\left\{ s^{-\frac{3}{2}}\ e^{-a\sqrt{s}} \right\} = 2\sqrt{\frac{t}{\pi}}\ e^{-\frac{a^2}{4t}}$$

$$-a\, \text{erfc}\left(\frac{a}{2\sqrt{t}}\right), \ a > o$$

$$\therefore \ U(x,t) = \frac{C\sqrt{k}}{K} \left[2\sqrt{\frac{t}{\pi}} \ e^{-\frac{x^2}{4kt}} - \frac{x}{\sqrt{k}} \ \text{erfc}\left(\frac{x}{2\sqrt{kt}}\right) \right]$$

$$\text{i.e } U(x,t) = \frac{C}{K} \left[2\sqrt{\frac{kt}{\pi}} \ e^{-\frac{x^2}{4kt}} - x\text{erfc}\left(\frac{x}{2\sqrt{kt}}\right) \right]$$

Example 6

A bar of length l is initially at temperature U_o. The end $x = o$ is insulated while the end $x = l$ is suddenly given the constant temperature U_1. Find the temperature at any point x of the bar and at any instant assuming the surface of the bar to be insulated.

Solution:

$$\frac{\partial U}{\partial t} = k \frac{\partial^2 U}{\partial x^2} \qquad (0 < x < l, t > 0)$$

Boundary conditions are

$$U(x,0) = U_o \quad , \quad U_x(0,t) = 0 \ , \ U(l,t) = U_1$$

$$su(x,s) - U(x,0) = ku_{xx} \ (x,s)$$

$$\text{i.e } \quad su(x,s) - U_o = ku_{xx} \ (x,s)$$

$$\frac{d^2u}{dx^2} - \frac{s}{k} u = -\frac{U_o}{k}$$

$$\therefore \ u(x,s) = A \cosh \sqrt{s/k} \ x + B \sinh \sqrt{s/k} \ x + \frac{U_o}{s}$$

Transforming $\quad U_x(0,t) = 0 \quad$ we get $\quad u_x(0,s) = 0$

$$\therefore \ B = 0 \quad \text{Hence} \quad u(x,s) = A \cosh \sqrt{s/k} \ x + \frac{U_o}{s}$$

Transforming $\quad U(l,t) = U_1 \quad$ we get $\quad u(l,s) = \frac{U_1}{s}$

$$\therefore \ \frac{U_1}{s} = A \cosh \sqrt{s/k} \ l + \frac{U_o}{s} \quad \therefore \ A = \frac{U_1 - U_o}{s \cosh (\sqrt{s/k} \ l)}$$

$$\therefore \ u(x,s) = \frac{U_o}{s} + (U_1 - U_0) \frac{\cosh (\sqrt{s/k}x)}{s \cosh(\sqrt{s/k} \ l)}$$

$$\therefore \ U(x,t) = U_0 + (U_1 - U_o) \ L^{-1} \left\{ \frac{\cosh(\sqrt{s/k} \ x)}{s \cosh (\sqrt{s/k} \ l)} \right\}$$

According to the theory of the inversion integral we have

$$L^{-1} \left\{ \frac{\cosh (\sqrt{s/k} \ x)}{s \cosh (\sqrt{s/k} \ l)} \right\} = \text{sum of residues of } \frac{e^{st} \cosh (\sqrt{s/k} \ x)}{s \cosh (\sqrt{s/k} \ l)}$$

262

at its poles. The poles are the zeros of $s \cosh (\sqrt{s/k}\, l)$ i.e $s = 0$ and zeros of $\cosh (\sqrt{s/k}\, l)$

Now consider $\cosh u = 0$ $\therefore$ $\dfrac{e^u + e^{-u}}{2} = 0$

$\therefore$ $e^u = - e^{-u}$ or $e^{2u} = -1 = e^{\pi i + 2n \pi i}$

where $n = 0,1,2,3, \cdots$

$\therefore$ $2u = (2n + 1)\pi i$ i.e $u = (n + \tfrac{1}{2})\pi i$

$\therefore$ $\cosh (\sqrt{s/k}\, l) = 0$ if $\sqrt{s/k}\, l = (n + \tfrac{1}{2})\pi i$

i.e $s = -\dfrac{(2n + 1)^2 \pi^2 k}{4\, l^2}$, $n = 0,1,2,3 \cdots$

which can also be written

$$s = -\frac{(2n - 1)^2 \pi^2 k}{4\, l^2} , \quad n = 1,2,3, \cdots$$

Hence zeros of $s \cosh (\sqrt{s/k}\, l)$ are

$$s = 0 , \quad s = -\frac{(2n - 1)^2 \pi^2 k}{4\, l^2} , \quad n = 1,2,3, \cdots$$

Residue at $s = 0$ is $\lim_{s \to 0} s \dfrac{e^{st} \cosh (\sqrt{s/k}\, x)}{s \cosh (\sqrt{s/k}\, l)} = 1$

Residue at $s = -\dfrac{(2n-1)^2 \pi^2 k}{4 l^2} = s_n$ is

$$\lim_{s \to s_n} (s - s_n) \left[\frac{e^{st} \cosh (\sqrt{s/k}\, x)}{s \cosh (\sqrt{s/k}\, l)} \right]$$

$$= \left[\lim_{s \to s_n} \frac{s - s_n}{\cosh (\sqrt{s/k}\, l)} \right] \left[\lim_{s \to s_n} \frac{e^{st} \cosh (\sqrt{s/k}\, x)}{s} \right]$$

$$= \left[\lim_{s \to s_n} \frac{1}{\sinh(\sqrt{s/k}\, l)\,(l/2\sqrt{ks})} \right] \left[\lim_{s \to s_n} \frac{e^{st} \cosh (\sqrt{s/k}\, x)}{s} \right]$$

$$= \frac{4(-1)^n}{(2n-1)\pi} e^{-\frac{(2n-1)^2 \pi^2 k t}{4 l^2}} \cos \frac{(2n-1)\pi x}{2l}$$

according to L'Hiospital's rule in evaluating indeterminate forms, and the fact that

$\cosh iu = \cos u$ and $\sinh iu = i \sin u$.

$$\therefore \ U(x,t) \ = \ U_1 \ +$$

$$\frac{4(U_1-U_0)}{\pi} \sum_{n=1}^{\infty} \frac{(-1)^n}{2n-1} \ e^{-\frac{(2n-1)^2\pi^2 kt}{4l^2}} \ \cos \ \frac{(2n-1)\pi x}{2l} \ .$$

Example 7

An infinitely long circular cylinder of radius unity is initially at temperature Uo. At time $t = 0$, a temperature 0°C is applied to the surface and is maintained. Determine the temperature distribution over the cylinder at any instant t.

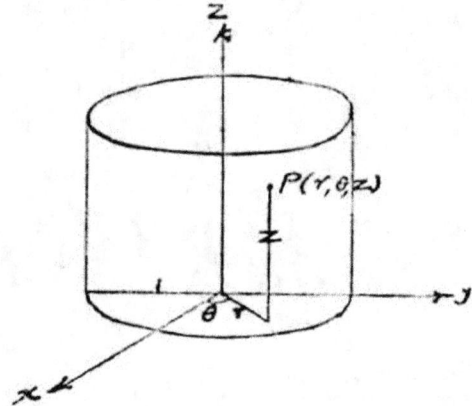

Solution:

The equation satisfied by the temperature U is

$$\frac{\partial U}{\partial t} \ = \ k \ \nabla^2 U$$

Now in cylindrical coordinates the Laplacian operator takes the form

$$\nabla^2 \ \equiv \ \frac{\partial^2}{\partial r^2} \ + \ \frac{1}{r} \ \frac{\partial}{\partial r} \ + \ \frac{1}{r^2} \ \frac{\partial^2}{\partial \theta^2} \ + \ \frac{\partial^2}{\partial z^2}$$

and hence the heat equation becomes

$$\frac{\partial U}{\partial t} \ = \ k \left[\ \frac{\partial^2 U}{\partial r^2} \ + \ \frac{1}{r} \ \frac{\partial U}{\partial r} \ + \ \frac{1}{r^2} \ \frac{\partial^2 U}{\partial \theta^2} \ + \ \frac{\partial^2 U}{\partial z^2} \ \right]$$

Since in this problem, due to symmetry, it is independent of θ and z, then

$$\frac{\partial U}{\partial t} \ = \ k \left(\frac{\partial^2 U}{\partial r^2} \ + \ \frac{1}{r} \ \frac{\partial U}{\partial r} \right) \qquad 0 < r < 1$$

Now it is more convenient to take this equation in the form

$$\frac{\partial U}{\partial t} \ = \ \frac{\partial^2 U}{\partial r^2} \ + \ \frac{1}{r} \ \frac{\partial U}{\partial r}$$

264

and then in the final result t is changed to kt.

The boundary conditions are

$$U(r,o) = U_o \ , \ U(1,t) = o \text{ and } U \text{ is bounded.}$$

Transform the heat equation w r.t. t

$$su(r,s) - U(r,o) = \frac{d^2u}{dr^2} + \frac{1}{r}\frac{du}{dr}$$

$$\therefore \frac{d^2u}{dr^2} + \frac{1}{r}\frac{du}{dr} - su = -U_o$$

$$\therefore u(r,s) = AJ_o(i\sqrt{s}\,r) + BY_o(i\sqrt{s}\,r) + \frac{U_o}{s}$$

Since u(r,s) is finite then B = o

$$\therefore u(r,s) = AJ_o(i\sqrt{s}\,r) + \frac{U_o}{s}$$

Transforming $U(1,t) = o$ we get $u(1,s) = o$

$$\therefore o = AJ_o(i\sqrt{s}) + \frac{U_o}{s} \qquad i.e \ A = -\frac{U_o}{sJ_o(i\sqrt{s})}$$

$$\therefore u(r,s) = \frac{U_o}{s} - \frac{U_o J_o(i\sqrt{s}\,r)}{sJ_o(i\sqrt{s})}$$

$$\therefore U(r,t) = U_o - U_o \ L^{-1}\left\{\frac{J_o(i\sqrt{s}\,r)}{sJ_o(i\sqrt{s})}\right\}$$

Now according to the inversion integral formula

$$L^{-1}\left\{\frac{J_o(i\sqrt{s}\,r)}{s\,J_o(i\sqrt{s})}\right\} = \text{sum of residues of } \frac{e^{st}J_o(i\sqrt{s}\,r)}{sJ_o(i\sqrt{s})} \text{ at its poles.}$$

The poles are zeros of $sJ_o(i\sqrt{s})$ i.e $s=o$ and zeros of $J_o(i\sqrt{s})$.

Now $J_o(i\sqrt{s})$ has simple zeros at $i\sqrt{s} = \lambda_1, \lambda_2, ..., \lambda_n, ...$

i.e $s = -\lambda_n^2$ where $n = 1,2,3,...$

Residue at $s=o$ is $\displaystyle\lim_{s\to o} s \frac{e^{st}J_o(i\sqrt{s}\,r)}{s\,J_o(i\sqrt{s})} = 1$

Residue at $s = -\lambda_n^2$ is

$$\lim_{s\to-\lambda_n^2} (s+\lambda_n^2) \frac{e^{st}J_o(i\sqrt{s}\,r)}{s\,J_o(i\sqrt{s})}$$

$$= \left[\lim_{s\to-\lambda_n^2}\frac{s+\lambda_n^2}{J_o(i\sqrt{s})}\right]\left[\lim_{s\to-\lambda_n^2}\frac{e^{st}J_o(i\sqrt{s}\,r)}{s}\right]$$

$$= \left[\lim_{s \to -\lambda_n^2} \frac{1}{J_0'(i\sqrt{s})i/2\sqrt{s}} \right] \left[\frac{e^{-\lambda_n^2 t} J_0(\lambda_n r)}{-\lambda_n^2} \right]$$

$$= -\frac{2e^{-\lambda_n^2 t} J_0(\lambda_n r)}{\lambda_n J_1(\lambda_n)}$$

according to L' Hospital's rule in evaluating indeterminate forms and the fact that

$$J_0'(x) = - J_1(x).$$

$$\therefore \quad U(r,t) = U_0 - U_0 \left[1 - \frac{2 \sum_{n=1}^{\infty} e^{-\lambda_n^2 t} J_0(\lambda_n r)}{\lambda_n J_1(\lambda_n)} \right]$$

Changing t into kt we get

$$U(r,t) = 2U_0 \sum_{n=1}^{\infty} \frac{e^{-k\lambda_n^2 t} J_0(\lambda_n r)}{\lambda_n J_1(\lambda_n)}$$

Example 8   A semi-infinite solid $x \geqslant 0$ has zero initial temperature.

The temperature at x=o is given by

$$U(o,t) = U_0 \qquad o < t < t_0$$

$$= o \qquad t > t_0$$

Determine the temperature at any point x of the solid and at any instant t.

Solution:

$$\frac{\partial U}{\partial t} = k \frac{\partial^2 U}{\partial x^2}$$

Boundary conditions are

$$U(x,o) = o , \qquad U(o,t) = U_0 \qquad o < t < t_0$$

$$= o \qquad t > t_0$$

$$su(x,s) - U(x,o) = ku_{xx}(x,s)$$

$$\frac{d^2 u}{dx^2} - \frac{s}{k} u = o \quad \therefore \quad u(x,s) = Ae^{\sqrt{s/k}\, x} + Be^{-\sqrt{s/k}\, x}$$

Since U is bounded and consequently u, then A = o

$$u(x,s) = Be^{-\sqrt{s/k}\, x}$$

Since $u(o,s) = \frac{U_0}{s} \left(1 - e^{-t_0 s} \right)$

$$\therefore \quad B = \frac{U_0}{s} \left(1 - e^{-t_0 s} \right)$$

$$u(x,s) = \frac{U_o}{s}\left(1-e^{-t_o s}\right) e^{-\sqrt{s/k}\; x}$$

i.e $u(x,s) = U_o\left(1-e^{-t_o s}\right) \cdot \frac{1}{s} e^{-\sqrt{s/k}\; x}$

Now $L^{-1}\left\{\frac{1}{s} e^{-a\sqrt{s}}\right\} = \text{erfc} \frac{a}{2\sqrt{t}}$, $a > 0$

$\therefore\; U(x,t) = U_o \text{ erfc} \frac{x}{2\sqrt{kt}}$ $\quad o < t < t_o$

$\qquad = U_o \text{ erfc} \frac{x}{2\sqrt{kt}} - U_o \text{ erfc} \frac{x}{2\sqrt{k(t-t_o)}}$ $\quad t > t_o$

11.6.5. Exercise on using Laplace Transformation in solving Linear Partial Differential Equations

(1) Solve the partial differential equation

$$U_{xx}(x,t) + U_{tx}(x,t) - 2U_{tt}(x,t) = o \qquad (x>o, t>o)$$

subject to the following boundary conditions

$U(x,o) = U_t(x,o) = o$. $\lim\limits_{x \to \infty} U(x,t) = o$, $U(o,t) = F(t)$.

[Ans. $\quad U(x,t) = o \qquad o < t < 2x$

$\qquad\qquad = F(t-2x) \qquad t > 2x$]

(2) Solve the following boundary value problem $\quad Y_x(x,t) + x\, Y_t(x,t) = o$

$\quad Y(x,o) = o,\quad Y(o,t) = t$

$\left[\text{Ans.} \quad Y(x,t) = o \qquad o < t < \frac{x^2}{2} \right.$

$\left. \qquad\qquad = t - \frac{x^2}{2} \quad t > \frac{x^2}{2} \right]$

(3) Solve the partial differential equation $\quad Y_{tt}(x,t) = a^2 Y_{xx}(x,t) \quad x>o\,, t>o$

with the following boundary conditions

$Y(x,o) = o,\; Y_t(x,o) = -uo,\; Y(o,t) = o,\; \lim\limits_{x \to \infty} Y_x(x,t) = o$

$\left[\text{Ans.} \quad Y(x,t) = -u_o t \qquad o < t < \frac{x}{a} \right.$

$\left. \qquad\qquad = -u_o \frac{x}{a} \qquad t > \frac{x}{a} \right]$

(4) Find a bounded solution of

$$x \frac{\partial U}{\partial x} + \frac{\partial U}{\partial y} = xe^{-y} \qquad 0 < x < 1, \; y > 0$$

which satisfies $U(x,o) = x \qquad 0 < x < 1$

[Ans. $U(x,y) = xe^{-y}(1+y)]$

(5) Solve the equation $\dfrac{\partial U}{\partial t} + x \dfrac{\partial U}{\partial x} + U = x \qquad x > 0, \; t > 0$

with the boundary conditions $U(o,t) = o, \quad U(x,o) = o$

[Ans. $U(x,t) = \frac{1}{2} x \, (1 - e^{-2t})]$

(6) Show that *in* solving the partial differential equation
$$\frac{\partial^2 Y}{\partial t^2} = a^2 \frac{\partial^2 Y}{\partial x^2} \quad \text{we can solve the equation} \quad \frac{\partial^2 Y}{\partial x^2} = \frac{\partial^2 Y}{\partial t^2}$$
and then in the result we replace t by at.

(7) Solve the equation $\dfrac{\partial^2 Y}{\partial x^2} = 16 \dfrac{\partial^2 Y}{\partial t^2} \qquad x > 0, \; t > 0$

subject to the following boundary conditions

$Y(x,o) = o, \; Y_t (x,o) = -1, \; Y(o,t) = t^2, \; \lim\limits_{x \to \infty} \; Y(x,t)$ exists

for fixed $t > 0$

[Ans. $\quad Y(x,t) = -t \qquad\qquad 0 \leqslant t \leqslant 4x$

$\qquad\qquad\quad = (t-4x)^2 - 4x \qquad t \geqslant 4x \;]$

(8) Solve the equation $\quad \dfrac{\partial Y}{\partial x} + 4 \dfrac{\partial Y}{\partial t} = - 8t \qquad x > 0, \; t > 0$

$Y(x,o) = o, \; Y(o,t) = 2t^2$

[Ans. $\quad Y(x,t) = - t^2 \qquad\qquad 0 \leqslant t \leqslant 4x$

$\qquad\qquad\quad = - t^2 + 3(t-4x)^2 \qquad t \geqslant 4x \;]$

(9) Solve the equation $\dfrac{\partial Y}{\partial x} + 2 \dfrac{\partial Y}{\partial t} = 4t \qquad x > 0, \; t > 0$

subject to the boundary conditions

$Y(x,o) = o, \qquad Y(o,t) = 2t^3$

[Ans. $Y(x,t) = t^2$ $0 \leqslant t \leqslant 2x$

            $= t^2 + 2 (t \quad 2x)^3 - (t-2x)^2$ $t \geqslant 2x$]

(10) $\dfrac{\partial^2 Y}{\partial t^2} = 4 \dfrac{\partial^2 Y}{\partial x^2}$ $x > 0,\ t > 0$

$Y(x,o) = o,\ Y_t(x,o) = 2,\ Y(o,t) = \sin t$

   $\displaystyle\lim_{x \longrightarrow \infty}\ Y(x,t)$ exists for $t > 0$.

[Ans. $Y(x,t) = 2t$ $0 \leqslant t \leqslant \dfrac{x}{2}$

            $= 2t + \sin \left(t - \dfrac{x}{2} \right) - 2 \left(t - \dfrac{x}{2} \right)$,

            $t \geqslant \dfrac{x}{2}$] .

12.1. Ordinary differential equations

21-10-1971

Ordinary Differential Equation.

d. eq^n.

Any relation connecting the independent variable x, the dependent variable y, and one or more of the derived function $\frac{dy}{dx}$, $\frac{d^2y}{dx^2}$ ..... is called an ordinary d. eq^n.

ex: $\frac{d^3y}{dx^3} + 5x\frac{d^2y}{dx^2} - 3\frac{dy}{dx} + 4y = e^x + \cos x$. ——(1)

It is called an ordinary d. eq^n. to distinguish it from a *partial* d. eq^n. which <u>contains</u> partial <u>derivatives</u> w. r. t. more than one independent variable.

ex: the wave equation: $\frac{\partial^2 y}{\partial t^2} = c^2 \frac{\partial^2 y}{\partial x^2}$.

We shall consider in what follows d. eq^n. of the ordinary type:-

In order to eliminate A and B we need three eq^{ns} from the above three eq^{ns} we see that.

$\ddot{x} = -n^2 x$.

Thus the result of eliminating two const. is a d. eq^n of order two.

In General: suppose we have n arbitrary const. in the eqs. in the arbitrary const.

$$f(x, y, c_1, c_2 \ldots c_n) \qquad \text{——} \quad \textcircled{1}$$

To eliminate these n constants we need $n+1$ eq^{ns}. eq^n ① is one of these eq^{ns}. Hence we need more eq^n these are obtained by successive differentiation (n) times.

Thus when we eliminate $c_1, c_2 \ldots c_n$ we obtain a d. eq^n of the form

$$f\left(x, y, \frac{dy}{dx}, \frac{d^2x}{dx^2} \ldots \frac{d^n y}{dx^n}\right) = 0$$

Thus the result of eliminating n arbitrary const. is a d. eq^n of the nth order.

In practice we are usually confronted by the converse problem namely we are given eq^n ② and is required to obtain eq^n ① thus the general solution of the d. eq^n of the nth order contains n arbitrary const.

Example (12.1.1): $y' = y^2 + x$

Picard's $\textcircled{Ex}$

$$y' = y^2 + x$$

$$y = -1 \implies x = 1$$

evaluate $y \implies x = 1.1$

$$x = X + x_0 = X + 1$$
$$y = Y + y_0 = Y - 1$$

$$\frac{dY}{dX} = -(Y-1)^2 + X + 1$$
$$= -Y^2 + 2Y - 1 + X + 1$$

$$\frac{X}{X} = -Y^2 + 2Y + X$$

$$\int_{X=0} dY = \int_{X=0} (-Y^2 + 2Y + X)\, dX$$

$$Y_1 = 0 + \int_0^X (X + 2\cdot 0 - 0)\, dX$$
$$= \frac{X^2}{2}$$

$$Y_2 = \int_0^X (X + X^2 - \frac{X^4}{4})\, dX$$
$$= \frac{X^2}{2} + \frac{X^3}{3} - \frac{X^5}{20}$$

12.1.1. First Order differential equations

An ordinary first- order differential equation may be represented as follows:

$$F\left(x, y, \frac{dy}{dx}\right) = 0$$

Or

$$\frac{dy}{dx} = f(x, y) \qquad \cdots \cdots (1)$$

Sometimes it is expedient to write this equation in the form

$$M(x, y)\, dx + N(x, y)\, dy = 0.$$

The differential equation (1) establishes a relation between the coordinates of a point and the slope of the tangent dy/dx to the graph of the solution at that point. Knowing x and y, it is possible to calculate dy/dx..

Hence, a differential equation of the type under consideration defines a direction field and the problem of integrating the differential equation consists in finding the curves, called integral curves, the direction of the tangents to which at each point coincides with the direction of the field.

Diff. Eqns. of the 1st order and 1st degree.

General form $M + N\frac{dy}{dx} = 0$ where M & N are in general f^n of (x, y) the above eq^n is sometimes written in the symmetrical form:—

$$M dx + N dy = 0$$

or;
$$\frac{dy}{dx} = -\frac{M}{N} = f(x, y)$$

Diff. eq^ns of the 1st order & 1st degree simple as they are cannot be solved in all cases in a finite form of known element-ary fn^s.

Solvable types are usually classified as follows:—

I. Eq^ns solvable by separation of variables.

II. Homogeneous eq^ns

III. Exact eq^ns.

IV. Linear diff. eq^ns of the 1st order.

28 \ 10 \ 1911

IV. Linear d. eq^ns of of the 1st order:—

General form is $\frac{dy}{dx} + Py = Q \longrightarrow ①$

where P and Q are f^ns of x only.
Linear means that y and its derivatives enter to the first degree only.

To solve the eq^n ① multiply by an <u>Integrating Factor</u> (I.F.) μ.

$$\mu \frac{dy}{dx} + P\mu y = \mu Q \longrightarrow ②$$

$$\frac{d}{dx}(\mu y) - y \frac{d\mu}{dx} + P\mu y = \mu Q \longrightarrow ③$$

$$\frac{d}{dx}(\mu y) + y\left(P\mu - \frac{d\mu}{dx}\right) = \mu Q \longrightarrow ④$$

Such choose μ $\mu P - \frac{d\mu}{dx} = 0 \longrightarrow ⑤$

$$\int \frac{d\mu}{\mu} = \int P \, dx \qquad\qquad P \text{ may not be integrated because}$$
$$\qquad\qquad\qquad\qquad\qquad\qquad\qquad\qquad it \text{ is a } f^n \text{ of } x.$$
$$\log_e \mu = \int P X$$
$$\boxed{\mu = e^{\int P \, dx}}$$

The eq^n now reduced to

$$\frac{d}{dx}(\mu y) = \mu Q \longrightarrow ⑥$$

$$\therefore \mu y = \int \mu Q \, dx + C$$

$$y e^{\int P \, dx} = \int Q e^{\int P \, dx} \, dx + C \longrightarrow ⑦$$

The solution is considered in two steps.
1.   I.F. $= e^{\int P \, dx} \longrightarrow 1^{\underline{st}}$
2.   The solution is given by $y \,(I.F.) =$

$$y\,(I.F.) = \int (R.H.S)(I.F.) + C \longrightarrow 2^{\underline{nd}}$$

<u>ex. 1</u>
$$\cos x \frac{dy}{dx} + y \sin x = 1$$

$$\frac{dy}{dx} + y \tan x = \sec x$$

$$①st \quad I.F. = e^{\int P \, dx} = e^{\int \tan x \, dx} = e^{\log \sec x} = \sec x$$

<u>NOTE</u>: $e^{-\log x} = e^{\log \frac{1}{x}} = \frac{1}{x} \longrightarrow ①$
$$e^{2 \log x} = e^{\log x^2} = x^2 \longrightarrow ②$$

$$\therefore \underline{\text{Solution}} \quad y\,(I.F.) = \int R.H.S\,(I.F.) + c$$

$$y \sec x = \int \sec x \cdot \sec x \, dx + c$$

$$y \sec x = \tan x + c$$

$$\boxed{y = \sin x + c \cos x}$$

12.1.1.1. Separable Equations

Differential equations of the form

Y (y) dy = X (x) (2)

are called equations with separated variables. The functions X (x) and Y (y) will be considered continuous. Integrating both sides of (2) we get

$$\int Y\,(y)\,dy = \int X\,(x)\,dx + C \qquad \cdots\cdots (3)$$

where C is an arbitrary constant. Thus, equation (3) is the complete (general) integral of equation (2).

<u>Example</u> 1   **x dx + y dy = 0**

The variables are separated since the coefficient of dx is a function of x alone, whereas the coefficient of dy is a function of y alone. Integrating we obtain

$x^2 + y^2 = C^2$

This is the equation of a family of concentric circles with centre at the coordinate origin and radius C.

<u>Example</u> 2:   **x (1 + y^2) dx — y (1 + x^2) dy = 0**

Separate the variables and integrate:

$$\frac{y}{1+y^2}\,dy = \frac{x}{1+x^2}\,dx$$

$$\int \frac{2y}{1+y^2}\,dy = \int \frac{2x}{1+x^2}\,dx + C_1$$

$\therefore$ *ln* $(1 + y^2)$ = *ln* $(1 + x^2)$ + *ln* **C**

Or

$1 + y^2$ = C (1 + x^2) $\cdots$ **(4)**

It should be noted that the general solution of a first – order differential equation contains only one arbitrary constant. In some problems it is required to specify a certain solution which satisfies a given condition. In such case nr initial condition will be given in the form y = yo when, x = xo. Substituting this condition into the general solution we obtain an equation for the determination of the arbitrary constant.

275

In the last Example let the initial condition be, y = 0 when x = 0. Substitution in the general solution (4) yields

$$1 + y^2 = 1 + x^2$$

or

$$y = \pm x$$

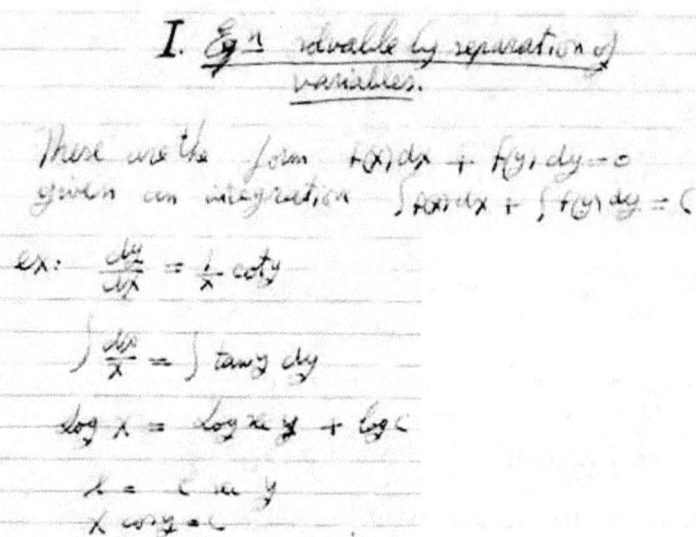

I. Eq ᵅ solvable by separation of variables.

These are the form f(x) dx + f(y) dy = o
given an integration ∫ f(x) dx + ∫ f(y) dy = C

ex: $\frac{dy}{dx} = \frac{1}{x} \cot y$

∫ $\frac{dx}{x}$ = ∫ tan y dy

log x = log sec y + log c

x = c sec y
x cos y = c

12.1.1.2. Homogeneous Equations

Differential equations of the form

$$\frac{dy}{dx} = f\left(\frac{y}{x}\right) \qquad \cdots \ (5)$$

are called homogeneous differential equations. This type can be reduced to equations with variables separation as follows.

Make the substitution v = y/x, we get

$$\frac{dy}{dx} = x\frac{dv}{dx} + v$$

Substitution in (5) yields

$$x\frac{dv}{dx} + v = f(v)$$

and the separation of the variables gives

$$\frac{dv}{f(v) - v} = \frac{dx}{x}$$

and hence

$$\int \frac{dv}{f(v) - v} = \ln x + \ln c = \ln cx$$

Example 1:

$$\frac{dy}{dx} = \frac{y}{x} + \tan \frac{y}{x}$$

276

Putting $y = x v$; $\dfrac{dy}{dx} = x\dfrac{dv}{dx} + v$

and substituting into the initial equation we have

$$x\frac{dv}{dx} + v = v + \tan v$$

or

$$\cot v \, dv = \frac{dx}{x}$$

which gives by integration

$$\ln(\sin v) = \ln x + \ln c,$$
$$\sin v = c x,$$
$$\sin \frac{y}{x} = c x.$$

Example 2: $\qquad \dfrac{dy}{dx} = \dfrac{x\,y}{x^2 - y^2}$

This equation can be rewritten as

$$\frac{dy}{dx} = \frac{\dfrac{y}{x}}{1 - \left(\dfrac{y}{x}\right)^2}$$

which is a homogeneous equation. Making the substitution $v = y/x$ we have

$$y = x v \; ; \quad \frac{dy}{dx} = v + x\frac{dv}{dx} \; ;$$

$$v + x\frac{dv}{dx} = \frac{v}{1 - v^2} \; ;$$

$$x\frac{dv}{dx} = \frac{v^3}{1 - v^2}$$

Separating variables we obtain

$$\frac{1 - v^2}{v^3} \, dv = \frac{dx}{x} \; ;$$

$$\left(\frac{1}{v^3} - \frac{1}{v}\right) dv = \frac{dx}{x}$$

Whence, integrating, we find

$$-\frac{1}{2v^2} - \ln v = \ln x + \ln C$$

or

277

$$- \frac{1}{2v^2} = \ln C \, x \, v$$

Substituting $v = y/x$ we get the general solution of the original equation

$$- \frac{x^2}{2y^2} = \ln C \, y.$$

It should be noted that there are other types of first – order differential equations which are reducible to homogeneous equations. Equations of the form

$$\frac{dy}{dx} = f\left(\frac{a_1 x + b_1 y + c_1}{a_2 x + b_2 y + c_2}\right) \quad \cdots \quad \cdots \quad \cdots \quad (6)$$

are converted into homogeneous equations by translating the origin of coordinates to the point of intersection (h, k) of the straight lines

$$a_1 x + b_1 y + c_1 = 0 \text{ and } a_2 x + b_2 y + C_2 = 0$$

Indeed, the constant terms in the equations of these lines in the new coordinates
$X = x - h, Y = y - k$
will be zero, the coefficients of the running coordinates remain unchanged, while

$$\frac{dy}{dx} = \frac{dY}{dX}.$$

The equation (6) is transformed to

$$\frac{dY}{dX} = f\left(\frac{a_1 X + b_1 Y}{a_2 X + b_2 Y}\right)$$

Or

$$\frac{dY}{dX} = f\left(\frac{a_1 + b_1 \frac{Y}{X}}{a_2 + b_2 \frac{Y}{X}}\right) = \phi\left(\frac{Y}{X}\right)$$

and is now a homogeneous equation.

This method cannot be used only when the lines

$$a_1 x + b_1 y + c_1 = 0 \text{ and } a_2 x + b_2 y + c_2 = 0$$

are parallel. But in this case the coefficients of the running coordinates are proportional:

$$\frac{a_2}{a_1} = \frac{b_2}{b_1} = k$$

and (6) may be written as

$$\frac{dy}{dx} = f\left(\frac{a_1 x + b_1 y + c_1}{k(a_1 x + b_1 y) + c_2}\right) = F(a_1 x + b_1 y),$$

and consequently the change of variables $z = a_1 x + b_1 y$ transforms the equation under consideration into an equation with variables separable.

Example 3: $\quad \dfrac{dy}{dx} = \dfrac{x - y + 1}{x + y - 3}$

Solving the system of equations

278

$x - y + 1 = 0$, $x + y - 3 = 0$, we get $h = 1$, $k = 2$. Putting $x = X + 1$, $y = Y + 2$,

we have $\quad \dfrac{dY}{dX} = \dfrac{X - Y}{X + Y}$.

The change of variables

$$v = \frac{Y}{X} \quad \text{or} \quad Y = v\, X$$

leads to the separable equation:

$$v + X\,\frac{du}{dX} = \frac{1 - v}{1 + v},$$

$$\frac{(1 + v)\, dv}{1 - 2v - v^2} = \frac{d X}{X},$$

$$-\tfrac{1}{2}\, ln\, (1 - 2v - v^2) = ln\, X - \tfrac{1}{2}\, ln\, C_1$$

$$(1 - 2v - v^2)\, X^2 = C_1$$

Or $\quad X^2 - 2\, X\, Y - Y^2 = C_1$

Or $\quad x^2 - 2\, x\, y - y^2 + 2\, x + 6\, y = C.$

Example 4: $\quad \dfrac{dy}{dx} = \dfrac{2x + y - 1}{4x + 2y + 5}$

In this case the straight lines
$2x + y - 1 = 0$ and $4x + 2y + 5 = 0$ are parallel since they have the some slope. Putting $z = 2x + y$ and

$$\frac{dy}{dx} = \frac{dz}{dx} - 2$$

then, the differential equation reduces to

$$\frac{dz}{dx} - 2 = \frac{z - 1}{2z + 5}$$

Or

$$\frac{dz}{dx} = \frac{5z + 9}{2z + 5}$$

Separating the Variables we get

$$\frac{2z + 5}{5z + 9}\, dz = dx$$

which gives *by* integration

$$\frac{2}{5}\, z + \frac{7}{25}\, ln\, (5z + 9) = x + C_1.$$

Since $z = 2x + y$, we obtain the final solution of the initial equation in the form

$$\frac{2}{5}(2x+y) + \frac{7}{2}\, ln\,(10x + 5y + 9) = x + C_1$$

Or

$$10y - 5x + 7\, ln\,(10x + 5y + 9) = C$$

II. Homogeneous eqns.

These are the form $\dfrac{dy}{dx} = f\left(\dfrac{y}{x}\right)$ —(1)

put $y = vx$

... v are fn in x.

$$\frac{dy}{dx} = v + x\frac{dv}{dx} \quad —(2)$$

By differentiation of eqn (2)

$$\therefore v + x\frac{dv}{dx} = f(v)$$

$$x\frac{dv}{dx} = f(v) - v$$

$$\int \frac{dv}{f(v) - v} = \int \frac{dx}{x}$$

Example 5: Reducing non-homogeneous diff. eq. into homogeneous by substitution

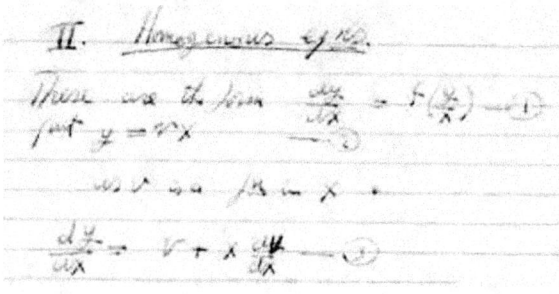

$$ex\quad \frac{dy}{dx} = \frac{xx}{x^2 - y^2} = \frac{\frac{y}{x}}{1 - \left(\frac{y}{x}\right)^2}$$

put $y = vx$

$$v + x\frac{dv}{dx} = \frac{v}{1 - v^2}$$

$$x\frac{dv}{dx} = \frac{v}{1 - v^2} - v = \frac{v - v + v^3}{1 - v^2} = \frac{v^3}{1 - v^2}$$

$$\frac{1 - v^2}{v^3}\, dv = \frac{dx}{x}$$

$$\int \left(\frac{1}{v^3} - \frac{1}{v}\right) dv = \int \frac{dx}{x}$$

$$-\frac{1}{2v^2} - \log v = \log x + c$$

$$-\frac{1}{2v^2} - \log v = \log x + c$$

$$-\frac{x^2}{2y^2} = \log\ cx \qquad \therefore x^2 = \log cy \qquad \therefore y = \log cy.$$

Example 6: Reducing non-homogeneous diff. eq. into homogeneous by transfer of axis

Eqns reducible to the homogeneous form.

ex: $\dfrac{dy}{dx} = \dfrac{y - x + 1}{y + x + 5}$

the point of intersection of the two lines:

$$\left.\begin{array}{r} y - x + 1 = 0 \\ y + x + 5 = 0 \end{array}\right\} \quad (-2, \ 3)$$

transfer to parallel axis through the point $(-2, 3)$ put $\begin{array}{l} x = X - 2 \\ y = Y - 3 \end{array}$

$dx = dX$
$dy = dY$

$\dfrac{dY}{dX} = \dfrac{Y - 3 - X + 2 + 1}{Y - 3 + X - 2 + 5} = \dfrac{Y - X}{Y + X}$

$\dfrac{dY}{dX} = \dfrac{Y - X}{Y + X}$ which is the Homogeneous form

12.1.1.3. Homogeneous linear differential equations (using D- operator)

Homogeneous linear Diff^l eqns

They are given by $\left[a_0 x^n D^n + a_1 x^{n-1} D^{n-1} + \cdots + a_n x D\right] y = f(x)$

to transform this eq^n to an eq^n in which the coeff^s are constants we make the substitution $x = e^t$

$$\begin{cases} x = e^t & \dfrac{dt}{dx} = e^{-t} = \dfrac{1}{x} \\ \dfrac{dx}{dt} = e^t & \\ \dfrac{dx}{dt} = e^t \end{cases}$$

let $\theta = \dfrac{d}{dt}$ $\qquad \dfrac{dy}{dt} = \dfrac{dt}{dx} = \dfrac{dy}{dx} e^t = \dfrac{1}{x} \dfrac{dy}{dt}$

$$\dfrac{1}{x} \dfrac{dy}{dt} = \dfrac{dy}{dx}$$

$$\boxed{x D y = \theta y} \quad \text{—} \quad \text{①}$$

$\dfrac{d^2 y}{dx^2} = \dfrac{d}{dx}\left(\dfrac{dy}{dx}\right) = \dfrac{d}{dx} \cdot \dfrac{y}{dt} e^{-t} = \dfrac{d}{dt}\left(\dfrac{dy}{dt} e^{-t}\right) \dfrac{dt}{dx}$

$\qquad = \left(-e^{-t} \dfrac{dy}{dt} + e^{-t} \dfrac{d^2 y}{dt^2}\right) e^{-t}$

281

$$\frac{d^2y}{dx^2} = \left(\frac{d^2y}{dt^2} - \frac{dy}{dt}\right)e^{-2t} = \frac{1}{x^2}\left(\frac{d^2y}{dt^2} - \frac{dy}{dt}\right) \text{—③}$$

$$\therefore \boxed{x^2 \frac{d^2y}{dx^2} = \frac{d^2y}{dt^2} - \frac{dy}{dt}}$$

$$x^2 D^2 y = (\theta^2 - \theta) y$$

similarly $\boxed{x^3 D^3 y = \theta(\theta-1)(\theta-2)y}$ —③

In General:

$$x^n D^n y = \theta(\theta-1)(\theta-2)(\dots)(\theta - n+1) y$$

$$x^n D^n y = \theta(\theta-1)(\theta-2)(\dots)(\theta-n+1) y$$

Example 1: $x^2 y'' + 9 x y' + 25 y = 50$

ex.① solve the eq^n $x^2\frac{d^2y}{dx^2} + 9x\frac{dy}{dx} + 25y = 50$

put $x = e^t$

$\theta = \frac{d}{dt}$

$$\left(\theta(\theta-1) + 9\theta + 25\right) y = 50$$

$$\left(\theta^2 + 8\theta + 25\right) y = 50$$

$$\theta = \frac{-8 \pm \sqrt{64 - 100}}{2}$$

$$\theta = -4 \pm 3i$$

c.F $e^{-4t}[A\cos 3t + B\cos 3t]$

$= e^{-4}[A\cos(3\log x) + B\cos(3\log x)]$

R.I $= y = \frac{50}{25} = 2.$

Example 2:

ex.2. $(x^3 D^3 + 3x^2 D^2 + x D) y = 24 x^2$

put $e^t = x$ $\theta = \frac{d}{dt}$

$\left(\theta(\theta-1)(\theta-2) + 3\theta(\theta-1) + \theta\right) y = 24 e^{2t}$

$(\theta^3 - 3\theta^2 + 2\theta + 3\theta^2 - 3\theta + \theta) y = 24 e^{2t}$

$\theta^3 y = 24 e^{2t}$

282

$$\text{C.F.} = (A + Bt + ct^2)e^{\alpha t} = A + Bt + ct^2$$

$$\text{P.I} \quad y = \frac{1}{\theta^3} \, 2t e^{2t} = 3 e^{2t}$$

Solution $\quad y = A + Bt + Ct^2 + 3e^{2t}$

$\qquad = A + B \log x + C(\log x)^2 + 3x^2$

12.1.1.4. The solvability of a system of linear homogeneous differential equations

Linear homogeneous system

$$\frac{dx_1}{dt} = a_{11} x_1 + a_{12} x_2 + a_{13} x_3$$

$$\frac{dX}{dt} = AX$$

$$AX = \begin{pmatrix} a_{11} & a_{12} \cdots a_{1} \\ & & \\ & a_{nn} \end{pmatrix} \begin{pmatrix} x_1 \\ x_2 \\ x_3 \end{pmatrix}$$

$$(A - \lambda I)c = 0 \qquad \text{②}$$

$$\text{if} \quad \frac{dX}{dt} = \lambda e^{\lambda t} c$$

$$X = c e^{\lambda t}$$

$$\frac{dX}{dt} = A(t) \, X(t) \qquad \text{①}$$
$$\qquad \qquad \qquad t \in I$$

Theorem

The set of all solns of ① on I form an n-dimensional vector space over the complex field (The set of all solns is sometimes form the general soln)

Proof

If ϕ_1, ϕ_2 are soln of ① , c_1, c_2 are 2 complex nos

$c_1 \phi_1 + c_2 \phi_2$ is again a soln. ie a soln forms a vector space

283

To show space is n dimensional, a set of n linearly *independent* vectors $\phi_1, \phi_2, \dots \phi_n$

must be expected : *each other sol'n of* (1) is linear combination (with complex coeffts) of these ϕ

Let ξ_i, $i=1,2,\dots n$ be a linearly indep

set in n-dimensional vector space (e.g. ξ may be taken as a vector with all components zero except the ith, which is by the existence theorem

$$\forall \ r \in I, \ \exists \ n \ \text{sol}^{ns} \ \phi_i, \ i=1,2\dots n$$

of (1) :

$$\phi_i(r) = \xi_i$$

This sol'n satisfies the required condt:

If ϕ_i are linearly dependent, this

$\exists$ n complex n^{os} not all zeros

$$\sum_{i=1}^{n} c_i \phi_i(t) = 0$$

$$t \in I \implies \sum_{i=1}^{n} c_i \phi_i(r) = \sum_{i=1}^{n} c_i \xi_i$$

This contradicts the assumption that ξ_i are *linearly independent*

If $\phi(t)$ is any sol'n of (1) on I

$\therefore \phi(r) = \xi$ whatever when for constant $\xi = \sum_{i=1}^{n} c_i \xi_i$ basis of sp

hence the fn $\sum c_i \phi_i$ is a sol'n of (1) on I & by uniqueness theorem which assumes value ξ at r this must be ϕ

i.e $\phi = \sum c_i \phi_i$

∴ so ϕ is linear combination of ϕ_i.
& this proves theorem.

Theorem 2

Let A be $n \cdot n$ matrix with continuous elements on an interval I

: $a \le t \le b$ & suppose ϕ
is a matrix of f on I satisfying

$$\phi'(t) = A(t)\phi(t) \quad t \in I_1$$

Then $\det(\phi)$ satisfies on I, the 1st order eqn

$$(\det I)' = (tr A)(\det \phi) \quad ②$$

$$\text{trace } A = tr A = \frac{?}{?} = \sum_{i=1}^{n} a_{ii}(t)$$

Proof

Let ϕ_{ij}, a_{ij} be the elements

i rows
j columns   of ϕ & A

$$\phi'_{ij} = \sum_{k=1}^{n} a_{i_k}(t)\phi_{kj}(t) \quad ③$$

$$\begin{bmatrix} \phi'_{11} & \phi'_{12} & \phi'_{1n} \\ \phi'_{n1} & \phi'_{n2} & \phi'_{nn} \end{bmatrix} = \begin{bmatrix} a_{11} & a_{1n} \\ a_{n1} & a_{nn} \end{bmatrix}\begin{bmatrix} \phi_{11} & \phi_{12} \\ \end{bmatrix}$$

∴ $\phi'_{11} = a_{11}\phi_{11} + a_{12}\phi_{21} + \cdots + a_{1n}\phi_{n1}$

$$\phi'_{11} = \sum_{i=1}^{n} a_{1i}\phi_{i1}$$

The derivative of $\det \phi$ is a sum of n matrix determinants

$$(\det \phi)' = \begin{vmatrix} \phi_{11}' & \phi_{12}' & \phi_{1n}' \\ \phi_{21} & \phi_{22} & \phi_{2n} \\ & \vdots & \end{vmatrix} + \begin{vmatrix} \phi_{11} & \phi_{12} & \phi_{1n} \\ \phi_{21}' & \phi_{22}' & \phi_{2n}' \\ & \vdots & \end{vmatrix} + \cdots$$

$$\cdots + \begin{vmatrix} \phi_{11} & \phi_{12} \\ \phi_{n1}' & \phi_{nn}' \end{vmatrix}$$

using ③ the 1st det on the RHS ①
set
$$\phi_{11}' = \sum_{k=1}^{n} a_{1k}\, \phi_{k1}$$

$$\phi_{12}' = \sum_{k=1}^{n} a_{1k}\, \phi_{k2}$$

$$\left| \det \phi \right| = \begin{vmatrix} \sum_{k=1}^{n} a_{1k}\phi_{k1} & \sum_{k=1}^{n} a_{1k}\phi_{k2} & \\ \phi_{21} & \phi_{22} & \phi_{2n} \\ \phi_{n1} & \phi_{n2} & \phi_{nn} \end{vmatrix}$$

& this det. is unchanged if ① subtract from the 1st row a_{12} times the 2nd row now plus a_{13}, the 3rd row and up to a_{1n} times nth row

$$\left| \det \phi \right|' = \begin{vmatrix} a_{11}\phi_{11} & a_{11}\phi_{12} & a_{11}\phi_{1n} \\ \phi_{21} & \phi_{22} & \phi_{2n} \\ \phi_{n1} & \phi_{n2} & \phi_{nn} \end{vmatrix}$$

$$= a_{11} \det \phi$$

Carrying out a similar procedure with remaining dets

we obtain finally ③ Proof

12.1.1.5.. Bernoulli's Equation

Numerous differential equations can be reduced to linear equations by means of a change of variables. Consider an equation of the fern:

$$\frac{dy}{dx} + P(x) y = Q(x) y^n,$$

where $P(x)$ and $Q(x)$ are continuous functions of x, and n is neither equal to 0 or 1 (otherwise we would have a linear equation). This equation is called Bernoulli's equation and reduces to a linear equation by the following transformation.

Dividing all terms of the equation by y^n, we get

$$y^{-n} \frac{dy}{dx} + P y^{n+1} = Q.$$

Making the substitution

$$z = y^{-n+1}$$

we have

$$\frac{dz}{dx} = (-n + 1) y^{-n} \frac{dy}{dx}.$$

Substituting we get

$$\frac{dz}{dx} + (-n + 1) P z = (-n + 1) Q.$$

This is a linear equation. Finding its complete integral and substituting the expression y^{-n+1} for z we get the complete integral of the Bernoulli's equation.

Example   Solve the equation $\dfrac{dy}{dx} + x y = x^3 y^3$.

Solution:

Dividing all terms by y^3, we have

$$y^{-3}\frac{dy}{dx} + x\,y^{-2} = x^3.$$

Introducing the new function $z = y^{-2}$,

we get

$$\frac{dz}{dx} = -2\,y^{-3}\frac{dy}{dx}.$$

Substituting, we obtain

$$\frac{dz}{dx} - 2\,x\,z = -2\,x^3.$$

This is a linear equation. Let us find its complete solution

$$z = u\,v \; ; \; \frac{dz}{dx} = u\frac{dv}{dx} + v\frac{du}{dx}.$$

Put expressions z and dz/dx in the above

$$u\frac{dv}{dx} + v\frac{du}{dx} - 2\,x\,u\,v = -2\,x^3$$

Equate to zero the expression in the brackets

$$\frac{dv}{dx} - 2\,x\,v = 0 \; ; \; \frac{dv}{v} = 2\,x\,dx \; ;$$

$$ln\;v = x^2 \; ; \; v = e^{x^2}.$$

For u we get the equation

$$e^{x^2}\frac{du}{dx} = -2\,x^3.$$

Separating variables, we have

$$du = -2\,e^{-x^2}\,x^3\,dx\,,$$

$$u = -2\int e^{-x^2}\,x^3\,dx\;+\;C.$$

Integrating by parts, we find

$$u = x^2\,e^{-x^2} + e^{-x^2} + C,$$

$$z = u\,v = x^2 + 1 + C\,e^{x^2}.$$

Consequently, the complete integral of the given equation is

$$y^{-2} = x^2 + 1 + C\,e^{x^2}$$

Or

$$y = \frac{1}{\sqrt{x^2 + 1 + C\,e^{x^2}}}.$$

Bernolli's eq^n :-

$$\frac{dy}{dx} + Py = Qy^n \qquad \text{non linear}$$

divide by y^n

$$y^{-n}\frac{dy}{dx} + Py^{-n+1} = Q \longrightarrow \textcircled{1}$$

$$\frac{1}{-n+1}\frac{d}{dx}(y^{-n+1}) + Py^{-n+1} = Q \longrightarrow \textcircled{2}$$

Put $y^{-n+1} = z$

$$\frac{1}{-n+1}\frac{dz}{dx} + Pz = Q \longrightarrow \textcircled{3}$$

which is of linear form.

Differentiation of the 1st order but not of the first degree.

General form is $L_0 P^n + L_1 P^{n-1} + L_2 P^{n-2} + \cdots + L_{n-1}P + L_n = 0$

where $p = \frac{dy}{dx}$

$L_0, L_1 \cdots L_n$
are in general fns of $x \,\&\, y$.

Solvable types d.e :-

I. Equation solvable for P :-

ex. $P^2 - 2P\cosh x + 1 = 0 \qquad$ as $P = \frac{dy}{dx}$

$$P = \frac{2P\cosh x \pm \sqrt{4\cosh^2 x - 4}}{2}$$

$$p = \cosh x \pm \sinh x = e^{\pm x} \longrightarrow \textcircled{1}$$

II. Eq^{ns} solvable for y :-

ex. $y = \frac{dy}{dx} + \left(\frac{dy}{dx}\right)^3$

i.e. $y = P + P^3 \qquad\qquad p = \frac{dy}{dx}$

Differentiation both sides w.r.t x.

$$P = (1 + 3P^2) \frac{dP}{dy}$$

$$\int dx = \int \left(\frac{1}{P} + 3P \right) dP$$

$$\left. \begin{array}{l} x = \log P + \frac{3}{2} P + C \\ y = P + P^3 \end{array} \right\}$$

which is the required Sol^n in the parametric form, The parameter is P (parameter).

III Eqns solvable for x

ex.
$$x = 4P^2 + 4P^3 \qquad P = \frac{dy}{dx}$$

Diff^{iate} w.r.t. y.

$$\frac{1}{P} = (4 + 12 P^2) \frac{dP}{dy}$$

$$dy = (4P + 12P^3) dP$$

$$\left. \begin{array}{l} y = 2P^2 + 3P^4 + C \\ x = 4P^2 + 4P^3 \end{array} \right\}$$

12.1.1.6. Riccati's Equation

The equation $\quad \dfrac{dy}{dx} + P(x)\,y + Q(x)\,y^2 = R(x)$.

where P (x), Q (x) and R (x) are given functions of x, is called Riccati's equation. In the general form it can not easily be integrated, but may be transformed into Bernoulli's equation by a change of variable if a single particular solution y, (N) of this equation is known. Indeed, assuming $y = y_1 + z$,

we get

$$y_1' + z' + P(x)\,(y_1 + z) + Q(x)\,(y_1 + z)^2 = R(x)$$

or, since

$$y_1' + P(x)\,y_1 + Q(x)\,y_1^2 \equiv R(x),$$

we will have the Bernoulli's equation

$$z' + [\,P(x) + 2Q(x)\,y_1\,]\,z = -\,Q(x)\,z^2$$

for the determination of z.

Example : Solve the differential equation $\quad \dfrac{dy}{dx} = y^2 - \dfrac{2}{x^2}$.

Solution:
In this Example it, is easy to choose a particular solution $\quad y_1 = \dfrac{1}{x}$

290

Putting

$$y = z + \frac{1}{x},$$

we get

$$y' = z' - \frac{1}{x^2},$$

and hence

$$z' - \frac{1}{x^2} = (z + \frac{1}{x})^2 - \frac{2}{x^2}$$

Or

$$z' - \frac{2}{x} z = z^2,$$

which is Bernoulli's equation. Dividing by z^2 we have

$$z^{-2} z' - \frac{2}{x} z^{-1} = 1$$

which on substituting

$$u = z^{-1}, \quad \frac{du}{dx} = -z^{-2} z'$$

reduces to

$$\frac{du}{dx} + \frac{2}{x} u = -1$$

which is a linear equation. Its integrating factor is

$$\mu = e^{\int \frac{2}{x} dx} = e^{2 \ln x} = x^2$$

and hence the solution is given from

$$x^2 u = \int -x^2 dx + C = -\frac{1}{3} x^3 + C_1$$

Or

$$u = -\frac{1}{3} x + \frac{C_1}{x^2}$$

and since z = 1/u we have

$$z = \frac{3 x^2}{C - x^3}.$$

Hence. the complete solution of the initial equation is

$$y = \frac{1}{x} + z,$$

$$y = \frac{1}{x} + \frac{3 x^2}{C - x^3}$$

12.1.2. Differential equations of the second order

Diff. eqns of the 2nd order.

We shall start by considering some types of d. eqns of the 2nd order of frequent occurrence in practical applications.—

12.1.2.1. Equations of the form y″ = f(x)

I. Eqns of the form $\frac{d^2y}{dx^2} = f(x)$.

$$\frac{dy}{dx} = \int f(x)\,dx + A$$

$$y = \int \left\{ \int f(x)\,dx \right\}\,dx + Ax + B.$$

e.g.

$$\frac{d^2y}{dx^2} = 6X \quad \text{—①} \quad \text{given that } X = 0$$
$$y = 0 \quad \frac{dy}{dx} = 0$$

$$\frac{dy}{dx} = 3X + A \quad \text{—②}$$
$$X = 0 \qquad y' = 0 \qquad \therefore A = 0$$

$$\frac{dy}{dx} = 3X$$

$$y = X^2 + B \quad \text{—③}$$
$$X = 0 \qquad y = 0 \qquad B = 0$$

$$y = X^3 \quad \text{—④}$$

In structures $B.M.F = EI\frac{d^2y}{dx^2} = F(x)$

12.1.2.2. Equations of the form y″ = f(y)

II Eqns of the form:

$$\frac{d^2y}{dx^2} = f(y) \qquad \text{put } P = \frac{dy}{dx}$$

$$\frac{d^2y}{dx^2} = \frac{dP}{dx} = \frac{dP}{dy}\cdot\frac{dy}{dx} = P\frac{dP}{dy}$$

$$\therefore \ P \frac{dP}{dy} = f(y)$$

$$\int P \, dP = \int f(y) \, dy$$

$$\frac{1}{2} P^2 = \int f(y) \, dy + A$$

$$P^2 = 2 \int f(y) \, dy + 2A$$

$$\frac{dy}{dx} = \pm \sqrt{2 \int f(y) \, dy + 2A}$$

$$\int \frac{dy}{\sqrt{2 \int f(y) \, dy + 2A}} = \int \pm \, dx$$

$$\int \frac{dy}{\sqrt{2 \int f(y) \, dy + 2A}} = \pm x + B$$

ex. $\dfrac{d^2y}{dx^2} = y$

$$P \frac{dP}{dy} = y \qquad\qquad \frac{dy}{dx} = \pm \sqrt{y^2 + A}$$

$$\int P \, dP = \int y \, dy \qquad \int \frac{dy}{\sqrt{y^2 + A}} = \int \pm \, dx = \pm x + B$$

$$\tfrac{1}{2} P^2 = \tfrac{1}{2} y^2 + c$$

$$P^2 = y^2 + A \qquad\qquad \therefore \log\left(y + \sqrt{y^2 + A}\right) = \pm x + B$$

in Mechanics $\ddot{x} = \mu x$

12.1.2.3. Equations of the form of inexplicit y (e.g., $xy'' + y' + c = b$)

III. Eq^{ns} in which y does not occur explicitly:—

There are eq^{ns} of the form $\phi\left(\frac{d^2y}{dx^2}, \frac{dy}{dx}, x\right) = 0$
y is not found clearly
put $\frac{dy}{dx} = p$

$$\therefore \frac{d^2y}{dx^2} = \frac{dP}{dx}.$$

the eq^n reduces to $\phi\left(\frac{dP}{dx}, P, x\right) = 0$ which is an d. eq^n of the 1st order with x as the independent variable and P the dependent variable.

ex:

ex: $\quad x \frac{d^2y}{dx^2} + \frac{dy}{dx} + 1 = 0$

i.e. y is absent and hence we write the eq^n in the form $x \frac{dP}{dx} + P + 1 = 0$

$$\int \frac{dP}{P+1} = \int \frac{-dx}{x}$$

$$\log(P+1) = -\log x + \log A$$

$$\frac{dy}{dx} = \frac{A}{x} - 1$$

$$y = A \log x - x + B$$

Example 1: Second order differential equation with singular points

$2\,x\,y'' + (x + 1)\,y' + 3\,y = 0$

Example
singular pts

$2X y'' + (x+1) y' + 3y = 0$

$y'' + \frac{x+1}{2x} y' + \frac{3y}{x} = 0$

$\lim\limits_{x \to 0} x P(x) = $ finite

$\lim\limits_{x \to 0} x^2 Q(x) = $ finite

295

$$y = \sum_{n=0}^{\infty} a_n x^{n+c}$$

$$y' = (n+c)a_n x^{n+c-1}$$

$$y'' = (n+c)(n+c-1)a_n x^{n+c-2}$$

$$2(n+c)(n+c-1)a_n x^{n+c-1} + (x+1)(n+c)a_n x^{n+c-1}$$
$$+3\, a_n x^{n+c} = 0$$

$\underline{\text{Coeff of } x^{n+c}}$

$$2(n+c+1)(n+c)a_{n+1} + (n+c)a_n$$
$$+ (n+1+c)a_{n+1} + 3 a_n = 0$$

$$a_{n+1}\left[(n+1+c)(2n+2c+1)\right] + a_n\left[n+c+3\right] =$$

$\underline{\text{Coeff of } x^{c-1}}$ NO (n-1)

$$2c(c-1)a_0 + c a_0 = 0$$

$$2c(c-1) + c = 0$$

$$c[2c - 2 + 1] = 0$$

$$c[2c - 1] = 0$$

$$\boxed{\begin{array}{l} c = 0 \\ c = +\tfrac{1}{2} \end{array}}$$

$\boxed{c = 0}$

$\underline{\text{coeff of } x^n}$

$$2(n+1)(n)a_{n+1} + n a_n + (n+1)a_{n+1}$$
$$+3 a_n = 0$$

$$a_{n+1}(n+1)(2n+1) + a_n(n+3) = 0$$

$$a_{n+1} = -a_n \frac{n+3}{(n+1)(2n+1)}$$

$$\boxed{a_0 = 4\lambda}$$

$$a_1 = -a_0 \frac{3}{1 \cdot 1}$$

$$a_2 = -a_1 \cdot \frac{4}{2 \cdot 3} = a_0 \frac{4 \cdot 3}{3 \cdot 2 \cdot 1}$$

$$a_3 = -a_2 \frac{5}{3 \cdot 5} = -a_0 \frac{5 \cdot 4 \cdot 3}{5 \cdot 3 \cdot 3 \cdot 4 \cdot 4}$$

12.1.2.4. Equations of the form of inexplicit x (e.g., y'' + y' + c = b)

IV. Eqns in which x does not occur explicitly.

These are the form of $\left(\frac{d^2y}{dx^2}, \frac{dy}{dx}, y \right) = c$

$$f\left(P \frac{dP}{dy}, P, y \right) = c$$

which is a eqn of the 1st order with y as the independent variable.

ex. $y \frac{d^2y}{dx^2} = 2 \left(\frac{dy}{dx} \right)^2$

$y P \frac{dP}{dy} = 2P^2$

$y \frac{dP}{dy} = 2P$, $\int \frac{dP}{P} = \int \frac{2 \, dy}{y}$

$\log P = 2 \log y + \log A$

$P = \frac{dy}{dx} = A y^2$

$\int \frac{dy}{y^2} = \int P \, dx$

$-\frac{1}{y} = A x + B$

$$y = \frac{-1}{A x + B} \qquad \boxed{y = \frac{1}{C x + D}}$$

V. Eqns of the form $f\left(\frac{d^2y}{dx^2}, \frac{dy}{dx} \right) = 0$

in which both x & y are absent the eqn can be treated by any one of above the methods.

12.1.2.5. General form of the linear differential equations of the second order

Linear diff^l of the second order:-

General form is $\dfrac{d^2y}{dx^2} + P\dfrac{dy}{dx} + Qy = R$ ——①

where $P \& Q \& R$ are f^{ns} of x only.

The solution of ① is usually written in the form:-

$$y = u + v \longrightarrow ②$$

where u is the general solution of the eq^n:-

$$\dfrac{d^2y}{dx^2} + P\dfrac{dy}{dx} + Qy = 0 \longrightarrow ③$$

Containing two arbitrary const^t & v is any particular solution of eq^n ①. The part u of a solution is called the complementary f^{n} (C.F.) ، الجزء

The part v of a solution is called the particular integral (P.I.)

الجزء

Substitute from ② in ①:-

$$\dfrac{d^2}{dx^2}(u+v) + P\dfrac{d}{dx}(u+v) + Q(u+v) - R = 0$$

$$= \left(\dfrac{d^2u}{dx^2} + P\dfrac{du}{dx} + Qu\right) + \left(\dfrac{d^2v}{dx^2} + P\dfrac{dv}{dx} + Qv - R\right) = 0$$

  1st expression   2nd expression

The 1st expression bet. brackets vanishes since u is the sol. of ③. Now suppose that u_1, u_2 are solutions of:-
& the 2nd expression vanishes since v is the sol^n of eq^n ①

$$\dfrac{d^2y}{dx^2} + P\dfrac{dy}{dx} + Qy = 0$$

we shall now show that $y = c_1 u_1 + c_2 u_2$.
is also a sol. of the eq^n.

$$\therefore \dfrac{d^2}{dx^2}(c_1 u_1 + c_2 u_2) + P\dfrac{d}{dx}(c_1 u_1 + c_2 u_2) + Q(c_1 u_1 + c_2 u_2)$$

$$= c_1\left(\frac{d^2u_1}{dx_1^2} + P\frac{du_1}{dx} + Qu_1\right) + c_2\left(\frac{d^2u_2}{dx_2^2} + \frac{du_2}{dx} + Qu_2\right) = 0$$

Since u_1 & u_2 satisfies the eqⁿ.

c_1 c_2 being arbitrary const. This Theory is useful in finding complementary fⁿ (C.F.)

12.1.2.6. Polynomial solution of linear differential equations of the second order

<u>Linear Eqⁿ of 2nd Order</u>

$$\frac{d^2y}{dx^2} + P(x)\frac{dy}{dx} + q(x)y = 0$$

$$P(x) = P_0 + \sum_{i=1}^{\infty} P_i x^{-i}$$

$$q(x) = q_0 + \sum_{j=1}^{\infty} q_j x^{-j}$$

$q(x)$, $P(x)$ are finite at ∞

substitute

$$\begin{cases} y = e^{\lambda x}V & \text{——①} \\ y' = e^{\lambda x}V' + \lambda e^{\lambda x}V & \text{——②} \\ y'' = e^{\lambda x}V'' - 2\lambda e^{\lambda x}V' + \lambda^2 e^{\lambda x}V & \text{——③} \end{cases}$$

$$V'' + 2\lambda V' + \lambda^2 V + P(V' + \lambda V) + qV = 0$$

$$\boxed{V'' + V'[2\lambda + P] + V[\lambda^2 + \lambda P + q] = 0}$$

$$\boxed{y = e^{\lambda x}V}$$
$$\boxed{V = x^\sigma u(x)}$$

if λ is the root of the eqⁿ

$$\lambda^2 + \lambda P_0 + q_0 = 0$$ coeff.

the constant term in the eqⁿ of V is disappear

[because $\lambda^2 = -q_0 + \lambda P_0$

299

so
$$v\left[x^2 + \lambda P_1 + q\right] = v\left[(-q_0 - \lambda\ell) + \lambda\left[P_0 + \sum_{i=1}^{n} \frac{z}{x^i}\right] + q_0 + \sum_{j=1}^{m} \frac{z}{x^j}\right]$$
$$= v\left[\sum_{i=1}^{n} P_i x^{-i} + \sum_{j=1}^{m} q_j x^{-j}\right]$$

& the eq. become

$$\boxed{v'' + v'\left[\omega_0 + \omega_1 x^{-1} + \omega_2 x^{-2} + \cdots\right] + v\left[\beta_1 x^{-1} + \beta_2 x^{-2} + \cdots\right] = 0}$$

where $\begin{cases} \omega_0 = 2\lambda + P_0 \\ \beta_1 = P_1 + q_1 \end{cases}$

If we let $\begin{cases} v = x^{\sigma} U(x) \\ v' = \sigma x^{\sigma-1} U(x) + x^{\sigma} U'(x) \\ v'' = \sigma(\sigma-1) x^{\sigma-2} U(x) + 2\sigma x^{\sigma-1} U' + x^{\sigma} U'' \end{cases}$

So eq. become

$$\left[\sigma(\sigma-1) x^{\sigma-2} U + 2\sigma x^{\sigma-1} U' + x^{\sigma} U''\right]$$
$$+ \left[\sigma x^{\sigma-1} U + x^{\sigma} U'\right]\left[\omega_0 + \omega_1 x^{-1} + \cdots\right]$$
$$+ \left[x^{\sigma} U\right]\left[\beta_1 x^{-1} + \beta_2 x^{-2} + \cdots\right] = 0$$

Divide by x^{σ}

$$\left[\sigma(\sigma-1) x^{-2} U + 2\sigma x^{-1} U' + U''\right]$$
$$+ \left[\sigma x^{-1} U + U'\right]\left[\omega_0 + \omega_1 x^{-1} + \cdots\right]$$
$$+ U\left[\beta_1 x^{-1} + \beta_2 x^{-2} + \cdots\right] = 0$$

Or
$$U'' + U'\left[\frac{2\sigma}{x} + \left(\omega_0 + \omega_1 x^{-1} + \omega_2 x^{-2} + \cdots\right)\right]$$
$$+ U\left[\frac{\sigma(\sigma-1)}{x^2} + \left(\frac{\sigma}{x}\left(\omega_0 + \omega_1 x^{-1} + \omega_2 x^{-2} + \cdots\right)\right) + \left(\beta_1 x^{-1} + \beta_2 x^{-2} + \cdots\right)\right] = 0$$

Assume $\sigma \omega_0 + \beta_1 = 0 \Rightarrow$ coeff to $\int x^{-1}$

        & $\omega_0 = -1$.   in $U\{\ \}$

So

$$U'' + U'\left[2\sigma x^{-1} + \omega_0 + \omega_1 x^{-1} + \omega_2 x^{-2} + \dots\right]$$
$$+ U\left[\underline{\sigma(\sigma-1)x^{-2}} + \underline{\sigma\omega_1 x^{-2}} + \sigma\omega_2 x^{-3} + \dots\right.$$
$$\left. + \beta_2 x^{-2} + \beta_3 x^{-3} + \dots\right] = 0$$

or $\omega_0 = -1$

$$\underline{U'' + U'\left[-1 + a_1 x^{-1} + a_2 x^{-2} + \dots\right] + U\left[b_2 x^{-2} + b_3 x^{-3} \dots\right) = 0}$$

where
$$\begin{cases} a_1 = 2\sigma + \omega_1 \\ a_2 = \omega_2 \dots \\ b_2 = \sigma(\sigma-1) + \beta_2 + \sigma\omega_1 \\ b_3 = \sigma\omega_2 + \beta_3 \dots \end{cases}$$

We shall solve the eq^n by using approx. sol.

if $U_1(x) = y$ as $x \to \infty$

$$U'' - U' = -\left[a_1 x^{-1} + a_2 x^{-2} + \dots\right]U' - \left[b_2 x^{-2} + b_3 x^{-3} + \dots\right]U$$

$$\frac{d^2 U_2}{dx^2} - \frac{dU_2}{dx} = -\left\{a_1 x^{-1} + a_2 x^{-2} + \dots\right\}\frac{dU_1}{dx} - \left\{b_2 x^{-2} + b_3 x^{-3} + \dots\right\}_{U_1}$$

$$\frac{d^2 U_n}{dx^2} - \frac{dU_n}{dx} = -\left\{a_1 x^{-1} + a_2 x^{-2} + \dots\left[\frac{dU_{n-1}}{dx} - \left\{b_3 x^{-2} + \dots\right\}U_{n-1}\right.\right.$$

$$\therefore \left[U\right]_0^\infty = -\int_x^{(x-t)}\left(e - 1\right)\left\{a_1 t^{-1} + a_2 t^{-2} \dots\right\}\frac{dU_{n-1}}{dx}\,dt$$

$$-\int_x^\infty \left(e^{(x-t)}-1\right)\left\{b_2 t^{-2} + b_2 t^{-3}...\right\} u_{n-1}\, dt$$

$$U_n(x) = \eta + \int_x^\infty \left(e^{(x-t)}-1\right)\left(a_1 t^{-1} + a_1 t^{-2}...\right) dU_{n-1}$$

$$+ \int_x^\infty \left(e^{(t-t)}-1\right)\left(b_2 t^{-2} + b_3 t^{-3}...\right) u_{n-1}\, dt$$

Sol^n will be found which assumes the value η when $n \to \infty$. Let

$u_1 = \eta$ & define the sequence of f=

$$u_n = \eta + \int_x^\infty e^{(x-t)}\left(\frac{\alpha_1}{t} + \frac{\alpha_2}{t^2} + \frac{\alpha_3}{t}...\right) u_{n-1}$$

$$+ \int_x^\infty e^{x-t}\left(\frac{\beta_2}{t^2} + \frac{\beta_3}{t^3} + ...\right) u_{n-1}\, dt$$

where $\alpha_1, \alpha_2 ..., \beta_2, \beta_3 ...$ are expressed determined by $a_1, a_2 ... b_1, b_2 ...$
The we can get

$$U_{n-1} = \eta + \int_x^\infty e^{x-t}\left\{\frac{\alpha_i}{t^i}\right\} u_{n-2}\, dt + \int_x^\infty e^{-t}\left\{\frac{\beta_i}{t^i}\right\} u_{n-2}\, dt$$

The sequence
$$u_n = u_1 + (u_2 - u_1) + (u_3 - u_2) + ... + (u_n - u_{n-1})$$

$$u_n - u_{n-1} = \int_x^\infty e^{x-t}\left\{\frac{\alpha_i}{t^i}\right\}\left\{u_{n-1} - u_{n-2}\right\} dt$$

$$+ \int_x^\infty e^{x-t}\left\{\frac{\beta_i}{t^i}\right\}\left\{u_{n-1} - u_{n-2}\right\} dt$$

at $n = 2$

$$u_2 - u_1 = \int_x^\infty e^{x-t}\left\{\alpha's\right\}\left\{u_1 - u_0\right\} dt$$

$$+ \int_x^\infty e^{x-t}\left\{\beta's\right\}\left\{u_1 - u_0\right\} dt$$

302

so as $\lim_{n\to\infty} u_n = u \qquad \therefore u = ?$

$u_1 - u_0 = ?$

$u_2 - u_1 = ? \left[\left\{ \int_n^\infty e^{x-t} \right\} \alpha^3 \int dt \right.$

$\left. + \int_x^\infty e^{x-t} \right\} \beta \right\} dt \right]$

$= \dfrac{A_1'}{\cancel{x}} + \dfrac{A_2'}{x^2} + \cdots + \dfrac{A_{n-1}'}{x^{n-1}} + \dfrac{A_{0+6}'}{x^{n+6}}$

as $x \to \infty$

as like

$u_3 - u_2 = \dfrac{A_2^2}{x^2} + \dfrac{A_3^2}{x^3} + \cdots + \dfrac{A_{n-1}^2}{x^{n-1}} + \dfrac{A_n^2}{x^n}$

$\& \qquad \to 0 \text{ as } x \to 0$

$u_n = \dfrac{A_1'}{x} + \dfrac{A_2' + A_2^2}{x^2} + \dfrac{A_3' + A_3^2 + A_3^3}{x^3} + \cdots$

$= \eta + \dfrac{c_1}{x} + \dfrac{c_2}{x^2} + \cdots + \dfrac{c_n + \gamma_n}{x^n}$

$\gamma_n \to 0 \text{ as } x \to \infty$

$\therefore y = e^{\lambda x} v \quad , v = x^\sigma u$

$y = e^{\lambda x} x^\sigma u$

$$\boxed{ y = e^{\lambda x} x^\sigma \left\{ \eta + \dfrac{c_1}{x} + \dfrac{c_2}{x^2} + \cdots + \dfrac{c_n}{x^n} \right\} }$$

Which is the polynomial solution of the linear diff. eq. $y'' + P y' + Q y = 0$

Example 1:

303

Power Series

$$\frac{dy}{dx} = f(x) \qquad \text{—①}$$

$$y = y_0 \quad \text{at } x = x_0 = 0$$

$$y = A_0 + A_1 x + A_2 x^2 + \cdots \qquad \text{—②}$$

$$= \sum_{n=0}^{\infty} A_n x^n$$

B.C

$$x = 0 \qquad y = y_0 \qquad \boxed{A_0 = y_0}$$

Example 2:   y' – y + x² = 0

Example

$$y' - y + x^2 = 0 \qquad \text{—①}$$

$$y = y_0 \quad \text{at } x = 0$$

$$y = A_0 + A_1 x + A_2 x^2 + \cdots \qquad \text{—②}$$

$$\boxed{A_0 = y_0}$$

$$y' = A_1 + 2A_2 x + 3A_3 x^2 + \cdots$$

eq ① become $\qquad$ —③

$$A_1 + 2A_2 x + 3A_3 x^2 + 4A_4 x^3 - (A_0 + A_1 x + A_2 x^2$$

$$+ A_3 x^3 + \cdots) + x^2 = 0$$

$$(A_1 - A_0) + (2A_2 - A_1)x + (3A_3 - A_2 + 1)x^2 +$$

$$+ (4A_4 - A_3)x^3 + (5A_5 - A_4)x^4 = 0$$

un $x \neq 0$ so

304

$$\boxed{A_1 = A_0 = y_0}$$
$$2A_2 = A_1 = y_0 \qquad \boxed{A_2 = \frac{y_0}{2}}$$
$$A_3 = \frac{A_2 - 1}{3} \qquad \boxed{A_3 = \frac{y_0 - 2}{6}}$$
$$A_4 = \frac{A_3}{4}$$
$$A_5 = \frac{A_4}{5} = \frac{A_3}{5 \cdot 4} = \frac{(y_0 - 2)}{5 \cdot 4 \cdot 3 \cdot 2} = \frac{y_0 - 2}{5!}$$
$$A_6 = \frac{A_5}{6} = \frac{y_0 - 2}{6!}$$

$$y = y_0 + y_0 x + \frac{y_0}{2} x^2 + (y_0 - 2) \sum_{n=3}^{\infty} \frac{x^n}{n!}$$
$$= y_0 + y_0 x + \frac{y_0}{2} x^2 + (y_0 - 2)\left[\left(1 + x + \frac{x^2}{2}\right) - \left(1 + x + \frac{x^2}{2}\right)\right]$$
$$+ (y_0 - 2) \sum_{n=3}^{\infty} \frac{x^n}{n!}$$

$$= y_0 \left[1 + x + \frac{x^2}{2}\right] - (y_0 - 2)\left(1 + x + \frac{x^2}{2}\right) + (y_0 - 2) \sum_{n=0}^{\infty} \frac{x^n}{n!}$$
$$= \left(1 + x + \frac{x^2}{2}\right)\left(y_0 - y_0 + 2\right) + (y_0 - 2) e^x$$

$$\boxed{y = 2\left[1 + x + \frac{x^2}{2}\right] + (y_0 - 2) e^x}$$

$$y = P(x) + C e^x$$

$$\underline{\text{Polynomial}}$$
$$\boxed{P_0(x) \frac{d^2 y}{dx^2} + P_1(x)\frac{dy}{dx} + P_2(x) = 0}$$

* $x = a$ & $P_0(a) \neq 0$
  $\underline{\text{ordinary pt}}$
$$\boxed{y = \sum_{n=0}^{\infty} a_n (x - a)^n}$$

$x = a \Rightarrow P_0(a) = 0, \underset{\underset{x \to a}{lim}}{} x P(x) = 0$

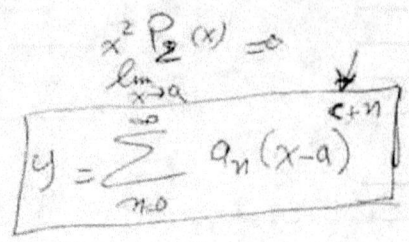

$$\underset{\underset{x \to a}{lim}}{} x^2 P_2(x) = 0$$

$$\boxed{y = \sum_{n=0}^{\infty} a_n (x-a)^{c+n}}$$

$$P_0 \frac{d^2 y}{dx^2} + P_1 \frac{dy}{dx} + P_2 y = 0$$

or

$$\frac{d^2 y}{dx^2} + \frac{P_1}{P_0} \frac{dy}{dx} + \frac{P_2}{P_0} y = 0$$

$$\boxed{\frac{d^2 y}{dx^2} + P \frac{dy}{dx} + Q\, y = 0}$$

ordinary pt $\boxed{x = a}$

$$\underset{x \to a}{lim} P(x) = \text{finite}$$

$$\underset{x \to a}{lim} Q(x) = \text{finite}$$

Regular pt $x = a$

$$\underset{x \to a}{lim} x P(x) = \text{finite}$$

$$\underset{x \to a}{lim} x^2 Q(x) = \text{finite}$$

Example 3: $(1 + x^2)\, y'' + x\, y' - y = 0$

Example

$$(1 + x^2) y'' + x y' - y = 0 \quad —(1)$$

$$y'' + \frac{x}{1+x^2} y' - \frac{y}{1+x^2} = 0$$

about $x = x_0$

306

$$P(0) = 0 \quad \underset{\nu}{finite}$$
$$Q(0) = -1$$

$$y = \sum_{n=0}^{\infty} a_n x^n$$
$$y' = \sum n a_n x^{n-1} \left.\right\} \cdot (2$$
$$y'' = \sum n(n-1) a_n x^{n-2}$$

$$\sum n(n-1)(1 + x^2) x^{n-2} a_n + \left\{ a_n nx^n - \sum a_n x^n = 0 \right.$$

$$\underline{coeff^{ts} \wedge x^n}$$

$$n(n-1) a_n + (n+2)(n+1) a_{n+2} + a_n n$$

$$a_n \left[n(n-1) + n - 1 \right] + a_{n+2} \left[(n+2)(n+1) \right] = 0$$
$$a_n \left[(n-1)(n+1) \right] + a_{n+2} \left[(n+1)(n+1) \right] = 0$$

$$\boxed{a_{n+2} = -a_n \frac{n-1}{n+2}}$$

$$a_0 = y_0$$

$n=0 \qquad a_2 = -y_0 \frac{-1}{2} = \frac{y_0}{2}$

$n=1 \qquad a_1 = 0$

$n=2 \qquad a_4 = -a_2 \cdot \frac{1}{4} = -\frac{y_0}{4 \cdot 2}$

$n=4 \qquad a_6 = -a_4 \frac{3}{6} = + \frac{3 y_0}{6 \cdot 4 \cdot 2}$

$n=6 \qquad a_8 = -a_6 \frac{5}{8} = -\frac{5 \cdot 3 y_0}{8 \cdot 6 \cdot 4 \cdot 2}$

$$n+2 = 2K$$
$$n = 2(K-1)$$

$$a_{2k} = -a_{2(k-1)} \frac{2k-2-1}{2k-2+2} = -$$

$$= -a_{2(k-1)} \frac{2k-3}{2k}$$

$$a_{2(k+1)} = -a_{2k} \frac{2k+2-3}{2k+2}$$

$$= +a_{2k-2} \frac{(2k-3)(2k-1)}{2^2 k(k+1)}$$

$$\boxed{a_{2k} = (-1)^k \frac{(2k-3)!!}{2^k k!}}$$

Example 4: $y'' - 2x^2 y' + 4xy = x^2 + 2x + 2$

$$\underline{\text{Example}}$$

$$y'' - 2x^2 y' + 4xy = x^2 + 2x + 2$$

$x = 0 \qquad P(0) = 0$
$\qquad\qquad Q(0) = 0 \quad$ finite

$$y = \sum_{n=0}^{\infty} a_n x^n$$

$$y' = \sum n a_n x^{n-1}$$
$$y'' = \sum n(n-1) a_n x^{n-2}$$
$$\sum n(n-1) a_n x^{n-2} - 2n a_n x^{n+1} + 4 a_n x^{n+1}$$

$$\text{coeffts of } x^0 = x^2 + 2x + 2$$

$$2(2-1) a_2 = 2$$
$$\boxed{a_2 = 1}$$

$\underline{x^1} \quad 3(3-1) a_3 + 4 a_0 = 2$

$\qquad a_3 = \frac{-2a_0 + 1}{3} = \boxed{-\frac{1}{3}}$

$\underline{x^2} \quad 4(3) a_4 - 2a_1 + 4a_1 = 1$

$\qquad a_4 = \frac{1 - 2a_1}{6} = \frac{1 - a_1}{}$

308

$\text{coeff}^t A \underline{x}^n$

$(n+2)(n+1)a_{n+2} - 2(n-1)a_{n-1} + 4a_{n-3}$

$(n+2)(n+1)a_{n+2} + a_{n-1}[6-2n] = 0$

$$a_{n+2} = a_{n-1}\frac{2(n-3)}{(n+2)(n+1)}$$

$a_3 = \dfrac{-a \cdot a_0}{3 \cdot 2} = -\dfrac{n}{3}a_0$

$\text{valid for } n \geq 2$

$a_5 = a_2 \dfrac{0}{2} = 0$

$a_6 = a_3 \dfrac{2}{2} = -\dfrac{2}{6 \cdot 5 \cdot 3} \cdot \dfrac{1}{}$

Example 5:

series Sol =
Example

solve $y' - y + x^2 = 0$ —①

$y = y_0 \quad x = 0$

$y = \sum_{k=0}^{n} A_k x^k = A_0 + A_1 x + A_2 x^2 + \cdots$

$y' = A_1 + 2A_2 x + 3A_3 x^2 + \cdots$

substitute in ①

$A_1 + 2A_2 x + 3A_3 x^2 + \cdots - (A_0 + A_1 x + A_2 x^2 + \cdots) + x^2 = 0$

$(A_1 - A_0) + (2A_2 - A_1)x + (3A_3 - A_2)x^2 + \cdots + (nA_n - A_{n-1})x^{n-1} = 0$

so $A_1 - A_0 = 0$

i.e $A_1 = A_0 = y_0$

or $\sum_{k=0}^{n} kA_k x^{k-1} - \sum_{k=0}^{n} A_k x^k + x^2 = 0$

$\text{coeffts of } x^2 \qquad A_0 = y_0 \quad \text{given}$

$3A_3 - A_2 + 1 = 0.$

Coeff. of x^0

$$A_3 = \frac{A_2 - 1}{3}$$

$$A_1 - A_0 = 0 \qquad A_1 = A_0$$

Coeff. of x^k

$$(k+1) A_{k+1} - A_k = 0$$

$$A_{k+1} = \frac{A_k}{k+1}$$

of x^1

$$2A_2 - A_1 = 0 \qquad A_2 = \frac{A_1}{2}$$

$$A_1 = A_0 = y_0$$

$$A_2 = \frac{y_0}{2}$$

$$A_3 = \frac{\frac{y_0}{2} - 1}{3} = \frac{y_0}{6} - \frac{1}{3}$$

$$A_4 = \frac{\left(\frac{y_0}{6} - \frac{1}{3}\right)}{4}$$

$$A_5 = \frac{\frac{y_0}{6} - \frac{1}{3}}{4 \cdot 5} , \quad A_J = \frac{\frac{y_0}{6} - \frac{1}{3}}{\frac{J!}{3!}}$$

$$y = A_0\left(1 + x + \frac{x^2}{2}\right) + \frac{1}{3}\left(\frac{A_0}{2} - 1\right)\left[x^3 + \frac{x^4}{4} + \frac{x^5}{4 \cdot 5}\right.$$
$$\left. + \cdots + \frac{3! \, x^n}{n!} \right]$$

$$= A_0\left(1 + x + \frac{x^2}{2}\right) + \frac{3!}{3}\left(\frac{A_0}{2} - 1\right)\left[\frac{x^3}{3!} + \frac{x^4}{4!} + \frac{x^5}{5!}\right.$$
$$\left. + \cdots \frac{x^n}{n!} \right\}$$

$$= A_0\left(1 + x + \frac{x^2}{2}\right) + \frac{3!}{3}\left(\frac{A_0}{2} - 1\right)\left\{\left[1 + x + \frac{x^2}{2!} \cdots\right.\right.$$
$$\left.\left. \right] - \left(1 + x + \frac{x^2}{2}\right)\right]$$

$$a) \quad e^x = 1 + x + \frac{x^2}{2!} + \frac{x^3}{3!} + \dots$$

$$so$$

$$y = A_0 \left(1 + x + \frac{x^2}{2}\right) + \frac{3!}{3}(A_0 - 1)\left[e^x - (1 + x + \frac{x^2}{2})\right]$$
$$= A_0 \left(1 + x + \frac{x^2}{2}\right) + (A_0 - 1)\left\{e^x - (1 + x + \frac{x^2}{2})\right\}$$

$$= A_0 e^x - 2e^x + 2 + 2x + x^2$$

$$\boxed{y = \quad x^2 + 2x + 2 + e^x(A_0 - 2)}$$

check $\quad y' - y + x^2 = 0$

$$(D - 1)y = -x^2$$

C.F $\quad y = B e^x$

P.f $\quad y = \frac{-1}{D-1} x^2$

$$= +(1 + D + D^2 + \dots)x^2$$

$$= x^2 + 2x + 2$$

soln $\quad \boxed{y = B e^x + x^2 + 2x + 2}$

B $\quad$ at $\quad x = 0 \quad y = 0$

$$A_0 = B_0 + 2 \qquad A_0 = 2 - 2$$

$$y = (A_0 - 2)e^x + x^2 + 2x + 2$$

12.1.2.7. Solution of linear differential equations of the second order by definite integrals

<u>Soln by definite intgl</u>

$$(a_0 D^n + a_1 D^{n-1} + \dots + a_n)y +$$

$$(b_0 x D^n + b_1 x D^{n-1} + \dots + b_n x)y = 0 \quad \text{①}$$

or $\quad (a_0 + b_0 x)D^n y + (a_1 + b_1 x)D^{n-1}y + \dots + (a_n + b_n x)y$

or $\quad \underline{\psi(D)y} + \times \underline{\phi(D)}\, y = 0 \quad$ (2)

suppose
$$\boxed{y = \int_{\alpha}^{\beta} e^{xr} R(r)\, dr}$$ (3)

We can show that

$$Dy = \int_{\alpha}^{\beta} x e^{xr} R(r)\, dr$$
$$D^s y = \int_{\alpha}^{\beta} x^s e^{xr} R(r)\, dr \quad (4)$$

from 2, 2, 4 we get
$$\int_{\alpha}^{\beta} \psi(x) e^{xr} R(r)\, dr + \int_{\alpha}^{\beta} x\,\phi(x) e^{xr} R(r)\, dr$$
$$= \int_{\alpha}^{\beta} \psi(r) e^{xr} R(r)\, dr + \left[\phi(r) R(r) e^{xr} \right]_{\alpha}^{\beta}$$
$$\qquad\qquad = 0$$
$$- \int_{\alpha}^{\beta} e^{xr} \frac{d}{dr}\left(\phi(r) R\, dr \right) = 0$$

$$\int_{\alpha}^{\beta} e^{xr} \left[\psi(r) R(r) - \frac{d}{dr}\, \phi(r) R(r) \right] dr$$
$$+ \left[\phi(r) R(r) e^{xr} \right]_{\alpha}^{\beta} = 0$$

The last eqn will be satisfied
if $\int_{\alpha}^{\beta} e^{xr} \left[\psi(r) R(r) - \frac{d}{dr}\left(\phi(r) R(r) \right) \right] dr = 0$

or $\quad \dfrac{d}{dr}\left\{ \phi(r) R(r) \right\} = \psi(r) R(r)$

or $\dfrac{d}{dr}\left\{ \phi(r) R(r) \right\} = \dfrac{\psi(r)}{\phi(r)} \phi(r) R(r)$

so
$\qquad \phi(r) R(r) = A\, e^{\int \frac{\psi(r)}{\phi(r)} dr}$

$$\boxed{R(r) = \dfrac{A}{\phi(r)}\, e^{\int \frac{\psi(r)}{\phi(r)} dr}}$$

$$y = A \int \frac{1}{\phi_{(r)}} e^{(xr + \int \frac{\phi}{\phi} \, dr)} \, dr$$

$$\boxed{y = A \int \frac{\exp[xr + \int \frac{\phi}{\phi} \, dr]}{\phi_{(r)}}}$$

Which is the definite integral solution of the linear diff. eq. of the second order.

The limits of integrat α, β are determined s.t

$$\left[e^{xr} \phi_{(r)} R_{(r)} \right]_{\alpha}^{\beta} = 0$$

Example 1:

Ex

$$\frac{d^n y}{dx^n} - xy = a \qquad \textcircled{1}$$

$$\text{let} \quad y = \int_{\alpha}^{\beta} e^{xr} R_{(r)} \, dr \qquad \textcircled{2}$$

$$D^n y = \int_{\alpha}^{\beta} r^n e^{xr} R_{(r)} \, dr \qquad \textcircled{3}$$

So eqn $\textcircled{1}$ become

$$\int_{\alpha}^{\beta} r^n e^{xr} R_{(r)} \, dr - \int_{\alpha}^{\beta} x e^{xr} R_{(r)} \, dr =$$

$$\int_{\alpha}^{\beta} r^n e^{xr} R \, dr - \int_{\alpha}^{\beta} \frac{x}{x} R \, de^{xr} =$$

$$\int_{\alpha}^{\beta} r^n e^{xr} R \, dr - \left[R e^{xr} \right]_{\alpha}^{\beta} + \int_{\alpha}^{\beta} e^{xr} \frac{dR}{dr}$$

$$\int_{\alpha}^{\beta} e^{xr} \left[\frac{dR}{dr} + r^n R \right] dr - \left[e^{xr} R \right]_{\alpha}^{\beta} =$$

$$\therefore \quad \frac{dR}{dr} + r^n R = a \qquad \textcircled{4}$$

313

$$\& \left[e^{xr} R(r) \right]_\alpha = a \qquad \text{(5)}$$

$$\text{so} \quad R = A e^{-\frac{r^{n+1}}{n+1}} \qquad \text{(6)}$$

so (5) becomes

$$\left[A e^{\left(xr - \frac{r^{n+1}}{n+1} \right)} \right]_\alpha^\beta = a$$

or

$$A e^{xr} e^{-\frac{r^{n+1}}{n+1}} = -\mu$$

$$r \to \infty$$
$$\mu = a, \quad r = a, \quad \mu = -A$$

$$y = A \int_0^\infty e^{xr - \frac{r^{n+1}}{n+1}} \, dr$$

If $r = \omega r$

$$y = A \int_0^\infty e^{\left(x\omega r - \frac{\omega^{n+1} r^{n+1}}{n+1} \right)} \, d\omega r$$

$$\therefore \frac{d^n y}{dr^n} - xy = a$$

$$\sum A \int_0^\infty r^n e^{xr - \frac{r^{n+1}}{n+1}} \, dr - \sum A \int x e^{xr - \frac{r^{n+1}}{n+1}} = a$$

if we replace $r = r\omega$

$$\sum A_i \omega \int_0^\infty (\omega)^n r^n e^{x\omega r} e^{-\frac{r^{n+1}}{n+1}} \, dr$$

$$- \sum \omega A \int_0^\infty x e^{x\omega r} e^{-\frac{r^{n+1} \cdot \omega^{n+1}}{n+1}} \, dr = a$$

$$\sum A_i \int_0^\infty r^n e^{x\omega r - \frac{r^{n+1}}{n+1}} \, dr - \sum A_i x \int_0^\infty e^{\frac{r^{n+1}}{n+1}} \, d = a$$

$$\sum A_i \int_0^\infty r^n e^{x\omega r - \frac{r^{n+1}}{n+1}} \, dr - \left[\sum A_i e^{x\omega r - \frac{r^{n+1}}{n+1}} \right]_0^\infty - \int \sum A_i r^n e^{\cdots} = a$$

314

$$\& \quad -\sum A_i e^{\left. \omega r - \frac{r}{\omega r}\right\rbrace_b} = a$$

$$i.e \ \sum A_i = a$$

Example 2:

$$EX$$

$$x\frac{d^2y}{dx^2} + a\frac{dy}{dx} - b^2xy = 0 \qquad (1)$$

$$y = \int_\alpha^\beta e^{xr} R(r)\,dr \qquad (2)$$

$$x\int_\alpha^\beta r^2 e^{xr} R\,dr + a\int_\alpha^\beta r e^{xr} R\,dr - bx\int_\alpha^\beta e^{xr} R_r$$
$$= 0$$
$$\int_\alpha^\beta x(r^2-b^2)e^{xr} R\,dr + a\int_\alpha^\beta r e^{xr} R\,dr = 0$$

$$\int_\alpha^\beta (r^2-b^2)R\,d e^{xr} + a\int_\alpha^\beta r e^{xr} R\,dr = 0$$
$$\left[e^{xr}(r^2-b^2)R\right]_\alpha^\beta - \int_\alpha^\beta e^{xr}\frac{d}{dr}(r^2-b^2)R\,dr$$
$$+ \int_\alpha^\beta a r e^{xr} R\,dr = 0$$

$$Put \ \left[e^{xr}(r^2-b^2)R\right]_\alpha^\beta = 0 \qquad (3)$$
$$so \ \int_\alpha^\beta e^{xr}\left[\frac{d}{dr}(r^2-b^2)R - arR\right]dr = 0$$

$$\therefore \quad \frac{d}{dr}[r^2-b^2]R - arR = 0$$
$$(r^2-b^2)\frac{dR}{dr} + 2rR - arR = 0$$
$$\frac{dR}{R} = \frac{(a-2)r}{r^2-b^2}\,dr$$

315

$$\log R = \tfrac{1}{2}(a-1)\log(r^2-b^2)$$

$$\boxed{R = A\,(r^2-b^2)^{\frac{a-2}{2}}}$$

$$\text{in 3}\quad \left[e^{xr}(r^2-b^2)^{\frac{a-2}{2}+1}\right]_{\alpha}^{\beta} = C$$

$$\left[(r^2-b^2)^{\frac{a}{2}}e^{xr}\right]_{\alpha}^{r=\beta} = C$$

if $r \pm b \Rightarrow C \to 0,\ -\infty,$

The sol^n is

$$\boxed{\begin{aligned} y &= A \int_{b}^{-b} e^{xr}(r^2-b^2)^{\frac{a}{2}-1}\,dr \\ &\quad + B \int_{-b}^{\infty} e^{xr}(r^2-b^2)^{\frac{a}{2}-1}\,dr \end{aligned}}$$

Example 3:

Example

$$\boxed{\begin{aligned} (D^2+n^2)\,y &= fx \\ \frac{d^2y}{dx^2} + n^2 &= f(x) \end{aligned}}$$

$$y = \frac{1}{D^2+n^2}\, f(x)$$

$$= \frac{1}{(D-in)(D+in)}\, f(x)$$

$$y = \frac{1}{2ni}\left[\frac{1}{D-ni}f(x) - \frac{1}{D+ni}f(x)\right]$$

$$= \frac{1}{2ni}\left[\frac{1}{D-ni}e^{inx}e^{-inx}f(x) - \frac{1}{D+ni}e^{inx}e^{-inx}f(x)\right]$$

$$= \frac{1}{2ni}\left[e^{inx}\frac{1}{D}e^{-inx}f(x) - e^{-inx}\frac{1}{D}e^{inx}f(x)\right]$$

316

$$= \frac{1}{2\pi i}\left[\int_0^x e^{in(x-t)} f(t)\,dt - \int_0^x e^{-in(x-t)} f(t)\,dt\right]$$

$$= \frac{1}{2\pi i}\left[\int_0^x \left[e^{in(x-t)} - e^{-in(x-t)}\right] f(t)\,dt\right]$$

$$y = \frac{1}{2\pi i}\left[\int_0^x 2i\,\sin(n(x-t))\, f(t)\,dt\right]$$

$$\boxed{y = \frac{1}{n}\int_0^x \sin n(x-t)\, f(t)\,dt}$$

$$\left(D^2 + n^2\right)^m y = \cos n x$$

$$y = \frac{1}{\left(D^2 + n^2\right)^m}\cos n x$$

$$= \frac{1}{\left(D - in\right)^m\left(D + in\right)^m}\cos n x$$

$$= \frac{1}{\left(D - in\right)^m\left(D + in\right)^m}\, e^{inx}\, e^{-inx}\,\cos n x$$

$$= \frac{1}{\left(D - in\right)^m\, D^m}\, e^{-inx}\, e^{inx}\,\cos n x$$

$$= \frac{1}{\left(D - in\right)^m\left(D + in\right)^m}\left(\frac{e^{inx} + e^{-inx}}{2}\right)$$

$$= \frac{1}{\left(D - in\right)^m\left(D + in\right)^m}\frac{e^{inx}}{2} + \frac{1}{\left(D - in\right)^m\left(D + in\right)^m}\frac{e^{-inx}}{2}$$

$$= \frac{1}{\left(D - in\right)^m}\frac{e^{inx}}{(2in)^m\, 2}$$

317

$$= \frac{e^{inx}}{2(in)^n} \overline{D^{-n}} (1) + \frac{1}{(D+in)^n} \frac{e^{-inx}}{2(-2in)^n}$$

$$= \frac{e^{inx}}{2(2in)^n} \frac{1}{\overline{D^n}} + \frac{e^{-inx}}{2(in)^n} \frac{1}{\overline{D^n}}$$

$$= \frac{1}{2} \left[\frac{e^{inx} + e^{-inx}}{(-1)^n} \right] \frac{1}{(2in)^n} \frac{1}{\overline{D^n}}$$

$$= \frac{1}{2(2in)^n} \left[e^{inx} + \frac{e^{-inx}}{(-1)^n} \right] \left[A_0 + A_1 x + A_2 x^2 + \dots + A_n x^n \right]$$

12.1.2.8. The linear differential equation of second order with constant coefficients

The linear diff^l eq^n of second order with const. coeff

$$a \frac{d^2 y}{dx^2} + b \frac{dy}{dx} + cy = f(x).$$

This eq^n is a fundamental in the study of vibrations whether mechanical, electrical, acoustical or ..... etc.

The complementary f^n :–

This of a general solution of:

$$a \frac{d^2 y}{dx^2} + b \frac{dy}{dx} + c \cdot y = 0 \quad \text{——①}$$

Containing two arbitrary const^s we assume a trial sol^n

$$y = Ae^{mx} \quad \text{——②}$$

Because e^{my} remains unchanged only m is multiplied.

$$\frac{dy}{dx} = A m e^{mx} \quad \text{——③}$$

$$\frac{d^2 y}{dx^2} = A m^2 e^{mx} \quad \text{——④}$$

Substituting in eq^n ① :–

$$\therefore a m^2 + b m + c = 0$$

Roots of this auxiliary eq^n are sol^n :–

$$\therefore \quad y = Ae^{m_1 x} + Be^{m_2 x} \quad \text{——⑤}$$

providing that m_1 & m_2 are distinct.

12.1.2.8.1. Real and distinct coefficients

Case I. m_1, m_2 are real and distinct.

Ex.1. Solve the eq^n $\dfrac{d^2y}{dx^2} + 7\dfrac{dy}{dx} + 12y = 0$

$$m^2 + 7m + 12 = 0$$
$$(m+3)(m+4) = 0$$
$$m = -3, -4$$

Sol^n
$$y = A e^{-3x} + B e^{-4x}.$$

Ex. 2.

$$\frac{d^2y}{dx^2} + 4\frac{dy}{dx} + 2y = 0$$

Auxiliary eq^n
$$m^2 + 4m + 2 = 0$$
$$m = -\frac{4 \pm \sqrt{16-8}}{2} = -2 \pm \sqrt{2}$$

Sol^n
$$y = A e^{(-2+\sqrt{2})x} + B e^{(-2-\sqrt{2})x} \longrightarrow ①$$
$$y = e^{-2x}\left(A e^{\sqrt{2}x} + B e^{-\sqrt{2}x}\right) \longrightarrow ②$$

$$= e^{-2x}\left[A(\cosh \sqrt{2}x + \sinh \sqrt{2}x) + B(\cosh \sqrt{2}x - \sinh \sqrt{2}x)\right]$$
$$y = e^{-2x}\left[C\cosh \sqrt{2}x + D\sinh \sqrt{2}x\right] \longrightarrow ③$$

12.1.2.8.2. Equal roots of auxiliary equation

Case II Roots of aux. eq^n are equal.

Let m, m aux. eq^n $am^2 + bm + c = 0 \longrightarrow ①$
$$\therefore m + m = -\frac{b}{a}$$
$$2ma + b = 0$$

we now assumed as a sol^n of the eq^n if
$$y = e^{mx} \cdot u \longrightarrow ②$$
$$\therefore \frac{dy}{dx} = e^{mx}\frac{du}{dx} + m e^{mx} \cdot u \longrightarrow ③$$
$$\frac{d^2y}{dx^2} = m e^{mx}\frac{du}{dx} + e^{mx}\frac{d^2u}{dx^2} + m e^{mx}\frac{du}{dx} + m^2 e^{mx} u$$

319

$$\cdot \quad \frac{d^2y}{dx^2} = e^{mx}\frac{d^2u}{dx^2} + 2ae^{mx}\frac{du}{dx} + m^2e^{mx}u \quad -\textcircled{3}$$

Subt. in the d. eq^n.

$$a\left[\frac{d^2u}{dx^2} + 2m\frac{du}{dx} + m^2u\right] + b\left[\frac{du}{dx} + mu\right] + cu =$$

$$= a\frac{d^2u}{dx^2} + (2am+b)\frac{du}{dx} + (am^2 + bm + c)u = 0$$

$$\therefore \frac{d^2u}{dx^2} = 0 \qquad \therefore u = A + Bx$$

$$\text{sol}^n \qquad y = e^{mx}(A + Bx)$$

Ex. $\qquad \dfrac{d^2y}{dx^2} + 4\dfrac{dy}{dx} + 4y = 0$

aux. eq^n = $m^2 + 4m + 4 = 0$

$$(m+2)^2 = 0 \qquad m = -2, -2$$

Sol^n. $\qquad y = e^{-2x}(A + Bx)$.

12.1.2.8.3. Complex imaginary roots of auxiliary equation

Case III. Roots of aux. eq^n are complex imaginary :-

Roots $\qquad \alpha \pm i\beta$

The sol^n

$$y = Ae^{(\alpha + i\beta)x} + Be^{(\alpha - i\beta)x}$$

$$= e^{\alpha x}\left(Ae^{i\beta x} + Be^{-i\beta x}\right)$$

$$= e^{\alpha x}\left[A(\cos\beta x + i\sin\beta x) + B(\cos\ldots\right.$$

$$y = e^{\alpha x}[C\cos\beta x + iR\sin\beta x].$$

This sol^n can also be written

$$y = Re^{\alpha x}\sin(\beta x + \varepsilon)$$

Ex.-. $\qquad \dfrac{d^2x}{dt^2} + 2k\dfrac{dx}{dt} + (n^2 + k^2)x$

aux. eq^n $\qquad m^2 + 2km + (n^2 + k^2)$

$$m = \frac{-2k \pm \sqrt{4k^2 - 4n^2 - 4k^2}}{2}$$

$$m = -k \pm ni$$

$$x = e^{-kt}[A\cos nt + B\sin nt]$$

or $\qquad x = Re^{-kt}\sin(nt + \varepsilon)$

$\qquad\qquad\qquad\qquad$ *PHASE ANGLE.*

Extension of the above methods to eq^n of order higher than 2:-

Ex. 1 $\dfrac{d^3y}{dx^3} - 6\dfrac{d^2y}{dx^2} + 11\dfrac{dy}{dx} - 6y = 0$

aux. eq^n $m^3 - 6m^2 + 11m - 6 = 0$
        $m = 1, 2, 3$

soln $y = Ae^x + Be^{2x} + Ce^{3x}$.

Ex. 2. $\dfrac{d^3y}{dx^3} - 4\dfrac{d^2y}{dx^2} + 5\dfrac{dy}{dx} - 2y = 0$

        $m^3 - 4m^2 + 5m - 2 = 0$
        $(m-2)(m-1)^2 = 0$
        $m = 2, 1, 1$

soln $y = A e^{2x} + e^x(B + Cx)$.

Ex. 3. $\dfrac{d^3y}{dx^3} - 8y = 0$

aux. eq^n $m^3 - 8 = 0$
        $(m-2)(m^2 + 2m + 4) = 0$

        $m = 2 \quad 6 \quad \dfrac{-2 \pm \sqrt{4 - 16}}{2}$

        $m = 2 \quad 6 \quad -1 \pm \sqrt{3}\,i$

$\therefore y = A e^{2x} + Re^{-x} \sin(\sqrt{3}x + \epsilon)$

Ex. 4. $(D-2)^4 y = 0$ Roots $2, 2, 2, 2$
        $y = e^{2x}[A + Bx + Cx^2 + Dx^3]$

12.1.2.9. Uniqueness and existence of solution of second order differential equations

2nd Order eqns

$f(t, y, z)$ 3 variables

a, b, β constants $\beta > 0$

$$\left\{ \begin{array}{l} y'' = F(t, y, y') \\ y(0) = a, \quad y'(0) = b \end{array} \right\} \quad -\text{(a)}$$

$$0 \leq t \leq \beta$$

y unknown fn of t

(a) $y'' = F(t, y, y')$
$$y(0) = a, \quad y'(0) = b$$
$$\beta > 0$$

(b) **Hypothesis**

$\forall \ 0 \leq t \leq \beta$, all y & all z the
$f = F(t, y, z)$ is assumed to

- exist
- continuous int
- partially differentiable in y,

& $|F_y(t, y, z)| \leq k, |F_z(t, y, z)| \leq \ell$

k, ℓ constants.

__Theorem A__   Existence:

Under (b) system a has at least
one sol.

__Theorem B__   Uniqueness

Under (b) system a has at
most one sol.

__Theorem C__   Approximat.

Under (b) system a can be
solved by successive approxim.

322

c) $\begin{cases} y_2(t) = a + bt \\ y''_{n+1}(t) = F\left(t, x_n(t), y_1(t)\right) \\ y_{m+1}(0) = a, \; y'_{n+1}(0) = b \end{cases}$

Proof

For any 5 $n \geq 0$ t, y_1, y_2, z_1, z_2
$0 \leq t \leq \beta$

$|F(t, y_1, z_1) - F(t, y_2, z_2)| \leq k |y_2 - y_1|$
$\qquad\qquad\qquad\qquad + \ell |z_2 - z_1|$ $\quad\text{①}$

by mean value theorem

step ①

Our choice of approx $=$
is given in Theorem 5

step ②

we transform ⓒ into more convenient eqs

$C \; \begin{cases} y_0(t) = a + bt \\ y''_{n+1}(t) = F\left(t, y_n(t), y_n'(t)\right) \\ y_{m+1}(0) = a, \; y'_{n+1}(0) = b \end{cases}$

$z = y', \quad u = y''$

Then $u_0(t) = 0$
$\qquad y_n(t) = a + \int^t z_n(s)\,ds$
$\qquad z_n(t) = b + \int^t u_n(s)\,ds$

$u_{n+1}(t) = F\left(t, y_n(t), z_n(t)\right)$

This gives in particular $z_0(t) = b$
$\qquad y_0(t) = a + bt$
$\qquad u_1(t) = F(t, y_0, z_0) = F(t, a + bt, b)$

323

<u>3 ci step</u>

We convert the sequence $\{U_n(t)\}$,
$\{Z_n(t)\}$, $\{y_n(t)\}$ into series

$$e \begin{cases} U_N(t) = u_1(t) + \sum_{n=1}^{N} \left[U_n(t) - U_{n-1}(t) \right] \\ Z_N(t) = b + \sum_{n=1}^{N} \left(Z_n - Z_{n-1} \right) \\ y_n(t) = a + bt + \sum_{n=1}^{N} \left(y_n - y_{n-1} \right) \end{cases}$$

in equation test items

$$f \begin{cases} \left| Z_m - Z_{m-1} \right| \le \int_0^t \left| U_{n-1}^{(s)} - U_{n-1}^{(s)} \right| ds \\ \left| y_m - y_{m-1} \right| \le \int_0^t \left| Z_n - Z_{m-1} \right| ds \\ \left| U_{m+1} - U_m \right| \le k \left| y_m - y_{m-1} \right| + \ell \left| Z_n - Z_{m-1} \right| \end{cases}$$

Comparison series, let h, K

$|U_1(t)| \le h$, $K = k \frac{\beta}{2} + \ell$

as $U_1(t) = F(t, a+bt, b)$ continous
                            bounded

⑨ <u>Prove</u> $\left| U_{m+1}(t) - U_n(t) \right| \le \frac{h K^n t}{n!}$

Ⅰ. if $n = 0$
$$\left| U_1(t) - 0 \right| \le h$$

which is true by def of h

Ⅱ. let m be   by ⑥ ⑦
$$\left| Z_{m+1}(t) - Z_m(t) \right| \le \frac{h K^m}{m!} \int_0^t s^m ds$$
$$\le \frac{h K^m t^{m+1}}{(m+1)!}$$

324

assuming integrability

$$\left| \left[b + \int_0^t U(s)ds \right] - Z_n(t) \right| = \left| \int_0^t [z_n - u_n(s)] ds \right|$$

$$\leq h R_n \beta$$

$$\left| \left[a + \int_0^t Z_n(s) \right] - Y_n(t) \right| = \left| \int_0^t [z_n - z_n(s)] ds \right|$$

$$\leq \frac{h}{k} R_{n+1} \beta$$

$$\left| f(t, Y(t), Z(t)) - U_{n+1} \right| =$$

$$\left| f(t, y, z) - f(t, y_n, z_n) \right| \leq$$

$$\leq k |y - y_n| + \ell |z - z_n| \leq$$

$$\leq \frac{k}{k} h R_{n+1} + \frac{\ell h}{k} R_{n+1} \quad \xrightarrow[n \to \infty]{}$$

step 6

prove uniqueness

If $\overline{Y}(t), \overline{Y}(t)$ are 2 solutions @

then Y, Z, U & $\overline{Y}, \overline{Z}, \overline{U}$ are solutions

(I) $Y_{(t)} - \overline{Y}_{(t)} = \int_0^t \left\{ Z_{(s)} - \overline{Z}_{(s)} \right\} ds$

$Z_{(t)} - \overline{Z}_{(t)} = \int_0^t \left(U_{(s)} - \overline{U}_{(s)} \right) ds$

$U_{(t)} - \overline{U}_{(t)} = f(t, Y_{(t)}, Z_{(t)}) - f(t, \overline{Y}_{(t)}, \overline{Z}_{(t)})$

assume f:

$|U_{(t)} - \overline{U}_{(t)}| \leq h \quad \forall s \leq t \leq \beta$

we can prove

$$\left| U(t) - \overline{U}(t) \right| < h \frac{k^n t^n}{n!}$$

hence $U(t) = \overline{U}(t)$

& $Y(t) = \overline{Y}(t)$

we prove existence & uniqueness

12.1.2.10. Differential equations of second order with variable coefficients

D. E of 2nd O. with variable coeffts

$$\frac{d^2 y}{dx^2} + P(x) \frac{dy}{dx} + Q(x) y = R(x) \quad ①$$

sol^n

$$y = UV \quad ②$$

$$\frac{dy}{dx} = U \frac{dV}{dx} + V \frac{dU}{dx}$$

$$\frac{d^2 y}{dx^2} = U \frac{d^2 V}{dx^2} + 2 \frac{dV}{dx} \frac{dU}{dx} + V \frac{d^2 U}{dx^2}$$

eq ① become

$$U \frac{d^2 V}{dx^2} + 2 \frac{dV}{dx} \frac{dU}{dx} + V \frac{d^2 U}{dx^2} + P(x) \left[U \frac{dV}{dx} + V \frac{dU}{dx} \right]$$
$$+ Q(x) UV = R(x)$$

so
$$V \left[\frac{d^2 U}{dx^2} + P(x) \frac{dU}{dx} + Q(x) U \right] + U \frac{d^2 V}{dx^2} + 2 \frac{dV}{dx} \frac{dU}{dx}$$
$$+ P(x) U \frac{dV}{dx} = R(x) \quad ③$$

326

if U satisfies

i.e $\boxed{\dfrac{d^2U}{dx^2} + P\dfrac{dU}{dx} + Q_{(x)}U = 0}$ —— ④

So $U\dfrac{d^2V}{dx^2} + 2\dfrac{dV}{dx}\dfrac{dU}{dx} + UP_{(x)}\dfrac{dV}{dx} = R_{(x)}$

or $\dfrac{d^2V}{dx^2} + \dfrac{dV}{dx}\left[\dfrac{2}{U}\dfrac{dU}{dx} + P_{(x)}\right] = \dfrac{R_{(x)}}{U}$ ⑤

denote $P_1{(x)} = \dfrac{2}{U}\dfrac{dU}{dx} + P_{(x)}$

     & $R_1 = \dfrac{R_{(x)}}{U}$

so ⑤ become

$\boxed{\dfrac{d^2V}{dx^2} + P_1\dfrac{dV}{dx} = R_1}$ ⑥

Take $W = \dfrac{dV}{dx}$, $\dfrac{dW}{dx} = \dfrac{d^2V}{dx^2}$

Eq: ⑥ become

$\boxed{\dfrac{dW}{dx} + P_1 W = R_1}$ ⑦

Example

$x\,y'' - (\,2x+1\,)\,y' + (\,x+1)\,y = 0$

Example

Prove that $e^x = y$ is a sol of

$x\dfrac{d^2y}{dx^2} - (2x+1)\dfrac{dy}{dx} + (x+1)y = 0$ ①

hence find comple sol of

$x\dfrac{d^2y}{dx^2} - (2x+1)\dfrac{dy}{dx} + (x+1)y = (x+1)$

     ②

$$\underline{\text{soln}}$$

$$y' = e^x, \quad y'' = e^x$$

so ① become

$$xe^x - (2x+1)e^x + (x+1)e^x$$
$$= 0$$

$\therefore$ $y = e^x$ is a soln of eqn ①

suppose soln $y = uV = ue^x$ —③

$$\frac{dy}{dx} = ue^x + \frac{du}{dx}e^x \qquad = uV$$

$$\frac{dy}{dx} = u\frac{dV}{dx} + V\frac{du}{dx}$$

$$\frac{d^2y}{dx^2} = u\frac{d^2V}{dx^2} + 2\frac{du}{dx}\frac{dV}{dx} + V\frac{d^2u}{dx^2}$$

$$\left(u\frac{d^2V}{dx^2} - 2\frac{du}{dx}\frac{dV}{dx} + V\frac{d^2u}{dx^2}\right) - \frac{2x+1}{x}\left(u\frac{dV}{dx} + V\frac{du}{dx}\right)$$
$$+ \frac{(x+1)}{x}uV = \frac{x^2+x-1}{x}e^x$$

$$U\left[\frac{d^2V}{dx^2} + P\frac{dV}{dx} + QV\right] + V\frac{d^2u}{dx^2} + \frac{du}{dx}\left[\frac{2dV}{dx} + P\right]$$
$$0 \qquad = R(x)$$

$$P = -\frac{(2x+1)}{x} \qquad V = e^x$$
$$Q = \frac{x+1}{x} \qquad \frac{dV}{dx} = e^x$$
$$R = \frac{x^2+x-1}{x}e^x$$

$$\frac{d^2u}{dx^2} + \frac{du}{dx}\left[\frac{2}{V}\frac{dV}{dx} + P\right] = \frac{R}{V} \quad -④$$

$$P_1 = \frac{2}{V}\frac{dV}{dx} + P = \frac{2}{e^x}e^x - \frac{2x+1}{x}$$

$$\boxed{P_1 = -\frac{1}{x}}$$

$$R_1 = \frac{R}{V} = \frac{x^2+x-1}{x}\cdot\frac{e^x}{e^x}$$

$$W = \frac{du}{dx} \;,\quad \frac{dW}{dx} = \frac{d^2u}{dx^2}$$

so

$$\frac{dW}{dx} + P_1 W = R_1$$

$$\int P_1\,dx$$

$$dWe + eP_1 W\,dx = R_1\,dx\,e^{\int P_1 dx}$$

$$d\left(We^{\int P_1 dx}\right) = R_1 e^{\int P_1 dx}\,dx$$

$$We^{-\log x} = \int \frac{x^2+x-1}{x}\,e^{-\log x}\,dx + C$$

$$\frac{W}{x} = \int \frac{x^2+x-1}{x}\,dx + C$$

$$\frac{W}{x} = x + \log x + \frac{1}{x} + C$$

$$W = x^2 + x\log x + C_1 x$$

$$u = \int W\,dx = \frac{x^3}{3} + C_1\frac{x^2}{2} + \int x\log x\,dx + C_2$$

$$\int x\log x\,dx = \int \log x\,\frac{dx^2}{2} = \left(\frac{x}{2}\log x\right) - \frac{1}{2}\int x^2\,d\log x$$

$$= \frac{x^2}{2}\log x - \frac{1}{2}\int x\,dx$$

$$= \frac{x^2}{2}\log x - \frac{1}{2}\frac{x^2}{2} + C$$

329

$$U = \frac{x^3}{2} + q\frac{x^2}{2} + \frac{x^2}{2}\log x - \frac{1}{2}\frac{x^2}{2} + C$$

$$= \frac{x^2}{2}\left[x + C_1 + \log x - \frac{1}{2}\right]$$

$$y = UV = e^x \frac{x^2}{2}\left[x + \log x + C\right]$$

II. Normal form

$$y'' + P_{(x)}y' + Q_{(x)}y = R_{(x)} \quad ①$$

$$y = UV \quad —② $$
$$\frac{dy}{dx} = uv' + vu' \quad \Big\} ③$$
$$\therefore \frac{d^2y}{dx^2} = uv'' + 2u'v' + vu''$$

eq ① become

$$uv'' + 2u'v' + vu'' + P(uv' + vu') + Quv = R$$

$$v'' + v'\left[\frac{2u'}{u} + P\right] + v\left[\frac{u'' + Pu' + Qu}{u}\right] = \frac{R}{u} \quad ④$$

Denote
$$\begin{cases} P_1 = \frac{2u'}{u} + P \quad \text{⑤} \\[2mm] Q_1 = \frac{1}{u}\left[u'' + Pu' + Qu\right] \\[2mm] R_1 = \frac{R}{u} \end{cases}$$

Put $P_1 = 0$

so $\frac{2}{u}\frac{du}{dx} = -P$

$$\frac{du}{u} = -\frac{P}{2}dx$$

$$\boxed{u = e^{-\int \frac{P}{2}dx}} \quad ⑥$$

330

$$u' = -\frac{P}{2} e^{-\int \frac{P}{2} dx} \quad \textcircled{7}$$

$$u'' = \frac{P^2}{4} e^{-\int \frac{P}{2} dx} - \frac{dP}{dx} \frac{e^{-\int \frac{P}{2}d}}{2} \longrightarrow \textcircled{8}$$

so

$$Q_1 = \frac{P^2}{4} - \frac{1}{2}\frac{dP}{dx} + \left(\frac{-P^2}{2}\right) + Q$$

$$\boxed{Q_1 = Q - \frac{P^2}{4} - \frac{1}{2}\frac{dP}{dx}} \longrightarrow \textcircled{9}$$

$$V'' + Q_1 V = R_1 \longrightarrow \textcircled{10}$$

consider $W = V'$

Example

Transforming to Normal form solve

$$\frac{d^2y}{dx^2} + 2x\frac{dy}{dx} + (x^2-8)\, y = 81 x^2 e^{\frac{x^2}{2}}$$

Soln

$$y'' + P y' + Q y = R$$

$$y = uv$$

$$\frac{dy}{dx} = u\frac{dv}{dx} + v\frac{du}{dx}$$

$$\frac{d^2y}{dx^2} = u\frac{d^2v}{dx^2} + 2\frac{du}{dx}\cdot\frac{dv}{dx} + v\frac{d^2u}{dx^2}$$

$$(uv'' + 2u'v' + vu'') + P(uv' + vu') + Quv = R$$

$$v'' + v'\left[\frac{2u' + Pu}{u}\right] + v\left[\frac{u'' + Pu' + Qu}{u}\right] = \frac{R}{u}$$

$$v'' + P_1 v' + Q_1 v = R_1 \longrightarrow \textcircled{1}$$

$$P_1 = \frac{2u'}{u} + P = 0$$

331

$$u = e^{-\int \frac{P}{2}dx} = e^{-\int \frac{x}{2}dx} = \boxed{e^{-\frac{x^2}{2}}}$$

$$u^\prime = e^{-\frac{x^2}{2}}(-x)$$

$$u^{\prime\prime} = x^2 e^{-\frac{x^2}{2}} - e^{-\frac{x^2}{2}} = e^{-\frac{x^2}{2}}[x^2-1]$$

$$R_1 = \frac{81 x^2 e^{-x^2/2}}{e^{-x^2/2}} = 81x^2 \quad\text{———}\boxed{2}$$

$$Q_1 = (x^2-1) + 2x(-x) + (x^2-8)$$

$$= -9 \quad\text{———}\boxed{3}$$

so

$$V^{\prime\prime} - 9V = 81x^2$$

$$\boxed{(D^2 - 9)V = 81x^2} \quad\text{—}\boxed{4}$$

$\underline{C.F}$ $\quad V = Ae^{-3x} + Be^{3x}$

$\underline{P.I}$ $\quad V = \dfrac{1}{D^2-9} \, 81 x^2$

$$= \frac{81}{-9} \frac{1}{(1+\frac{D}{3})(1-\frac{D}{3})} x^2$$

$$= -9 \frac{1}{(1+\frac{D}{3})}(1 + \frac{D}{3} + \frac{D^2}{9} + \cdots)x^2$$

$$= -9 \frac{1}{1+\frac{D}{3}}\left[x^2 + \frac{2y}{3} + \frac{2}{9}\right]$$

$$= -9 \left[1 - \frac{D}{3} + \frac{D^2}{9}\right]\left(x^2 + \frac{2x}{3} + \frac{2}{9}\right)$$

$$= -9 \left[x^2 + \frac{2x}{3} + \frac{2}{9} - \frac{2}{3}x - \frac{2}{9} + \frac{2}{9}\right]$$

$$= -9x^2 - 2$$

$\underline{G.S}$ $\quad V = Ae^{-3x} + Be^{3x} - (9x^2+2)$

$$y = Vu = e^{-\frac{x^2}{2}}\left[Ae^{-3x} + Be^{3x} - 9x^2 - 1\right]$$

Solving by polynomial expansions

$$y = 3(e^t = 3X \ln x$$

$$\underline{C.S} \quad y = AX + \beta X \cos(\ln x) + \xi X \sin(\ln x)$$

$$+ 3X \ln x$$

12.1.2.11. Operational integration using the Differential Operator D = d/dx

The operator D stands for $\frac{d}{dx}$; D^2 for $\frac{d^2}{dx^2}$ ... $D^n \frac{d^n}{dx^n}$

Then the expression $a_o \frac{d^n y}{dx^n} + a_1 \frac{d^{n-1} y}{dx^{n-1}} \cdots + a_n \frac{dy}{dx} + a_n y$

can be written in the form $[a_o D^n + a_1 D^{n-1} + \cdots + a_{n-1} D + a_n]y$

Now let $u \approx v$ be functions of x & let "a" constant. Then

$$D(u + v) = Du + Dv$$

I i.e. D obeys the distributive law.

again $D a u = a D u$

Hence the operator D is commutative with constant multipliers.

II i.e. D obeys the commutative law

again we have $D^m \cdot D^n u = D^{m+n} u$

III i.e. D obeys the index law

From the above 3 it follows: The operator D is commutative with multipliers obeys the fundamental laws of ordinary Algebra so far as operating with D goes.

I. Let $\psi(D) = a_o D^n + a_1 D^{n-1} + a_{n-1} D + a_n$

$$\psi(D) e^{\lambda x} = a_o \lambda^n e^{\lambda x} + a_1 \lambda^{n-1} e^{\lambda x} \cdots$$

$$\therefore \psi(D) e^{\lambda x} = \psi(\lambda) e^{\lambda x}$$

i.e. λ takes place of D $\underline{\psi(D) e^{\lambda x} = \psi(\lambda) e^{\lambda x}}$

II with the same meaning of $\psi(D) e^{\lambda x} V = e^{\lambda x}(D.\lambda) V$

$$\psi(D) e^{\lambda x} \cdot V = e^{\lambda x}(D + \lambda) V.$$

e.g. $D^5 e^{2x} \sin 3x = 2 e^{2x} \sin 3x + e^{2x} D \sin 3x$

$$= e^{2x}(D + 2)^5 \sin 3x$$

$$D e^{\lambda x} V = e^{\lambda x} DV + \lambda e^{\lambda x} V.$$
$$= e^{\lambda x} (D + \lambda) V$$

$$D^2 e^{\lambda x} V = D(e^{\lambda x} DV + \lambda e^{\lambda x} V)$$
$$= (\lambda e^{\lambda x} DV + e^{\lambda x} D^2 V) +$$
$$\lambda^2 e^{\lambda x} V + \lambda e^{\lambda x} DV)$$

$$= e^{\lambda x} (D^2 + \lambda D) V + \lambda e^{\lambda x} (\lambda + D) V$$
$$= e^{\lambda x} (D^2 + \lambda D + \lambda^2 + \lambda D) V$$
$$= e^{\lambda x} (D + \lambda)^2 V$$

i.e $D + \lambda$ takes place of D $\quad \underline{D^n e^{\lambda x} V = e^{\lambda x} (D + \lambda)^n V}$
operating on V

III. $\quad \psi(D^2) \cos \lambda x = \psi(-\lambda) \cos \lambda x$

$\quad D \cos \lambda x = -\lambda \sin \lambda x.$
$\quad D^2 \cos \lambda x = -\lambda^2 \cos \lambda x$

i.e. $-\lambda^2$ takes place of D^2 $\quad \underline{\psi(D^2) \cos \lambda x = \psi(-\lambda^2)}_{\cos x}$

IV. $\quad \psi(D^2) \sin \lambda x = \psi(-\lambda^2) \sin \lambda x.$
i.e $-\lambda^2$ takes place of D^2
V. $\quad$ If $I y = \int y \, dx \quad \psi(D^2) \sin \lambda x = \psi(-\lambda^2) \sin x$
$\quad D I y = y \quad " \quad D I = 1 \quad " D = \frac{1}{I}$
$$\underline{\frac{1}{D} = I}$$
$\frac{1}{D} = I$ That is $\frac{1}{D}$ is a process of
Integration

The particular integral $(P.I)$ of y

This depends on the nature of the y's in the R.H.S.

① R.H.S = const.
$\quad$ ex. $\frac{d^2 y}{dx^2} + 5 \frac{dy}{dx} + 6 y = 12$

$\quad (D^2 + 5D + 6) y = 12.$
$\quad (D + 2)(D + 3) y = 12$

$\quad D = -2, -3 \quad$ Roots of Aux. eq$\underline{^n}$.

$\underline{C.F.} \qquad y = A e^{-2x} + B e^{-3x}$
Complementary fraction.

$\underline{P.I} = \quad 6 y = 12 \quad$ i.e $y = 2$

334

ex. $(D^2 + D)\,y = 8$

$$D(D+1)y = 8$$

$$D = 0, -1$$

C.F $\quad y = A\,e^{-x} + B\,e^{0x}$

$\quad\quad = A\,e^{-x} + B$

P.I $\quad Dy = 8 \quad\quad y = \frac{1}{D}\,8 = x8$

$\quad\quad y = 8x$

Solution $\quad y = u + v = A\,e^{-x} + B + 8x$

Example

② R.H.S is polynomial in x :—

ex. $\dfrac{d^2y}{dx^2} + 3\dfrac{dy}{dx} + 2y = x^2 + x$

$$(D^2 + 3D + 2)\,y = x^2 + x$$

$$(D+1)(D+2)\,y$$

$$D = -1, -2$$

C.F $\quad A\,e^{-x} + B\,e^{-2x}$

P.I $\quad$ we assume a solution of the form of the R.H.S

$\quad\quad y = ax^2 + bx + c \quad\quad$ i)

$\quad\quad \dfrac{dy}{dx} = 2ax + b \quad\quad$ ii)

$\quad\quad \dfrac{d^2y}{dx^2} = 2a \quad\quad$ iii)

$2a + 3(2ax+b) + 2(ax^2 + bx + c) = x^2 + x$

$\{$ · coeff of x^2 · $2a = 1 \quad\quad$ ∴ $a = \frac{1}{2}$

$\quad$ · of x · $6a + 2b = 1 \quad\quad$ ∴ $b = -1$

$\quad$ · absolute terms $2a + 3b + 2c = 0 \quad c = 1$

∴ $1 + 3(x - 1) + 2(\frac{x^2}{2} - x + 1) = x^2 + x$

∴

P.I $\quad y = \frac{1}{2}x^2 - x + 1$

$\boxed{\begin{array}{l} \dfrac{1}{(1-D)} = 1 + D + D^2 + \cdots \\[2mm] \dfrac{1}{(1+D)} = 1 - D + D^2 - D^3 + \cdots \end{array}}$

P.I $= -e^{2x}\dfrac{1}{D(1+D)}\,x = -e^{2x}\,\frac{1}{D}(1 + D + D^2)\,$

$$= -e^{2x}\left(\frac{1}{D} + 1 + D + \cdots\right)1$$

$$= -e^{2x}\left(x + \frac{1}{+1}\right)$$

Solution = C.F + P.I

$$y = Ae^{2x} + Be^{3x} - xe^{2x} - e^{2x}$$ Part of C.F

$$\therefore y = A'e^{2x} + B e^{3x} - xe^{2x}.$$

Example

Ex. $(D-2)^2 y = 3e^{2x}$

C.F $= e^{2x}(A + Bx)$

P.I $y = \frac{1}{(D-2)^2} 3e^{2x}$

$$= 3e^{2x} \frac{1}{(D+2-2)^2} \times 1 = 3e^{2x} \frac{1}{D^2} 1$$

$$= \frac{3}{2} e^{2x} x^2$$

Solution $y = e^{2x}(A + Bx) + \frac{3}{2}x^2 e^{2x}.$

VI R.H.S $\cos \lambda x$ or $\sin \lambda x$.

$$\psi(D^2)\cos \lambda x = \psi(-\lambda^2)\cos \lambda x$$
$$\psi(D^2)\sin \lambda x = \psi(-\lambda^2)\sin \lambda x.$$

Ex. $(D^2 + 3D + 2) y = \cos 2x$

$\quad (D+1)(D+2) y = \cos 2x$

C.F $= Ae^{-x} + Be^{-2x}$

P.I $y = \dfrac{1}{D^2 + 3D + 2} \cos 2x$

Square $= \dfrac{1}{-4 + 3D + 2} \cos 2x$

$$= \frac{1}{3D - 2} \cos 2x$$

$$= \frac{-(2 + 3D)}{4 - 9D^2} \cos 2x$$

$$= \frac{-(2 + 3D)}{4 + 36} \cos 2x$$

$$= \frac{-2\cos 2x + 6 \sin 2x}{40}$$

P.I $= -\dfrac{\cos 2x + 3\sin 2x}{20}$

336

The inverse operator $\frac{1}{\alpha + \beta D}$ operating on $\sin \omega x$ or $\cos \omega x$

$$\frac{1}{\alpha + \beta D} \sin \omega t = \frac{\alpha - \beta D}{\alpha^2 - \beta^2 D^2} \sin \omega t$$

$$= \frac{\alpha \sin \omega t - \beta \omega \sin \omega t}{\alpha^2 + \beta^2 \omega^2}$$

$$= \frac{1}{\sqrt{\alpha^2 + \beta^2 \omega^2}} \left[\frac{\alpha}{\sqrt{\alpha^2 + \beta^2 \omega^2}} \sin \omega x - \frac{\beta \omega}{\sqrt{\alpha^2 + \beta^2 \omega^2}} \cos \omega x \right]$$

$$= \frac{1}{\sqrt{\alpha^2 + \beta^2 \omega^2}} \sin(\omega x - \epsilon)$$

Similarly

$$\frac{1}{\alpha + \beta D} \cos \omega t = \frac{1}{\sqrt{\alpha^2 + \beta^2 \omega^2}} \cos(\omega x - \epsilon)$$

The diff eq^n $\ddot{x} + \omega^2 x = N \sin \omega t$ ①
This eq^n is of fundamental importance in the study of Resonance.

C.F. : $(D^2 + \omega^2) x = N \sin \omega t$
$\quad A \cos \omega t + B \sin \omega t$

P.I.
$$x = \frac{N \sin \omega t}{D^2 + \omega^2}$$

here we meet with the difficulty that if we replace D^2 by $-n^2$ we obtain a zero. for this reason we consider first
$\frac{1}{D^2 + \omega^2} N \sin \omega t$ where $\omega = n + \epsilon$
$\qquad \epsilon \to 0$

$\therefore x = \frac{1}{D^2 + n^2} N \sin(nt + \epsilon t) = \frac{1}{D^2 + n^2} N(\sin nt \cos \epsilon t + \cos nt \sin \epsilon t)$

$$= \frac{1}{D^2 + n^2} N(\sin nt + \epsilon t \cos nt)$$

$$= \frac{1}{-(n + \epsilon)^2 + n^2} N(\sin nt + \epsilon t \cos nt)$$

$$= \frac{1}{-n^2 - 2n\epsilon - \epsilon^2 + n^2} N(\sin nt + \epsilon t \cos nt)$$

$$= \frac{-N}{2n\epsilon} \sin nt - \frac{N t}{2n} \cos nt$$
part of C.F.

337

$$f \cdot x = -\frac{N}{2n} + const.$$

Solution

$$x = A \cos nt + B \sin nt - \frac{N}{2n}t \ const.$$

12.1.2.12. Bessel's differential equation

Bessel

$$x^2 y'' + x y' + (x^2 - k^2) y = 0 \qquad k > 0$$

$$x = 0 \quad \text{regular}$$

$$\begin{cases} y = \sum a_n x^{c+n} \\ y' = (c+n) a_n x^{n+c-1} \\ y'' = (c+n)(n+c-1) a_n x^{n+c-2} \end{cases}$$

$$(n+c)(n+c-1) a_n x^{n+c} + (c+n) a_n x^{n+c}$$
$$+ (x^2 - k^2) a_n x^{n+c} = 0$$

coeff of x^c

$$c(c-1) a_0 + c a_0 - k^2 a_0 = 0$$
$$c(c-1) + c - k^2 = 0$$
$$c^2 - k^2 = 0$$
$$(c-k)(c+k) = 0$$

$$\boxed{c = \pm k}$$

$$\frac{c = k}{\text{coeff of } x^{n+k}}$$

$$(n+k)(n+k-1) a_n + (k+n) a_n + a_{n-2}$$
$$- k^2 a_n = 0$$

$$a_n \left[n^2 + 2nk - n + k^2 - k + k + n - k^2 \right] + a_{n-2} = 0$$

$$a_n = - \frac{a_{n-2}}{n(n+2k)}$$

$$a_2 = - \frac{a_0}{2(2+2k)}$$

338

$$a_4 = -\frac{a_2}{4(4+2k)} = \frac{a_0}{4 \cdot 2 (4+2k)(2+2k)}$$

$$a_6 = -\frac{a_4}{6(6+2k)} = -\frac{a_0}{6 \cdot 4 \cdot 2 (6+2k)(4+2k)(2+2k)}$$

$$a_6 = -\frac{a_0}{2^3 \cdot 2^3 \cdot 3 \cdot 2 \cdot 1 (3+k)(2+k)(1+k)}$$

$$a_n = (-1)^{\frac{n}{2}} \frac{}{2^n \left(\frac{n}{2}\right)! \, (n+k)(n-1+k)(n-1)}$$

$$\boxed{J_n = \frac{1}{2^k k!} \, y_k}$$

$$a_0 \qquad 2^k k!$$

$$x^2 y'' + xy' + \overset{\text{senex}}{(x^2 - k^2)} y = 0$$

singular pt.

$$y = \sum a_n x^{n+c}$$
$$y' = (n+c) a_n x^{n+c-1}$$
$$y'' = (n+c)(n+c-1) a_n x^{n+c-2}$$

$$(n+c)(n+c-1) a_n x^{n+c} + (n+c) a_n x^{n+c}$$
$$+ (x^2 - k^2) a_n x^{n+c} = 0$$

coefts of x^c

$$c(c-1) a_0 + c a_0 - k^2 a_0 = 0$$

$$c^2 - c + c - k^2 = 0$$

$$\boxed{c = \pm k}$$

coeffts of x^{n+c}

$$(n+c)(n+c-1)a_n + (n+c)a_n + a_{n-2} - k^2 a_n = 0$$

$$a_n \left[n^2 + 2nc - n + c^2 - c + n + c - k^2 \right] + a_{n-2} = 0$$

$$a_n \left[(n+c)^2 - k^2 \right] = -a_{n-2}$$

$$\boxed{a_n = - \frac{a_{n-2}}{(n+c)^2 - k^2}}$$

$\boxed{c = k}$
$$a_2 = -\frac{a_0}{4 + 4k} = -\frac{a_0}{2^2(k+1)}$$

$$a_4 = -\frac{a_2}{16 + 8k} = -\frac{a_2}{2^3(2+k)}$$

$$= \frac{a_0}{2^5(2+k)(k+1)}$$

$$a_6 = -\frac{a_4}{36 + 12k} = -\frac{a_4}{4 \cdot 3(3+k)}$$

$$a_6 = -\frac{a_0}{2^6 \cdot 3 \cdot 2(k+3)(k+2)(k+1)}$$

so $\qquad n = 2r$

$$\boxed{a_{2r} = (-1)^r \frac{a_0 \quad k!}{2^{2r} \, r! \, (k+r)!}}$$

$$y_k = \sum_{r=0}^{\infty} a_{2r} x^{2r}$$

$$= a_0 \sum (-1)^r \frac{k! \; x^{2r+k}}{2^{2r} \, r! \, (k+r)!}$$

$$J_k = \frac{1}{2^k \, k!} \, y_k$$

340

$$y_k = \left(\frac{x}{2}\right)^k \sum (-1)^r \frac{x^{2r}}{2^{2r} \, r! \, (k+r)!}$$

$$\boxed{c = -k}$$

$$a_n = -\frac{a_{n-2}}{(n-c)^2 - k^2}$$

$$a_2 = -\frac{a_0}{4 - 4k} = -\frac{a_0}{2^2(1-k)}$$

$$a_4 = -\frac{a_2}{16 - 8k} = -\frac{a_2}{2^3(2-k)}$$

$$= \frac{a_0}{2^5(2-k)(1-k)}$$

$$a_6 = -\frac{a_4}{36 - 12k} = -\frac{a_0}{4\cdot3\cdot2^5(3-k)(2-k)(1-k)}$$

$$= \frac{a_6}{2^6 \cdot 3\cdot3(k-3)(k-2)(k-1)k}$$

$$\boxed{a_{2r} = (-1)^r \frac{a_0 \, k\,(k-r-1)!}{2^{2r} \, r! \, (k-1)!}}$$

12.1.3.. Euler's linear differential Equation

3) Euler's eq:

Solve

$$(x^3 D^3 + 2xD - 2)y = x^2 \ln x + 3x$$

let

$$x = e^t \qquad \frac{dx}{dt} = x$$

$$\frac{dt}{dx} = \frac{1}{x} \qquad Dy = \frac{dy}{dt}\frac{dt}{dx} = \frac{1}{x}\frac{dy}{dt}$$

$$\frac{d^2t}{dx^2} = -\frac{1}{x^2} \qquad D^2y = \frac{1}{x}\frac{d^2y}{dt^2}\frac{dt}{dx} - \frac{dy}{dt}\frac{1}{x^2}$$

$$\frac{d^3t}{dx^3} = \frac{2}{x^3} \qquad = \frac{1}{x^2}\left(\frac{d^2y}{dt^2} - \frac{dy}{dt}\right)$$

$$D^3y = \frac{1}{x^3}\,\theta(\theta-1)(\theta-1)$$

$$Dy = \frac{1}{x}\theta$$
$$D^2y = \frac{1}{x^2}\theta(\theta-1) \qquad \theta = \frac{d}{dt}$$
$$D^3y = \frac{1}{x^3}\theta(\theta-1)(\theta-2)$$

341

$$\left(\Theta(\Theta-1)(\Theta-3)+2\Theta-2\right)y = te^{2t}+3e^{t}$$
$$\left(\Theta^{3}-3\Theta^{2}+2\Theta+2\Theta-2\right)y = te^{2t}+3e^{t}$$
$$\left(\Theta^{3}-3\Theta^{2}+4\Theta-2\right)y = te^{2t}+3e^{t}$$

$$\left[\Theta\left(\Theta^{2}-3\Theta+4\right)-2\right]y =$$
$$\left[\Theta\left(\Theta^{2}-4\Theta+1+\Theta\right)-2\right]y = \left[\Theta(\Theta-1)^{2}+\Theta-2\right]y$$
$$(\Theta-1)\left(\Theta^{2}_{,}\Theta+2\right)y = (\Theta-1)\left[\Theta^{2}-2\Theta+\Theta+2\right]y$$

$$= (\Theta-1)\left[\Theta^{2}-\Theta+2\right]y$$
$$a = \frac{1\pm\sqrt{1-8}}{2} = \frac{1\pm i\sqrt{7}}{2}$$
$$\text{so } (\Theta-1)\left(\Theta-\frac{1+i\sqrt{7}}{2}\right)\left(\Theta-\frac{1-i\sqrt{7}}{2}\right)y$$

$$= te^{2t}+3e^{t}$$

$\underline{C.F}$ $\quad y = At^{2}+Bt^{\frac{1+i\sqrt{7}}{2}}+Ct^{\frac{1-i\sqrt{7}}{2}}$

$\underline{P.I}$ $\quad y = \dfrac{1}{(\Theta-1)\left(\Theta-\frac{1+i\sqrt{7}}{2}\right)\left(\Theta-\frac{1-i\sqrt{7}}{2}\right)}te^{2t}+3e^{t}$

$$1st \quad \frac{1}{(\Theta-1)}\cdot\frac{1}{\left(\Theta-\frac{1+i\sqrt{7}}{2}\right)\left(\Theta-\frac{1-i\sqrt{7}}{2}\right)}te^{2t}$$

$$= \frac{1}{(\Theta-1)}\cdot\frac{1}{\left(\Theta-\frac{1+i\sqrt{7}}{2}\right)}\cdot e^{2t}\frac{1}{\left(\Theta+2-\frac{1-i\sqrt{7}}{2}\right)}t$$

$$= e^{2t}\frac{1}{(\Theta+2-1)}\cdot\frac{1}{\left(\Theta+2-\frac{1+i\sqrt{7}}{2}\right)}\cdot\frac{1}{\Theta+1.5-\frac{i\sqrt{7}}{2}}t$$

$$= e^{2t}\frac{1}{\Theta}\cdot\frac{1}{\left(\Theta+1.5-\frac{i\sqrt{7}}{2}\right)}\cdot\frac{1}{\left(1.5-\frac{i\sqrt{7}}{2}\right)}\left(1-\frac{\Theta}{1.5-\frac{i\sqrt{7}}{2}}\right)t$$

$$= \frac{e^{2t}}{1.5+\frac{i\sqrt{7}}{2}}\cdot\frac{1}{\Theta\left(1.5-\frac{i\sqrt{7}}{2}\right)}\left(1-\frac{\Theta}{1.5-\frac{i\sqrt{7}}{2}}\right)\left(t - \frac{1}{1.5+\frac{i\sqrt{7}}{2}}\right)$$

$$= \frac{e^{2t}}{2.25-\frac{7}{4}}\cdot\frac{1}{\Theta}\left[t - \frac{1}{1.5+\frac{i\sqrt{7}}{2}} - \frac{1}{1.5-\frac{i\sqrt{7}}{2}}\right]$$

$$= \frac{e^{2t}}{2.25+\frac{7}{4}}\left[\frac{t^{2}}{2} - t\left(\frac{3}{2.25+\frac{7}{4}}\right)\right]$$

$$= \frac{e^{2t}}{4}\left[\frac{t^{2}}{2} - \frac{3t}{4}\right] \longrightarrow \boxed{1}$$

$$2 \times 1 \quad \frac{1}{(\theta - 3)\left(\theta - \frac{1+i\sqrt{7}}{2}\right)\left(\theta - \frac{1-i\sqrt{7}}{2}\right)} 3e^t$$

$$= 3e^t \frac{1}{(1-3)\left(1 - \frac{1+i\sqrt{7}}{2}\right)\left(1 - \frac{1-i\sqrt{7}}{2}\right)}$$

$$= 3e^t \frac{1}{(-1)\left(0.5 - i\sqrt{\frac{7}{2}}\right)\left(0.5 + i\sqrt{\frac{7}{2}}\right)}$$

$$= \frac{3e^t}{(-1)} \frac{1}{\left(0.25 + \frac{7}{4}\right)}$$

$$= -\frac{3}{2} e^t \quad \text{——} \quad ②$$

General sol = C.F + P.I

$$= At^2 + Bt^{\frac{1+i\sqrt{7}}{2}} + Ct^{\frac{1-i\sqrt{7}}{2}}$$

$$+ \frac{e^{2t}}{8}\left[t^2 - \frac{3}{2}t\right] - \frac{3}{2}e^t$$

$$x = e^t \qquad t = \ln x$$

$$y = A\ln^2 x + B(\ln x)^{\frac{1+i\sqrt{7}}{2}} + C(\ln x)^{\frac{1-i\sqrt{7}}{2}}$$

$$+ \frac{x^2}{8}\left[\ln^2 x - \frac{3}{2}\ln x\right] - \frac{3}{2}x$$

$$y = A\ln^2 x + B(\ln x)^{\frac{1+i\sqrt{7}}{2}} + C(\ln x)^{\frac{1-i\sqrt{7}}{2}}$$

$$+ \frac{x^2}{8}\left[\ln^2 x - \frac{3}{2}\ln x\right] - \frac{3}{2}x$$

<u>Euler Linear d. Eqn</u>

$$a_0 x^n \frac{d^n y}{dx^n} + a_1 x^{n-1}\frac{d^{n-1}y}{dx^{n-1}} + \cdots + a_{n-1} x y'$$

replace

$$x = e^t \, , \quad D = \frac{d}{dt} \, , \quad \frac{dx}{dt} = x$$

$$\frac{dy}{dx} = \frac{dy}{dt} \cdot \frac{dt}{dx} = \frac{1}{x} Dy$$

$$\frac{d^2y}{dx^2} = \frac{1}{x} D^2y \cdot \frac{dt}{dx} + Dy \left(-\frac{1}{x^2}\right)$$

$$= \frac{1}{x^2} D(D-1)y$$

$$\frac{d^3y}{dx^3} = \frac{1}{x^2} D^3y \cdot \frac{dt}{dx} - D^2y \cdot \frac{2}{x^3} - \frac{1}{x^2} D^2y$$

$$+ Dy \cdot \frac{2}{x^3}$$

$$= \frac{D}{x^3} \left[D^2 - 3D + 2\right]y$$

$$= \frac{D(D-1)(D-2)y}{x^3}$$

$$\frac{d^n y}{dx^n} = \frac{1}{x^n} D(D-1) \cdots (D-n+1)y$$

So eqn ① become

$$a_0 D(D-1)(D-2)\cdots(D-n+1)y \; +$$
$$a_1 D(D-1)(D-2)\cdots(D-n)y \; +$$

$$a_{n-1} Dy = 0$$

Example 1:

Example

Solve
$$(x^3 D^3 + 2xD - 2)y = x^2 \ln x + 3x$$

$$\begin{cases} x = e^t, \\ \dfrac{dt}{dx} = \dfrac{1}{x} \end{cases} \qquad \theta = \dfrac{d}{dt}, \ D = \dfrac{d}{dx}$$

$$\dfrac{dy}{dx} = Dy = \dfrac{dy}{dt} \cdot \dfrac{dt}{dx} = \dfrac{1}{x} \theta y$$

$$\dfrac{d^2y}{dx^2} = \dfrac{1}{x^2} \theta^2 y - \dfrac{1}{x^2} \theta y = \dfrac{1}{x^2} \theta(\theta-1) y$$

$$\dfrac{d^3y}{dx^3} = \dfrac{1}{x^3} \theta(\theta-1)(\theta-2) y$$

$$\left[\theta(\theta-1)(\theta-2) + 2\theta - 2 \right] y = t e^{2t} + 3 e^{t}$$

C.F

$$(\theta-1)\left[\theta^2 - 2\theta + 2\right] y = 0$$

$$\theta^2 - 2\theta + 2 = 0$$

$$\theta = \dfrac{2 \pm \sqrt{4-8}}{2} = 1 \pm \sqrt{1-2}$$

$$= 1 \pm i$$

$$y = A e^{t} + B e^{(1+i)t} + C e^{(1-i)t}$$
$$= Ax + Bx(\cos t + i \sin t) + C x(\cos t - i \sin t)$$

$$= Ax + Bx \cos t + i C x \sin t$$

$$\boxed{y = Ax + Bx \cos t + C x \sin t}$$

P.I

$$y = \dfrac{1}{(\theta-1)\left[\theta^2 - 2\theta + 2\right]} t e^{2t} + \dfrac{3}{(\theta-1)(\theta^2 - 2\theta + 2)} e^{t}$$

$$= e^{2t} \dfrac{1}{(\theta+1)(\theta^2 + 2\theta + 4 - 2\theta - 4 + 2)} t + \dfrac{e^{t} \, 3}{\theta(\theta^2 \cdot 2\theta + 1 - 2\theta)}$$

$$= e^{2t} \dfrac{1}{(\theta+1)(\theta^2 + 2)} t + 3 e^{t} \dfrac{1}{\theta(\theta^2 + 1)}$$

$1^{\underline{st}}$

$$\frac{1}{\theta(\theta^2+1)} = \frac{1}{(\theta^3+1)}\; t = (1-\theta^3+)t = t$$

2nd

$$\frac{1}{(\theta+1)(\theta^2+2)} \; t = \frac{1}{\theta^3+2\theta+\theta^2+2}\; t$$

$$= \frac{1}{2\left[1 + \theta\left(\frac{\theta+\theta^2+2}{2}\right)\right]}\; t$$

$$= \frac{1}{2}\left[1 - \theta\left(\frac{\theta+\theta^2+2}{2}\right)\right] t$$

$$= 0$$

12.1.3..1. Solution of differential equation by successive approximation

The sol^n of $1^{\underline{st}}$ Order D. Eqns
by the method of successive approximat^n
Method of iteration

find sol^n of

$$\frac{dy}{dx} = f(x, y)$$

at $y = y_0$ for $x = x_0$

$$y = y_0 + \int_{x_0}^{x} f\, dx$$

$$y_1(x) = y_0 + \int_{x_0}^{x} f(x, y)\, dx$$

$$y_2(x) = y_0 + \int_{x_0}^{x} f(s, y_1)\, ds$$

$$y_n(x) = y_0 + \int_{x_0}^{x} f[x, y_{n-1}(x)]\, dx$$

Existence of solution of a differential equation

Proof of the existence of a
Sol^n of a d.eq^n.

Error estimate in approximate sol^{ns}

Theorem   If the eq^n:
$$\frac{dy}{dx} = y' = f(x,y)$$

the f$_c$ f(x,y) & its partial derivatives w.r.t
y, $\frac{df}{dy}$, are continuous in some domain D

in an xy-plane, containing some p' (x_0, y_0).
Then $\exists$ a unique sol^n to this eq^n

$$y = \phi(x)$$

which satisfies the condition $y = y_0$ at $x = x_0$

Another statement
diff.l eq: $\quad \frac{dy}{dx} = f(x,y)$ $\quad$ ①

initial cond^n: $\quad y = y_0 \quad$ for $x = x_0$ ②

let f(x,y) & f(x,y) be continuous in a
closed domain D:

$D \begin{Bmatrix} x_0 - a \le x \le x_0 + a, \\ y_0 - b \le y \le y_0 + b \end{Bmatrix}$ $\quad$ ③

Then, a some interval
$$x_0 - l < x < x_0 + l \quad ④$$
$\exists$ a sol^n to eq ①

(i) which satisfies I.C₂ $\underline{\underline{②}}$
(ii) & the sol^n is unique

347

Proof

As $f(x,y)$ & $f_y(x,y)$ are continuous in Domain D,

$\therefore \exists \ M > 0 \ , \ N > 0 \ ,$

$$|f(x,y)| \leq M , \qquad (5)$$
$$|f_y(x,y)| \leq N \qquad (6)$$

let $\ell = \min(a, \frac{b}{M})$ $\qquad (7)$

For two arbitrary pt (x, y_1), $(x, y_2) \in D$

Lagrange Theorem give

$$f(x, y_2) - f(x, y_1) = f_y'(x, \eta)(y_2 - y_1)$$

where $y_1 < \eta < y_2$

from (6)

$$|f(x, y_2) - f(x, y_1)| \leq N |y_2 - y_1| \qquad (8)$$

But as

$$y_1 = y_0 + \int_{x_0}^{x} f(x, y_0) dx$$

so

$$|y_1 - y_0| = |\int_{x_0}^{x} f(x, y_0) dx|$$

& by eq (5)

$$|y_1 - y_0| \leq \int_{x_0}^{x} M dx = M|x - x_0| \leq M\ell \leq b \qquad (9)$$

&

$$y_2 = y_0 + \int_{x_0}^{x} f(x, y_1) dx$$

also

$$|y_2 - y_0| \leq M|x - x_0| \leq M\ell \leq b \qquad (10)$$

finally $|y_n - y_0| \leq b \qquad (11)$

for arbitrary n.

We will now prove that $\exists$ a limite

$$\lim_{n \to \infty} y_n(x) = y_\infty \qquad (12)$$

348

& $y_{(x)}$ satisfies d.eq.①& I.C ②

Proof Consider a series

$$y_0 + (y_1 - y_0) + (y_2 - y_1) + \dots + (y_{n-1} - y_{n-2}) + y_n - y_{n-1}$$

with general term $U_n = y_n - y_{n-1}$ with $u_0 = y_0$

The sum of $n+1$ term equal to

$$S_{n+1} = \sum_{i=0}^{n} u_i = y_n \qquad ⑬$$

Let us estimate the terms of series ⑬ in absolute value:

$$|y_1 - y_0| \le M|x - x_0|$$

$$|y_2 - y_1| = \left| \int_{x_0}^{x} [f(x y_1) - f(x y_0)] dx \right|$$

$$= \left| \int_{x_0}^{x} f_y'(x \, \eta)(y_1 - y_0) \, dx \right|$$

$$\le \pm N \int_{x_0}^{x} M|x - x_0| \, dx$$

$$= N\frac{M}{2}|x - x_0|^2 \qquad ⑯$$

similarly

$$|y_3 - y_2| = \left| \int_{x_0}^{x} [f(x y_2) - f(x y_1)] dx \right|$$

$$= \left| \int_{x_0}^{x} f_y'(x \, \eta)(y_2 - y_1) \, dx \right|$$

$$\le \pm N \int_{x_0}^{x} \frac{NM}{2} |x - x_0|^2 \, dx$$

$$= M\frac{N^2}{1 \cdot 2 \cdot 3}|x - x_0|^3 \qquad ⑰$$

So $|y_n - y_{n-1}| \le M \frac{N^{n-1}}{n!}|x - x_0|^n \cdot ⑱$

Continuing, for interval $|x - x_0| < \ell$ the series ⑬ of $f_{n}^{(n)}$ is dominated.

The corresponding numerical series with +ve or terms which exceed in absolute value the corresponding terms of series 13 is

$$y_0 + M\ell + \frac{MN\ell^2}{2!} + \frac{MN^2\ell^3}{3!} + \dots + \frac{MN^{n-1}\ell^n}{n!} + \dots$$

general term
$$v_n = \frac{MN^{n-1}\ell^n}{n!}$$

This series converges by $\underline{d'Alembert's\ test}$

$$\lim_{n\to\infty} \frac{v_n}{v_{n-1}} = \lim_{n\to\infty} \frac{\dfrac{MN^{n-1}\ell^n}{n!}}{\dfrac{MN^{n-2}\ell^{n-1}}{(n-1)!}}$$

$$= \lim_{n\to\infty} \frac{N\ell}{n} = 0 < 1$$

I.e. series 13 converges. Since its terms are continuous fs, it converges to a continuous fr $y(x)$,

$$\lim_{n\to\infty} S_{n-1} = \lim_{n\to\infty} y_n = y(x) \qquad ⑳$$

where y is a continuous fr satisfying the initial condit: for all n. $y_n(x) = y_0$

We will prove that the $f = y(x)$ thus obtained satisfies eqⁿ (1). We again write down the eqⁿ

$$y_n = y_0 + \int_{x_0}^{x} f(x, y_{n-1})\,dx \qquad ㉑$$

We will prove

$$\lim_{n\to\infty} \int_{x_0}^{x} f(x, y_{n-1}(x))\,dx = \int_{x_0}^{x} f(x, y)\,dy$$

where $y(x)$ from ⑳ $\qquad ㉒$

$\underline{1^{st}\ note}$ Since series ⑬ is dominated, it follows from ⑳ that $\forall \varepsilon > 0 \;\exists$ an n:

$$|y - y_n| < \varepsilon \qquad (23)$$

having regard (23) over entire interval ℓ

$$\left| \int_{x_0}^{x} f(x,y)\,dx - \int_{x_0}^{x} f(x,y_n)\,dx \right| \leq \pm \int_{0}^{x} |f(x,y) - f(x,y_n)|\,dx$$

$$\leq \pm \int_{x_0}^{x} N|y - y_n|\,dx \leq N\varepsilon|x - x_0|$$

But $\lim\limits_{n \to \infty} \varepsilon = 0$, hence

$$\lim_{n \to \infty} \left| \int_{x_0}^{x} f(x,y)\,dx - \int_{x_0}^{x} f(x,y_n)\,dx \right| = 0$$

so $y(x)$ of ② satisfies the eq.

$$|y - y_n| \leq \frac{N^n M}{(n+1)!} |x - x_0|^{n+1}$$

$$\leq \frac{N^n M \ell^{n+1}}{(n+1)!}$$

Ex

Find 4th approx: y_3

$$y' = x + y^2$$

$$y_0 = 1 \qquad x = 0$$

D $\left\{ -\frac{1}{2} \leq x \leq \frac{1}{2} \ , \ -1 \leq y \leq 1 \right\}$

Sol:

$a = \frac{1}{2}$, $b = 1$. Then $M = \frac{3}{2}$,

$N = 2$

$\ell = \min\left(a, \frac{b}{M}\right) = \min\left(\frac{1}{2}, \frac{2}{3}\right) = \frac{1}{2}$

$$|y - y_3| \leq \frac{M N^n \ell^{n+1}}{(n+1)!} = \frac{\frac{3}{2}(2^3)(\frac{1}{2})^6}{6!}$$

$$|y - y_3| < \frac{1}{960}$$

351

<u>Notice</u> if $F(y)$ satisfies

$$|F(y_2) - F(y_1)| \leq K |y_2 - y_1|$$

where $y_1, y_2 \in$ domain $\underline{K}$ is constant

condit^n : <u>Lipschitz condition</u>

If $f(x, y)$ & its derivative $\frac{\partial f}{\partial y}$ bounded in some domain, then its satisfies the <u>Lipschitz condit^o</u> in that domain.

<u>Converse</u> may <u>not</u> be true

12.1.3..2. Uniqueness of solution of a differential equation

Uniqueness Theorem of the sol^n of a d. Eq^n

<u>Theorem</u>
If a f^n $f(x, y)$ is continuous & has a continuous derivative $\frac{\partial f}{\partial y}$ in a domain D then the <u>solution</u> of the d. eq^n

$$\frac{dy}{dx} = f(x, y) \quad ①$$

is unique for I.C $y = y_o$ when $x = x_o$ i.e, through the pt (x_o, y_o) there passes a <u>unique integral</u> curve

$$\text{when } \begin{array}{c} y = y_o \\ x = x_o \end{array} \Big\} \quad ②$$

$$\frac{dy}{dx} = f(x,y) \quad ①$$

is unique for I.C $y = y_0$ when $x = x_0$
i.e through the pt (x_0, y_0) there passes
a unique integral curve

$$\text{when} \quad \begin{matrix} y = y_0 \\ x = x_0 \end{matrix} \Big\} \quad —②$$

Proof

Assuming there are 2 soln's of eqn ①
satisfies condition ② $y(x)$ & $z(x)$
so
$$y = y_0 + \int_{x_0}^{x} f(x,y)\,dx$$

$$\text{&} \quad z = y_0 + \int_{x_0}^{x} f(x,z)\,dx$$

The diff^{ce}
$$y(x) - z(x) = \int_{x_0}^{x} \left[f(x,y) - f(x,z) \right] dx \quad ③$$

Use Lagrange's theorem

$$f(x,y) - f(x,z) = \frac{\partial f(x,y)}{\partial y}(y-z) \quad ④$$

so
$$|f(x,y) - f(x,z)| \leqslant N\,|y-z| \quad ⑤$$

& $$|y-z| = \left| \int_{x_0}^{x} \frac{\partial f}{\partial y}(y-z) \right| \leqslant \pm N \int_{x_0}^{x} |y-z|\,dx \quad ⑥$$

To be definite consider $|x - x_0| < \frac{1}{N}$

assume on interval $x - x_0 < \frac{1}{N}$,

$|y-z|$ assumes a maximum value for
$x = x^*$ & it be equal to λ

i.e $|y-z| = \lambda$ maximum
so inequality ⑥ for x^*

353

$$|y - z| = \lambda = \left| \int_{x_0}^{x} \frac{\partial f}{\partial y} (y-z) dx \right| \le \pm N \int_{x_0}^{x} \lambda \, dx$$

$$= N\lambda(x^* - x) < N\lambda \cdot \frac{1}{N} < \lambda$$

or $\lambda < \lambda$

which is a contradiction because we assume
two __solns__

Theorems of D-Eqns

$t \in [a,b]$ the set of all fns having
k derivatives $C^k(I)$ where I is open
interval $-\infty < t < \infty$ $\quad t \in I$

if f & derivatives are continuous $f \in C(D)$

D is a domain of complex field
I " " " real field
f " " fn of complex field
if The __set__ of all derivatives of f is $C^k(I$
belong __to I__ only exist & continuous so

$$f \in C(D)$$

$\exists$ a sol: $\phi(t) \in D \quad \forall \, t \in I$
& derivatives of $\phi(t)$: $\phi', \phi'', \ldots, \phi^{(n)}$ exist on
real field I:

$$f[t, x, x', x'', \ldots, x^{(n)}] = 0$$

$$\boxed{x' = \frac{dx}{dt} = f(t,x) \longrightarrow \textcircled{1}}$$

$f \in C(D)$ " xt plane

$\phi(t)$ is a sol: of $\textcircled{1}$

i.e $\int \phi'(t) \, dt = \int f[t, \phi(t)] \longrightarrow \textcircled{2}$

assume ξ is init condit= for $\phi(t_o)$

$$\int_{t_o}^{t} \phi'(t)\,dt = \int_{t_o}^{t} f[t, \phi(t)]\,dt$$

$$\phi(t) = \phi(t_o) + \int_{t_o}^{t} f[s, \phi(s)]\,ds$$

$$\phi(t) = \xi + \int_{t_o}^{t} f(s, \phi(s))\,ds \Bigg| \quad \text{(3)}$$

Theorem

$$\phi'(t) = f[t, \phi(t)] \quad \text{(1)}$$

$$f \in C(D) \qquad \text{ie } f = \text{on domain } D$$
of continuous &
existing derivatives
on <u>real field</u> I

$$f \in \text{lip condit=} \quad \text{(5)}$$

i.e $\left| f(t, \phi_1) - f(t, \phi_2) \right| < \underline{k} \left| \phi_1 - \phi_2 \right| \Bigg|$ (4)

$$f \in (C, lip) \text{ in } D \quad \text{(6)}$$

<u>Theorem</u> Suppose $f \in (C, lip)$ in D with
lipshite constant K, let ϕ_1, ϕ_2 be,
ϵ_1 & ϵ_2, approximately sol= f eq=

$$\phi'(t) = f[t, \phi(t)] \quad \text{(1)}$$

on some interval $[a, b]$ satisfying for
some τ:
$$a < \tau < b$$

$$\left|\phi_1(\tau) - \phi_2(\tau)\right| \leq \delta \qquad (7)$$

when $\delta > 0$ if $\epsilon = \epsilon_1 + \epsilon_2$, then
$\forall \ t \in [a,b]$

$$\left|\phi_1(t) - \phi_2(t)\right| \leq \delta e^{k(t-\tau)} + \frac{\epsilon}{K}\left\{e^{k(t-\tau)} - 1\right\} \qquad (8)$$

Existance of sol^n

Proof

Consider the case when $\tau \leq t < b$ a corresponding holds for $a < t \leq \tau$

since ϕ_1, ϕ_2 & ϵ_1, ϵ_2 are approximate sol^{u}s ϕ (1)

$$\left|\dot{\phi}_i(t) - f[t, \phi_i(t)]\right| \leq \epsilon_i \qquad (9)$$
$$i = 1, 2, \ldots$$

integrating from τ to t for $\tau \leq t < b$

$$\left|\phi_i(t) - \phi_i(\tau) - \int_\tau^t f[s, \phi_i(s)] ds\right|$$
$$\leq \epsilon_i (t - \tau)$$

It is easy to see
$$\left|\phi_1(t) - \phi_1(\tau) - \int_\tau^t f[s, \phi_1(s)] ds\right| \leq \epsilon_1 (t-\tau)$$

& $\left|\phi_2(t) - \phi_2(\tau) - \int_\tau^t f[s, \phi_2(s)] ds\right| \leq \epsilon_2(t-\tau)$

Note they are minus
so
$$\left[\phi_1(t) - \phi_2(t)\right] - \left[\phi_1(\tau) - \phi_2(\tau)\right] - \int_\tau^t \left[f(s,\phi_1(s)) - f(s, \phi_2(s))\right] ds$$
$$< (\epsilon_1 + \epsilon_2)(t - \tau) = \epsilon(t-\tau)$$

356

let $r(t) = |\phi_1(t) - \phi_2(t)|$ defined on $[\gamma, b]$

The The proceeding inequality gives

$$\left| \phi_1(t) - \phi_2(t) \right| - \left| \phi_1(\gamma) - \phi_2(\gamma) \right|$$

$$- \int_\gamma^t \left| f(s, \phi_1(s)) - f(s, \phi_2(s)) \right| ds < \varepsilon(t - \gamma)$$

i.e

$$\left| \phi_1(t) - \phi_2(t) \right| < \left| \phi_1(\gamma) - \phi_2(\gamma) \right| + \int_\gamma^t \left| f(s, \phi_1(s)) - f(s, \phi_2(s)) \right| ds$$

$$+ \varepsilon(t - \gamma)$$

or

$$r(t) \leqslant r(\gamma) + \int_\gamma^t \left| f(s, \phi_1(s)) - f(s, \phi_2(s)) \right| ds$$

$$+ \varepsilon(t - \gamma)$$

utilising lipschitz conditions in D

i.e $\left| f(s, \phi_1(s)) - f(s, \phi_2(s)) \right| < k|\phi_1 - \phi_2|$

so

$$r(t) \leqslant r(\gamma) + k \int_\gamma^t r(s) \, ds + \varepsilon(t - \gamma) \qquad (10)$$

Define for $R(t)$:

$$R(t) = \int_\gamma^t r(s) \, ds$$

or $\dfrac{dR(t)}{dt} = r(t)$

& remember that $r(\gamma) \leq \delta$ ⑦
eq. ⑩ become

$$\frac{dR}{dt} \leq \delta + K R(t) + \varepsilon(t-\gamma)$$

or
$$\boxed{\frac{dR}{dt} - K R(t) \leq \delta + \varepsilon(t-\gamma)} \quad ⑪$$

multiply both sides by $e^{-K(t-\gamma)}$ & integrate
from γ to t

$$(dR)\cdot e^{-K(t-\gamma)} - K e^{-K(t-\gamma)} R\,dt \leq \left[\delta + (t-\gamma)\varepsilon\right] e^{-K(t-\gamma)} dt$$

$$d\left[R e^{-K(t-\gamma)}\right] \leq \delta e^{-K(t-\gamma)} dt \quad ⑫$$
$$+ \varepsilon(t-\gamma) e^{-K(t-\gamma)} dt$$

assume $\displaystyle\int_{\gamma}^{t} t e^{-K(t-\gamma)} dt = \int_{\gamma}^{t} \frac{t\, d e^{-K(t-\gamma)}}{-K}$

$$= \frac{1}{-K}\left\{ \left[t e^{-K(t-\gamma)}\right]_{\gamma}^{t} - \int_{\gamma}^{t} e^{-K(t-\gamma)} dt \right\}$$

$$= \frac{1}{K}\left\{ \left[t e^{-K(t-\gamma)} - \gamma\right] + \frac{1}{K}\left[e^{-K(t-\gamma)} - 1\right] \right\}$$

$$= e^{-K(t-\gamma)}\left[-\frac{t}{K} - \frac{1}{K^2}\right] + \frac{\gamma}{K} + \frac{1}{K^2}$$

eq. ⑫ become
$$\int_{\gamma}^{t} d\left(R e^{-K(t-\gamma)}\right) \leq \delta \int_{\gamma}^{t} e^{-K(t-\gamma)} dt$$
$$+ \varepsilon \int t e^{-K(t-\gamma)} dt - \varepsilon\gamma \int e^{-K(t-\gamma)} dt$$

358

$$R(t)e^{K(t-\tau)} - R(\tau) \leq -\frac{\delta}{K}\left(e^{-K(t-\tau)} - 1\right) + \frac{\varepsilon}{K}\left\{\tau + \frac{1}{K}\right.$$

$$\left(R(\tau) = \int_{\tau}^{t} \dot{x}(s)ds\cdots\right) + \frac{\varepsilon\tau}{K}\left\{e^{K(t-\tau)} - 1\right\} \cdot e^{-K(t-\tau)}\left.(t+\frac{1}{K})\right\}$$

so

$$R(t)e^{-K(t-\tau)} \leq -\frac{\delta}{K}\left\{e^{-K(t-\tau)} - 1\right\} + \frac{\varepsilon}{K}e^{-K(t-\tau)}\left\{\tau - t\right\}$$

$$+ \frac{\varepsilon}{K^2}\left\{1 - e^{-K(t-\tau)}\right\}$$

$$R(t) \leq -\frac{\delta}{K}\left[1 - e^{K(t-\tau)}\right] + \frac{\varepsilon}{K}\left\{\tau - t\right\}$$

$$+ \frac{\varepsilon}{K^2}\left[e^{K(t-\tau)} - 1\right]$$

$$\boxed{R(t) \leq \frac{\delta}{K}\left[e^{K(t-\tau)} - 1\right] - \frac{\varepsilon}{K^2}\left[1 - K(\tau - t)\right]}$$

$$+ \frac{\varepsilon}{K^2}e^{K(t-\tau)} \qquad (14)$$

differentiate w.r.t t

$$\frac{dR(t)}{dt} = r(t) = |\phi_1(t) - \phi_2(t)|$$

$$\frac{dR(t)}{dt} = r(t) \leq \delta e^{K(t-\tau)} - \frac{\varepsilon}{K} + \frac{\varepsilon}{K}e^{K(t-\tau)}$$

$$\boxed{|\phi_1(t) - \phi_2(t)| \leq \delta e^{K(t-\tau)} + \frac{\varepsilon}{K}\left[e^{K(t-\tau)} - 1\right]}$$

$$\phi'(t) = f(t, \phi(t))$$

Hence the proof

359

Theorem ② Let $f \in (c, \text{lip})$ in D
& $(\tau, \xi) \in D$ if ϕ_1, ϕ_2 are any two
solns of ① on, $a < \tau < b$:

$$\phi_1(\tau) = \phi_2(\tau) = \xi$$

Then $\phi_1(t) = \phi_2(t)$

Proof as $\varepsilon \to 0, \delta \to 0$

$$|\phi_1(t) - \phi_2(t)| = 0$$

& $\phi_1(t) = \phi_2(t)$

If $\phi_1 \neq \phi_2 (\tau), \tau_0$

$$|\phi_1 - \phi_2(\tau)| \leq \int_0^t |f(\tau, \phi_1) - f(t, \phi_2)| \, ds$$

$$\leq K \int (\phi_1 - \phi_2) \, ds$$

$= \lambda$

$$\lambda \leq K \lambda / t - \tau) \qquad |t - \tau| = \frac{1}{K}$$

$$\lambda < \lambda \qquad \text{contradict}$$

Theorem ③

Suppose $f \in (C, \text{lip})$ on the domain Rectangle

$$R : |t - \tau| \leq a, \quad |x - \xi| \leq b$$

$a, b > 0$ & let $M = \max |f(t, x)|$

and $\alpha = \min\left(a, \frac{b}{M}\right)$

then $\exists$ a unique solk $\phi \in C'$
of ① & $\overline{|t - \tau| \leq \alpha}$ for which

$$\phi(\tau) = \xi$$

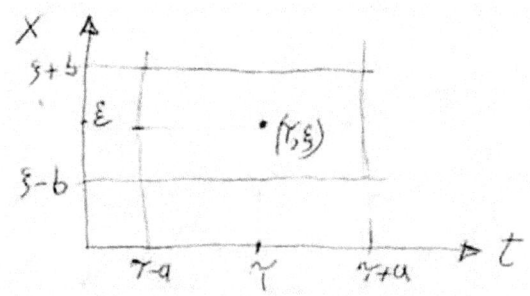

Proof

Let $\{\varepsilon_n\}$ a monotonic decreasing sequence of the realness $\varepsilon_n \xrightarrow{\text{tends}} \text{zero}$ as $n \to \infty$

choose for each $\varepsilon_n < \varepsilon_m$ approximate $\text{sol}^n \phi_n$:

$$\phi_n(t) = \xi + \int_\gamma^t \left[f(s, \phi_n(s)) + \Delta_n(s) \right] dt \quad \boxed{12}$$

where $\Delta_n(t) = \phi_n'(t) - f[t, \phi_n(t)]$

where ϕ_n' exist and $\Delta_n(t)$ tends to 0 as $n \to \infty$ uniformly on $|t - \gamma| \leq a$ as $f \in (C, \text{lip})$

$$|\phi_n - \phi_m| \leq \frac{\varepsilon_n + \varepsilon_m}{K} \left(e^{Ka} - 1 \right)$$

where K is lipschitz constant
The sequence $\{\phi_n\}$ is uniformly convergent on $|t - \gamma| < a$ & therefore $\exists$ a continuous limit f defined on the interval s.t. $\phi_n(t) \to \phi(t)$ as $n \to \infty$

361

uniformly on $[t, \gamma]$, $|t - \tau| \leq a$

This fact plus uniform continuity of f

$\Rightarrow$

$$f(t, \phi_n(t)) \rightarrow f(t, \phi(t))$$

$$\text{as } n \rightarrow \infty$$

uniformly on $|t - \tau| \leq a$

so

$$\lim_{n \to \infty} \int_\gamma^t \left\{ f(s, \phi_n(s)) + \Delta_n(s) \right\} ds$$

$$= \int_\gamma^t f(s, \phi(s)) ds$$

& from ⑫ on get

$$\phi(t) = \xi + \int_\gamma^t f(s, \phi(s)) ds$$

which proves the existence of sol.

$$\phi \in C \text{ of } ① \text{ on } |t - \tau| \leq a$$

& it is unique by Theorem ②

Theorem 4

If $f \in (C, \text{Lip})$ on D then the successive approximation ϕ_k exist on $|t - \tau| \leq a$ as continuous fn & converges uniformly on this interval to the unique sol ϕ_1 of eq ① s.t

$$\phi(\tau) = \xi$$

$$\phi'(t) = f(t, \phi(t)) \quad\text{---}\quad \textcircled{1}$$

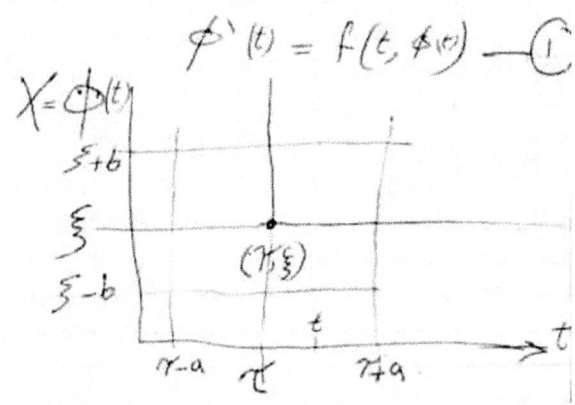

Proof

Consider the interval $(t, \tau+a)$
It will be shown that every ϕ_k existis on
$(t, \tau+a)$, $\phi \in C$ &

$$|\phi_k - \xi| \leq M(t-\tau) \qquad \textcircled{15}$$

observe that $\phi_0 = \xi$ constant satisfies
This condit= & ϕ_k does the same,
then $f(t, \phi_k(t))$ is defined &
continuous on $(t, \tau+a)$

$$|\phi_{k+1} - \xi| \leq M(t-\tau)$$

Geometrically this means that all ϕ_k start
at (τ, ξ) & stay within the Triangular
region bet= the lines $\phi_k - \xi = \pm M(t-\tau)$

It remains to prove the _convergence_ of ϕ_k.
let Δ_k be defined by

$$\Delta_k(t) = |\phi_{k+1}(t) - \phi_k(t)|$$

then as $\phi_{k+1} = \xi + \int_\tau^t f(s, \phi_k) ds$ & $f \in Lip$
so $\phi_k = \xi - \int_\xi^t f(s, \phi_{k-1}) ds$

363

the terms of the series

$$\sum_{K=0}^{\infty} \Delta_K \leq \frac{M}{K}\left(e^{Ka} - 1\right)$$

Therefore the series is continuous convergent on $(t, \gamma+a)$. Thus the series

$$\phi_0 + \sum_{K=0}^{\infty}(\phi_{K+1} - \phi_K) \quad \text{is uniformly convergent}$$

on $(t, \gamma+a)$, thus the partial sum

$$\phi_0 + \sum_{K=0}^{\infty}(\phi_{K+1} - \phi_K) = \phi_n \quad \text{constant limit f=}$$

We will prove that ϕ satisfies the eq=

$$\boxed{\phi(t) = \xi + \int_{\gamma}^{t} f(s, \phi(s)) ds}$$

& is hence a sol: of ① on $(t, \gamma+a)$

clearly

$$\left| \int_{\gamma}^{t} \left\{ f(s, \phi(s)) - f(s, \phi_K) \right\} ds \right|$$

$$\leq \int_{\gamma}^{t} \left| f(s, \phi) - f(s, \phi_K) \right| ds$$

$$\leq K \int_{\gamma}^{t} \left| \phi(s) - \phi_K(s) \right| ds$$

as $K \to \infty$ $\phi_K \to \phi$

so the sol ϕ is unique by theorem

② this completes proof

Notice

$$\phi_{K+1} = \xi + \int_{\gamma}^{t} f(s, \phi_K(s)) ds$$

$$\& \phi_K = \xi + \int_{\gamma}^{t} f(s, \phi_{K-1}(s)) ds$$

$$\phi_{K+1} - \phi_K = \int_{\gamma}^{t} \left[f(s, \phi_K(s)) - f(s, \phi_{K-1}(s)) \right] ds$$

364

$$|\phi_{k+1} - \phi_k| \leq \int_\tau^t |f(s,\phi_k) - f(s,\phi_{k-1})| \, ds$$

$$\leq K \int_\tau^t |\phi_k - \phi_{k-1}| \, ds$$

Due to <u>lipschitz</u> condition

as $\Delta_0 = |\phi_1 - \phi_0| \leq M(t - \tau)$

$$\Delta_1 \leq K \int_\tau^t \Delta_0 \, ds = K \int_\tau^t M(s-\tau) \, ds$$

$$\Delta_1 \leq KM \frac{(t-\tau)^2}{2}$$

& $\Delta_2 \leq K \int_\tau^t \Delta_1 \, ds$

$$\leq K \int_\tau^t \frac{KM(s-\tau)^2}{2} \, ds$$

$$\Delta_2 \leq MK^2 \frac{(t-\tau)^3}{3!}$$

finally

$$\Delta_n \leq \frac{M}{K} \frac{\left\{ K(t-\tau) \right\}^{n+1}}{(n+1)!}$$

& as

$$\phi_n = \phi_0 + \sum_{k=0}^{n-1} \left(\phi_{k+1} - \phi_k \right)$$

$$= \phi_0 + \sum_{k=1}^{n-1} \Delta k$$

so

$$|\phi_n| \leq |\phi_0| + \sum_{k=0}^{n-1} \Delta_k$$

Now $\sum_{k=0}^{n-1} \Delta_k = \dfrac{M}{K}\left\{ \dfrac{K(t-\gamma)}{1!} + \dfrac{K^2(t-\gamma)^2}{2!} + \right.$

$$\left. + \dfrac{K^n(t-\gamma)^n}{n!} \right\} + \dfrac{M}{K} - \dfrac{M}{K}$$

$$= \dfrac{M}{K}\left\{ 1 + \dfrac{K(t-\gamma)}{1!} + \dfrac{K^2(t-\gamma)^2}{2!} + \cdots \right\} - \dfrac{M}{K}$$

$$\lim_{n \to \infty} \sum \Delta_k = \dfrac{M}{K} e^{K(t-\gamma)} - \dfrac{M}{K}$$

$$= \dfrac{M}{K}\left[e^{K(t-\gamma)} - 1 \right]$$

12.1.4. Total Differential Equations

A differential equation of the form
$P(x, y, z)\, dx + Q(x, y, z)\, dy + R(x, y, z)\, dz = 0$ (1)

is called:
1 - A total differential equation in three variables
Or
2 - Pfaffian differential equation
Or.
3 - An exact differential equation in three variables

Suppose a solution of (1) can be written in the form

$\phi(x, y, z) = c$ (2)

Differentiating (2), we obtain the equation

$$\frac{\partial \phi}{\partial x}\, dx + \frac{\partial \phi}{\partial y}\, dy + \frac{\partial \phi}{\partial z}\, dz = 0 \qquad (3)$$

The left member of (3) is either i) identically the left member of (I) or ii) the left member of (3) differs from the left member of (1) by a factor $\mu(x, y, z)$ so that,

$$\frac{\partial \phi}{\partial x} = \mu P, \quad \frac{\partial \phi}{\partial y} = \mu Q, \quad \frac{\partial \phi}{\partial z} = \mu R \qquad (4)$$

where $\mu = 1$ in case i). Thus in case ii), $\mu \neq 1$ is an integrating factor for (1)

12.1.4.1.1. <u>First derivatives are functions of variables lacking integral terms</u>

Case (I): $\mu = 1$, so that $\dfrac{\partial \phi}{\partial x} = P$, $\dfrac{\partial \phi}{\partial y} = Q$, $\dfrac{\partial \phi}{\partial z} = R$

Differentiating each derivative w.r.t. other variables we get

$$\frac{\partial^2 \phi}{\partial y\, \partial x} = \frac{\partial P}{\partial y}, \quad \frac{\partial^2 \phi}{\partial x\, \partial y} = \frac{\partial Q}{\partial x},$$

$$\frac{\partial^2 \phi}{\partial z\, \partial x} = \frac{\partial P}{\partial z}, \quad \frac{\partial^2 \phi}{\partial x\, \partial z} = \frac{\partial R}{\partial x},$$

$$\frac{\partial^2 \phi}{\partial z\, \partial y} = \frac{\partial Q}{\partial z}, \quad \frac{\partial^2 \phi}{\partial y\, \partial z} = \frac{\partial R}{\partial y}$$

Hence we have $\dfrac{\partial Q}{\partial z} = \dfrac{\partial R}{\partial y}$, $\dfrac{\partial R}{\partial x} = \dfrac{\partial P}{\partial z}$, $\dfrac{\partial P}{\partial y} = \dfrac{\partial Q}{\partial x}$

In this case equation (1) is called exact since the left hand side is an exact differential.

12.1.4.1.2. _First derivatives are functions of variables have integral terms_

Case (ii): $\mu \neq 1$. In this case the left member of (1) has an integrating factor $\mu \neq 0$, so that

$$\mu P \, dx + \mu Q \, dy + \mu R \, dz = 0$$

is exact. Hence

$$\frac{\partial(\mu Q)}{\partial z} = \frac{\partial(\mu R)}{\partial y}, \quad \frac{\partial(\mu R)}{\partial x} = \frac{\partial(\mu P)}{\partial z}, \quad \frac{\partial(\mu P)}{\partial y} = \frac{\partial(\mu Q)}{\partial x} \qquad (6)$$

$$\mu \frac{\partial Q}{\partial z} + Q \frac{\partial \mu}{\partial z} = \mu \frac{\partial R}{\partial y} + R \frac{\partial \mu}{\partial y} \qquad (7)$$

$$\mu \frac{\partial R}{\partial x} + R \frac{\partial \mu}{\partial x} = \mu \frac{\partial P}{\partial z} + P \frac{\partial \mu}{\partial z} \qquad (8)$$

$$\mu \frac{\partial P}{\partial y} + P \frac{\partial \mu}{\partial y} = \mu \frac{\partial Q}{\partial x} + Q \frac{\partial \mu}{\partial x} \qquad (9)$$

multiplying (7), (8) and (9) by P, Q and R respectively and adding we get

$$P \left(\frac{\partial Q}{\partial z} - \frac{\partial R}{\partial y} \right) + Q \left(\frac{\partial R}{\partial x} - \frac{\partial P}{\partial z} \right) + R \left(\frac{\partial P}{\partial y} - \frac{\partial Q}{\partial x} \right) = 0 \qquad (10)$$

Equation (10) is often written in the operator form:

$$\begin{vmatrix} P & Q & R \\[6pt] \dfrac{\partial}{\partial x} & \dfrac{\partial}{\partial y} & \dfrac{\partial}{\partial z} \\[6pt] P & Q & R \end{vmatrix} = 0 \qquad (11)$$

Equation (11) gives the necessary condition that (1) have an integrating factor and is called the condition of integrability.

If we introduce the

$$\bar{U} = (P, Q, R) \quad \text{and} \quad dr = (dx, dy, dz),$$

we may write Pfaffian differential equation in three variables in the vector notation as

$$\bar{U} \cdot d\bar{r} = 0$$

and the condition of integrability takes the simple form

$$\bar{U} \cdot \operatorname{curl} \bar{U} = 0$$

12.1.4.2. Solution of Pfaffian differential equations in three variables

We shall now consider methods by which the solution of Pfaffian differential equations in three variables x, y, z may be derived.

12.1.4.2. (a) By inspection:
Once the condition of integrability has been verified it is often possible to derive the primitive of the equation by inspection.

In particular if the equation is such that $\operatorname{curl} \bar{U} = 0$, then $\bar{U}$ must be of the form grad v (since curl grad v = 0) and the equation $\bar{U} \cdot d\bar{r} = 0$ is equivalent to

$$\frac{\partial v}{\partial x} dx + \frac{\partial v}{\partial y} dy + \frac{\partial v}{\partial z} dz = 0$$

with primitive v (x, y, z) = c

Example 1   Solve the equation $(x^2 z - y^3) dx + 3x y^2 dy + x^3 dz = 0$

Solution:

Using the operator form in Equation (11), above, we get

$$\begin{vmatrix} x^2 z - y^3 & 3 x y^2 & x^3 \\ \frac{\partial}{\partial x} & \frac{\partial}{\partial y} & \frac{\partial}{\partial z} \\ x^2 z - y^3 & 3 x y^2 & x^3 \end{vmatrix}$$

$$= (x^2 z - y^3)(0 - 0) - 3xy^2 (3x^2 - x^2) + x^3 (3y^2 + 3y^2) = 0$$

Hence the equation is integrable. We may write the equation in the form

$$x^2 (z \, dx + x \, dz) - y^3 dx + 3x y^2 dy = 0$$

i-e $z \, dx + x \, dz - \dfrac{y^3}{x^2} dx + \dfrac{3y^2}{x} dy = 0$

i-e $d(xz) + d\left(\dfrac{y^3}{x}\right) = 0$

so that the primitive of the equation

$$x^2 z + y^3 = cx$$

12.1.4.2. (b) By variables separable:
In certain cases it is possible to write Pfaffian differential equation in the form

P (x) dx + Q (y) dy + R (z) dz = 0

in which case it is immediately obvious that the solution is given by

$$\int P(x)\ dx + \int Q(y)\ dy + \int R(z)\ dz = C$$

where C is a constant.

Example 2   Solve the equation $a^2 y^2 z^2\ dx + b^2 z^2 x^2\ dy + c^2 x^2 y^2\ dz = 0$

Solution:

If we divide both sides of this equation by $x^2 y^2 z^2$ we have
:

$$\frac{a^2}{x^2}\ dx + \frac{b^2}{y^2}\ dy + \frac{c^2}{z^2}\ dz = 0$$

showing that the primitive of the equation is

$$\frac{a^2}{x} + \frac{b^2}{y} + \frac{c^2}{z} = k$$

where k is a constant.

12.1.4.2. (c) One variable separable:
It may happen that one variable is separable, z say in which case the equation is of the form

$$P(x,y)\,dx + Q(x,y)\,dy + R(z)\,dz = 0 \qquad\qquad (1)$$

The condition of integrability implies that

$$R\left(\frac{\partial Q}{\partial x} - \frac{\partial P}{\partial y}\right) = 0$$

In other words, P dx + Q dy is an exact differential dv say and equation (1) reduces to

dv + R (z) dz = 0

with primitive

$$v(x,y) + \int R(z)\ dz = c$$

Example 3   Verify that the equation
$$x(y^2 - a^2)\ dx + y(x^2 - z^2)\ dy - z(y^2 - a^2)\ dz = 0$$
is exact and solve it.

Solution:

If we divide throughout by $(y^2 - a^2)(x^2 - z^2)$ we get

$$\frac{x}{x^2 - z^2}\, dx - \frac{z}{x^2 - z^2}\, dz + \frac{y}{y^2 - a^2}\, dy = 0$$

showing that it is separable in y and it is therefore integrable if

$$\frac{\partial P}{\partial z} = \frac{\partial R}{\partial x} \quad \text{which is true}$$

To determine the solution of the equation we note that it is

$$\tfrac{1}{2} \log (x^2 - z^2) + \tfrac{1}{2} \log (y^2 - a^2) = \tfrac{1}{2} \log c$$

show that the solution is $\quad (x^2 - z^2)(y^2 - a^2) = c$

12.1.4.2. (d) Homogeneous equations:

The equation

$$P(x, r, z)\, dx + Q(x, y, z)\, dy + P(x, y, z)\, dz = 0 \quad (1)$$

is said to be homogeneous if the functions P, Q, R are homogeneous in x, y, z of the same degree n. To derive the solution of such an equation we make the substitutions

$$y = ux, \ z = vx \quad (2)$$

substituting from (2) into (1) we see that equation (1) assumes the form

$$P(1, u, v,)\, dx + Q(1, u, v)(u\, dx + x\, du) + R(1, u, v)(x\, dv + v\, dx) = 0$$

a factor x^n canceling out. If we write

$$A(u, v) = \frac{Q(1, u, v)}{P(1, u, v) + u\, Q(1, u, v) + v\, R(1, u, v)},$$

$$B(u, v) = \frac{R(1, u, v)}{P(1, u, v) + u\, Q(1, u, v) + v\, R(1, u, v)}$$

we find that this equation is of the form

$$\frac{dx}{x} + A(u, v)\, du + B(u, v)\, dv = 0$$

and can be solved by method (c)

Example 4   Verify that the equation
$$yz(y + z)\, dx + xz(x + z)\, dy + xy(y + x)\, dz = 0$$
is integrable and find its solution

Solution:

Using the curl operator form as in Example 1, we get

370

$$\begin{vmatrix} yz(y+z) & xz(x+z) & xy(y+x) \\ \dfrac{\partial}{\partial x} & \dfrac{\partial}{\partial y} & \dfrac{\partial}{\partial z} \\ yz(y+z) & xz(x+z) & xy(y+x) \end{vmatrix}$$

$$= 2xyz(y^2 - z^2) + 2xyz(z^2 - x^2) + 2xyz(x^2 - y^2) = 0$$

The equation is homogeneous, it can be solved by putting $y = ux$, $z = vx$, we find that the equation satisfied by x, u, v, is

$$u\,v\,(u+v)\,dx + v\,(v+1)\,(u\,dx + x\,du)$$

$$+ u\,(u+1)\,(v\,dx + x\,dv) = 0$$

which reduces to

$$2\,\frac{dx}{x} + \frac{v+1}{u(1+u+v)}\,du + \frac{u+1}{v(1+u+v)}\,dv = 0$$

Or,

$$2\,\frac{dx}{x} + \left(\frac{1}{u} - \frac{1}{1+u+v}\right) du + \left(\frac{1}{v} - \frac{1}{1+u+v}\right) dv = 0$$

The solution of this equation is

$$2 \log x + \log \frac{u}{1+u+v} + \log v = \log c$$

Or,

$$\frac{x^2\,u\,v}{1+u+v} = c$$

Reverting to the original variables, we see that the solution of the given equation is $xyz = c\,(x + y + z)$

12.1.4.2. (e) The general method:
If the equation is integrable, we consider one of the variables, say z as a constant. Integrate the resulting equation, denoting the arbitrary constant of integration by $\varphi\,(z)$. Take the total differential of the integral just obtained and compare the coefficients of its differentials with those of the given differential equation, thus determining $\varphi\,(z)$.
This procedure is illustrated as follows:

Consider the equation
$$P\,dx + Q\,dy + R\,dz = 0$$
given that the condition of integrability is satisfied.

Consider one of the variables, say z as a constant for the moment and let the solution of the resulting equation
$P\,dx + Q\,dy = 0$ (1)
be
$v\,(x, y, z) = \varphi\,(z)$ (2)

From (2):
$$d\varphi = \frac{\partial v}{\partial x}\,dx + \frac{\partial v}{\partial y}\,dy + \frac{\partial v}{\partial z}\,dz \qquad (3)$$

Now, $\dfrac{\partial v}{\partial x} = \mu P, \dfrac{\partial v}{\partial y} = \mu Q$ where $\mu = \mu\,(x, y, z)$ is an

integrating factor of (1). Substituting in (3) we have

371

$$\mu P \; dx + \mu Q \; dy + \frac{\partial v}{\partial z} \; dz = d\phi$$

But from the given equation

$$\mu P \; dx + \mu Q \; dy + \mu R \; dz = 0$$

so that

$$d\phi = \frac{\partial v}{\partial z} \; dz - \mu R \; dz = \left(\frac{\partial v}{\partial z} - \mu R \right) dz$$

This relation is free from dx and dy and using (2) if necessary, can be written as a differential equation in z and ϕ. Solving the integral for ϕ and substituting in (2) we have the required solution.

Example 5 Verify that the equation
$$2yz \; dx + zx \; dy - xy \, (1 + z) \; dz = 0$$
is integrable and find its primitive.

Solution:

Using the curl operator form as in Example 1, we get

$$\begin{vmatrix} 2yz & zx & -xy\,(1+z) \\ \dfrac{\partial}{\partial x} & \dfrac{\partial}{\partial y} & \dfrac{\partial}{\partial z} \\ 2yz & zx & -xy\,(1+z) \end{vmatrix}$$

$$= 2yz\,(-2x - xz) + xz\,(3y + yz) + xyz\,(1 + z) = 0$$

Let z be a constant. Then we have

$$2yz \; dx + zx \; dy = 0$$

Or

$$2\frac{dx}{x} = -\frac{dy}{y}$$

$$\therefore \quad 2 \log x = -\log \frac{y}{c}$$

$$\therefore \quad x^2 y = c$$

or we can write

$$x^2 y = f(z)$$

Differentiating we have

$$2xy \; dx + x^2 \; dy - f'(z) \; dz = 0$$

and combining with the given differential equation

$$2yz \; dx + zx \; dy - xy \, (1 + z) \; dz = 0$$

we find:

$$\frac{2xy}{2yz} = \frac{x^2}{zx} = -\frac{f'(z)}{-xy(1+z)}$$

$$\therefore \quad \frac{x}{z} = \frac{f'(z)}{xy\,(1+z)}$$

$$\therefore \quad x^2 y = \frac{z\,f'(z)}{1+z}$$

$$\therefore \quad f(z) = \frac{z f'(z)}{1+z} \quad \text{or} \quad \frac{f'(z)}{f(z)} = 1 + \frac{1}{z}$$

Hence, $\log f(z) = z + \log z + c_1$

$$\therefore \quad \log x^2 y = z + \log z + c_1$$

Example 6   Solve the equation
$$(2x^2 + 2xy + 2xz^2 + 1)\,dx + dy + 2z\,dz = 0$$

Solution:

$$\begin{vmatrix} 2x^2 + 2xy + 2xz^2 + 1 & 1 & 2z \\[6pt] \dfrac{\partial}{\partial x} & \dfrac{\partial}{\partial y} & \dfrac{\partial}{\partial z} \\[6pt] 2x^2 + 2xy + 2xz^2 + 1 & 1 & 2z \end{vmatrix}$$

$$= (2x^2 + 2xy + 2xz^2 + 1) \times 0 - 1\,(0 - 4xz)$$

$$+ 2z\,(0 - 2x) = 0$$

Assume x = constant, then we have dy + 2z dz = 0

$$\therefore \quad y + z^2 = f(x)$$

Differentiating we have
$$dy + 2z\,dz = f'(x)\,dx$$
Or
$$f'(x)\,dx - dy - 2z\,dz = 0$$

Combining with the given differential equation
$$(2x^2 + 2xy + 2xz^2 + 1)\,dx + dy + 2z\,dz = 0$$

we find
$$\frac{f'(x)}{2x^2 + 2xy + 2xz^2 + 1} = -1$$

$$\therefore \quad -f'(x) = 2x^2 + 2x\,(y + z^2) + 1$$

Or
$$-f'(x) = 2x^2 + 2x\,f(x) + 1$$

$$\therefore \quad f'(x) + 2x\,f(x) = -(2x^2 + 1)$$

373

Integrating factor is $e^{\int 2x\,dx} = e^{x^2}$

$$\therefore \quad e^{x^2} f(x) = -\int (2x^2+1)\, e^{x^2}\, dx = -xe^{x^2} + c_1$$

$$\therefore \quad f(x) = -x + c_1\, e^{-x^2}$$

Hence, $\quad y + z^2 = -x + c_1\, e^{-x^2}$

12.1.4.2. (f) Natani's method:

As in the general method, we first treat the variable z as though it were constant, and solve the resulting differential equation $P\,dx + Q\,dy = 0$. Suppose we find that the solution of this equation is $\phi\,(x, y, z) = c_1$ which can be expressed in the form

$$\phi\,(x, y, z) = \psi\,(z) \tag{1}$$

where ψ is a function of z alone. To determine the function $\psi\,(z)$ we observe that if we give the variable x a fixed value α say, then $\phi\,(\alpha, y, z) = \psi\,(z)$ is a solution of the differential equation

$$Q\,(\alpha, y, z)\, dy + R\,(\alpha, y, z)\, dz = 0 \tag{2}$$

now we can find a solution of equation (2) in the form

$$F(y, z) = c \tag{3}$$

Since equations (1) and (3) represent general solutions of the same differential equation (2), they must be equivalent. Therefore if we eliminate the variable y between (1) and (3) we obtain an expression for the function $\psi\,(z)$. Substituting this expression in equation (1), we obtain the solution of the Pfaffian differential equation. The method is often simplified by choosing a value for x, such as 0 or 1, which makes the labor of solving the differential equation (2) as simple as possible.

Example 7   Verify that the equation

$$(e^x y + e^z)\, dx + (e^y z + e^x)\, dy + (e^y - e^x y - e^y z)\, dz = 0$$

is integrable and find its primitive

Solution:

We have

$$\begin{vmatrix} e^x y + e^z & e^y z + e^x & e^y - e^x y - e^y z \\[6pt] \dfrac{\partial}{\partial x} & \dfrac{\partial}{\partial y} & \dfrac{\partial}{\partial z} \\[6pt] e^x y + e^z & e^y z + e^x & e^y - e^x y - e^y z \end{vmatrix}$$

$$= (e^x y + e^z)\,[(e^y - e^x - e^y z) - (e^y)]$$

$$- (e^y z + e^x)\,[(-e^x y) - e^z]$$

$$+ (e^y - e^x y - e^y z)\,[(e^x) - (e^x)] = 0$$

374

assume $z = $ constant, then we have

$$(e^x y + e^z)\, dx + (e^y z + e^x)\, dy = 0$$

This equation is exact and its solution is

$$e^x y + x\, e^z + z\, e^y = c$$

or, we can write

$$e^x y + x\, e^z + z\, e^y = f(z) \qquad\qquad (1)$$

If we now let $y = 0$ in the original equation, we see that it reduces to the simple form

$$e^z\, dx + (1 - z)\, dz = 0$$

with solution

$$x = \int (z - 1)\, e^{-z}\, dz$$

Or

$$x = -z\, e^{-z} + c_1 \qquad\qquad (2)$$

This solution must be of the form assumed by (1) in the case $y = 0$, in other words, (2) must be equivalent to the relation

$$x\, e^z + z = f(z) \qquad\qquad (3)$$

Eliminating x between (2) and (3) we find that

$$(-z\, e^{-z} + c_1)\, e^z + z = f(z)$$

Or

$$f(z) = c_1\, e^z$$

substituting this expression in equation (1), we find that the solution of the equation is

$$e^x y + x\, e^z + z\, e^y = c_1\, e^z$$

Example 8
On the exactness of differential equation

$$\frac{\partial M}{\partial y} = \frac{\partial N}{\partial x}$$

This is the condition for exactness the solution can be obtain in the following way.
integrate M. w.r.t. x

as if y were const. and then integrate the terms of N which do not contain x w.r.t. y and the two equate to the const.

Ex. show that the eq^n

$$(3x^2 + 2xy + 2y - 3)dx + (x^2 - 4y + 2x + 1)dy$$

with M over the first bracket and N over the second

is exact and solve it.

$$\frac{\partial}{\partial y}(3x^2 + 2xy + 2y - 3) = 2x + 2y$$

$$\frac{\partial}{\partial x}(x^2 - 4y + 2x + 1) = 2x + 2$$

equal
∴ eqn is exact

Solution

$$x^3 + x^2 y + 2yx - 3x - 2y^2 + y = c.$$

Example 9
On linear simultaneous equations with constant coefficients

Linear simultaneous eqns with consts coeffs :-
consider these eqns

$$f(D)y + \phi(D)z = P(x) \quad —①$$
$$\psi(D)y + \psi(D)z = Q(x) \quad —②$$

1. **1st step** eliminate z bet ① and ② by operating on ① by ✗(3) operating by on② by ✗(D) and substract we thus obtain a diff. eq^n in y whose sol. gives y in terms of x.

2. **2nd step** we eliminate y and solve the remaining diff. eq^n in z.

3. **3rd step** we substitute from y and z in one of the eq^n ① ② the simpler the better and we adjust the consts such that the eq^n is satisfied.

ex. solve the eq^n.

$$(5D+4)y - (2D+1)z = e^{-x} \quad ①$$
$$(D+8)y - 3z = 5e^{-x}$$

1st eliminate z by multiplying 1st eq^n by (3) & operating on 2nd eq^n by (2D+1) & then substracting.

$$3(5D+4)y - 3(D+1)z = 3e^{-x}$$
$$(D+8)(2D+1)y - 3(2D+1)z = (2D+1)5e^{-x}$$

$$3(5D+4)y - (D+8)(2D+1)y = 3e^{-x} + 10e^{-x} - 5e^{-x}$$
$$(15D + 12 - 2D^2 - 8 - 17D)y = 8e^{-x}$$
$$(D^2 + D - 2)y = -4e^{-x}$$
$$(D+2)(D-1)y = -4e^{-x}$$

$$\underline{\underline{C.F}} \quad y = Ae^{x} + Be^{-2x}$$
$$\underline{P.I} \quad y = \frac{-4e^{-x}}{(D+2)(D-1)} = 2e^{-x}$$

$$\boxed{y = Ae^{x} + Be^{-2x} + 2e^{-x}} \quad ②$$

2nd eliminate y by operating on eqn ① by $D+8$
and on eqn ② by $5D+4$ & subtract.

$(D+8)(5D+4)y - (2D+1)(D+8)z = e^{-x}(D+8)$
$(D+8)(5D+4)y - 3(5D+4)z = (5D+4)5e^{-x}$

$(15D+12 - 2D^2 - 8 - 17D)z = -e^{-x} + 8e^{-x} + 15 \cdot e^{-x}$
$(D^2 + D - 2)z = -12 e^{-x} \cdot \frac{1}{2}$
$(D+2)(D-1)z = -6e^{-x}$
$\underline{\underline{C.F.}} \quad z = A \cdot E e^{x} + F e^{-2x}$

$P.I. \quad z = \dfrac{-6e^{-x}}{(D+2)(D-1)} = 3e^{-x}$

$\therefore \quad \boxed{z = E e^{x} + F e^{-2x} + 3e^{-x}}$

3rd Substituting in eqn ① ii)

$(D+8)(Ae^{x} + Be^{-2x} + 2e^{-x}) - 3(Ee^{x} + Fe^{-2x} + 3e^{-x})$
$\qquad = 5e^{-x}$

$\therefore Ae^{x} - 2Be^{-2x} - 2e^{-x} - 3Fe^{x} - 3Fe^{-2x}$
$+ 8Ae^{x} + 8Be^{-2x} + 16e^{-x}$
$\qquad -9e^{-x}$
$\qquad -5e^{-x} \quad = (9Ae^{x} + 6Be^{-2x}) - 3(Ee^{x} + Fe^{-2x})$

$\therefore \quad \boxed{(9A - 3E)e^{x} + (6B - 3F)e^{-2x} = 0}$

$\therefore \quad 9A = 3E \qquad 3A = E$
$\& \quad 6B = 3F \qquad 2B = F$

Hence the solution is

$$y = Ae^{x} + Be^{-2x} + 2e^{-x}$$

$$z = 3A e^{x} + 2Be^{-2x} + 3e^{-x}$$

In this ex. The sol⁼ can be simplified as follows
After obtaining y we substitute for y into eq 2 and
obtain z directly

$$(D+8)(Ae^x + Be^{-2x} + 2e^{-x}) - 3z = 3e^{-x}$$

$$9Ae^x + 6Be^{-2x} + 10e^{-x} - 3x = 5e^x$$

$$z = 3A\,e^x + 2Be^{-2x} + 3e^{-x}$$

Example 10: On the stability of the solution of differential function

Stability

Def= the sol⁼ $x = x(t)$, $y = y(t)$,
that satisfy eqs

$$\frac{dx}{dt} = f_1(t, x, y)$$

$$\frac{dy}{dt} = f_2(t, x, y)$$

& initial condit⁼

$$X_{t=0} = X_0$$

$$y_{t=0} = y_0$$

are stable as $t \to \infty$ if $\varepsilon > 0$ ∃
$\delta > 0$: ∀ $t > 0$
$$|\bar{x}(t) - x(t)| < \varepsilon$$
$$|\bar{y}(t) - y(t)| < \varepsilon$$

are stable as $t \to \infty$ if $\varepsilon > 0$ ∃
$\delta > 0$: ∀ $t > 0$
$$|\bar{x}(t) - x(t)| < \varepsilon$$
$$|\bar{y}(t) - y(t)| < \varepsilon$$

379

y $|\bar{x}_0 - x_0| < \delta$

$|\bar{y}_0 - y_0| < \delta$

where $\bar{x}$, $\bar{y}$ be sol^t satisfy

$\bar{x}_0$, $\bar{y}_{t_0}$

$$\frac{dx}{dt} = cx + gy$$

$$\frac{dy}{dt} = ax + by$$

$$A = \begin{pmatrix} c & g \\ a & b \end{pmatrix}$$

$$(A - \lambda I)x = 0$$

$$\begin{pmatrix} c & g \\ a & b \end{pmatrix} - \lambda \begin{pmatrix} 1 & 0 \\ 0 & 1 \end{pmatrix} = 0$$

$$\begin{vmatrix} c - \lambda & 0 \\ a & b - \lambda \end{vmatrix} = 0$$

$$(c - \lambda)(b - \lambda) = 0$$

$$\lambda_1 = c, \quad \lambda_2 = b$$

$$x = C_1 e^{ct} + C_2 e^{bt}$$

$$\frac{dx}{dy} = C_1 c e^{ct} + C_2 b e^{bt}$$

$$= c[C_1 e^{ct} + C_2 e^{bt}] + gy$$

$$y = \frac{1}{g}\left[C_2 e^{bt}(b - c)\right]$$

$$\frac{dx}{dt} = -1x$$

$$\frac{dy}{dt} = -2y$$

$$\begin{pmatrix} -1 & 0 \\ 0 & -2 \end{pmatrix} - \lambda \begin{pmatrix} 1 & 0 \\ 0 & 1 \end{pmatrix} = 0$$

$$\begin{pmatrix} -1-\lambda & 0 \\ -2 & -2-\lambda \end{pmatrix} = 0$$

$$(-1-\lambda)(-2-\lambda) = 0$$

$$(1+\lambda)(2+\lambda) = 0 \quad \lambda = -2$$

$$\lambda_2 = 1$$

$$x = C_1 e^{-2t}, \quad y = C_2 e^{-t}.$$

$$\frac{dx}{dt} = -x + y$$

$$\frac{dy}{dt} = -x - y$$

$$\begin{pmatrix} -1 & 1 \\ -1 & -1 \end{pmatrix} - \lambda \begin{pmatrix} 1 & 0 \\ 0 & 1 \end{pmatrix} = 0$$

$$\begin{pmatrix} -1-\lambda & 1 \\ -1 & -1-\lambda \end{pmatrix} = 0$$

$$(1+\lambda)(1+\lambda) + 1 = 0$$

$$\lambda^2 + 1 + 2\lambda + 1 = 0$$

$$\lambda + 2\lambda + 2 = 0$$

$$\lambda = -1 \pm \sqrt{1-2}$$

$$= -1 \pm i$$

$$x = C_1 e^{(-1+i)t} + C_2 e^{(-1-i)t}$$

$$= e^{-t}\left[C_1(\cos t + i\sin t) + C_2(\cos t - i\sin t)\right]$$

$$x = e^{-t}(A\cos t + B\sin t)$$

$$\frac{dx}{dt} = -e^{-t}(A\cos t + B\sin t)$$
$$+ e^{-t}(C\sin t + D\cos t)$$

$$x_1' = -x_1 + Kx_2 + 2x_3$$
$$x_2' = -x_1 - x_2 - x_3$$
$$x_3' = \qquad - x_2 - Kx_3$$

$$\begin{pmatrix} -1 & K & 2 \\ 0 & -1 & -1 \\ 0 & 1 & -K \end{pmatrix} - \lambda \begin{vmatrix} 1 & 0 & 0 \\ 0 & 1 & 0 \\ 0 & 0 & 1 \end{vmatrix}$$

$$\begin{vmatrix} -1-\lambda & K & 2 \\ 0 & -1-\lambda & 0 \\ 0 & 1 & -K-\lambda \end{vmatrix} = 0$$

$$(-1-\lambda)\left[(-1-\lambda)(-K-\lambda)\right] = 0$$
$$-(1+\lambda)(1+\lambda)(K+\lambda) = 0$$
$$\text{roots} \quad \lambda = -1$$
$$\lambda = K$$

$$x_2 = Ce^{-t}(Ax+B) + De^{K}$$

12.1.4.3. Simultaneous Total Differential Equations

382

Consider the following system of equations

$$\left.\begin{array}{l} P_1\ dx + Q_1\ dy + R_1\ dz = 0 \\ P_2\ dx + Q_2\ dy + R_2\ dz = 0 \end{array}\right\} \quad \dots \ (1)$$

where P, Q, R are functions of x, y and z. Equations (1) may be written as

$$\frac{dx}{P} = \frac{dy}{Q} = \frac{dz}{R} \qquad (2)$$

where

$$P = \begin{vmatrix} Q_1 & R_1 \\ Q_2 & R_2 \end{vmatrix}, \quad Q = \begin{vmatrix} R_1 & P_1 \\ R_2 & P_2 \end{vmatrix}, \quad R = \begin{vmatrix} P_1 & Q_1 \\ P_2 & Q_2 \end{vmatrix}$$

The general solution of (2) consists of two relations involving two arbitrary constants:

$$\left.\begin{array}{l} \phi_1\ (x,\ y,\ z,\ c_1) = 0, \\ \phi_2\ (x,\ y,\ z,\ c_2) = 0 \end{array}\right\} \quad \dots \ (3)$$

Equations (3) may be interpreted as representing a family of curves whose direction at (x, y, z) is given by dx: dy: dz, in (2), i.e., curves whose tangents at (x, y, z) have direction cosines proportional to P: Q: R.

We shall indicate certain methods of obtaining solutions to (2).

12.1.4.3.1. Method (1): Separating ordinary equations
It is sometimes possible to deduce from (2) two equations each of which contains only two variables and their differentials.

Example (8)   Consider $\dfrac{dx}{yz} = \dfrac{dy}{zx} = \dfrac{dz}{xy}$ \qquad (1)

Evidently,

$$\frac{dx}{z} = \frac{dz}{x}\ , \quad \frac{dy}{z} = \frac{dz}{y}$$

which have the solutions

$$x^2 = z^2 + c_1\ , \quad y^2 = z^2 + c_2 \qquad (2)$$

Equations (2) are solutions of (1).

12.1.4.3.2. Method (2): determining linear dependence

It may happen that but one equation containing only two variables and their differentials is readily obtained. In this case the solution of this equation may often be used to obtain another equation in two variables.

383

Example (9) Solve $\dfrac{dx}{y} = \dfrac{dy}{x+z} = \dfrac{dz}{y}$ (1)

Solution:
From the first and third members of (1) we find

$x = z + c_1$ (2)

substituting (2) in (1), we find

$$\dfrac{dy}{2z + c_1} = \dfrac{dz}{y}$$

whose solution is $\tfrac{1}{2} y^2 = z^2 + c_1 z + c_2$ (3)

The solution of (I) consists of (2) and (3) taken together.

Example (10) Solve $dx = \dfrac{dy}{2} = \dfrac{dz}{3z + \cos(y - 2x)}$ (1)

Solution:
From the first and second members of (1) we find

$y - 2x = c_1$ (2)

Substituting (2) in (1) we find

$$dx = \dfrac{dz}{3z + \cos c_1}$$

whose solution is $x = \tfrac{1}{3} \log (3z + \cos c_1) + c_2$ (3)

The solution of (1) consists of (2) and (3) taken together. The solution can be put in the form

$x = \tfrac{1}{3} \log \{ 3z + \cos (y - 2x) \} + c_2$

12.1.4.3.3. Method (3): detecting linear dependence by constant fractions

Each fraction in

$$\dfrac{dx}{P} = \dfrac{dy}{Q} = \dfrac{dz}{R}$$

is equal to $\dfrac{l\, dx + m\, dy + n\, dz}{l\, P + m\, Q + n\, R}$

where 1, m and n are arbitrary multipliers not necessarily constants. This fact may sometimes be used advantageously to obtain a zero denominator and a numerator that is an exact differential, or to obtain a non zero denominator for which the numerator is the differential

Example (11) Solve $\dfrac{dx}{y+z} = \dfrac{dy}{z+x} = \dfrac{dz}{x+y}$

384

Solution:

$$\frac{dx - dy}{y - x} = \frac{dy - dz}{z - y} = \frac{dx - dz}{z - x}$$

from which we find

$$y - x = c_1 \, (z - y) \; ; \; z - y = c_2 \, (z - x)$$

Example (12)   Solve

$$\frac{dx}{y + 2z} = \frac{dy}{x + y + 2z} = \frac{2dz}{-x} \tag{1}$$

Solution:

$$\frac{dx}{y + 2z} = \frac{dy}{x + y + 2z} = \frac{2dz}{-x} = \frac{dx - dy - 2dz}{0}$$

Hence, $dx - dy - 2\,dz = 0$
so that $x - y - 2z = c_1$ (2)

substituting (2) in the first fraction of (1) we find

$$\frac{dx}{x - c_1} = \frac{2dz}{-x}$$

$$\frac{dx}{x - c_1} = \frac{2dz}{-x}$$

The solution of (1) is (2) and (3) taken together.

Example (13) Solve the system:

$$\frac{dx}{4y - 3z} = \frac{dy}{4x - 2z} = \frac{dz}{2y - 3x} \tag{1}$$

Solution:
We seek multipliers 1, m, n such that

$$l \, (4y - 3z) + m \, (4x - 2z) + n \, (2y - 3x) = 0 \tag{i}$$

Rearranging (i) in the form

$$(4m - 3n) \, x + (4l + 2n) \, y + (- 3l - 2m) \, z = 0$$

we see that it will be satisfied when

$$4m - 3n = 0 \; , \quad 4l + 2n = 0 \; , \quad -3l - 2m = 0$$

Or $l : m : n = 2 : -3 : -4$ then

$$2\,dx - 3\,dy - 4\,dz = 0$$

Hence,

$$2x - 3y - 4z = c_1 \tag{2}$$

Using the arrangement:

$$4 \, (ly + mx) + 3 \, (- lz - nx) + 2 \, (ny - mz) = 0$$

and setting

$$ly + m x = 0 \, , \, - lz - nx = 0 \, , \, ny - mz = 0$$

we obtain

$$l : m : n = x : - y : - z$$

Then

$$x \, dx - y \, dy - z \, dz = 0,$$

hence, $\quad x^2 - y^2 - z^2 = c_2 \qquad\qquad (3)$

The solution of (1) is (2) and (3) taken together.

12.1.4.4. Geometrical interpretation of the equation:

$$P \, dx + Q \, dy + R \, dz = 0 \qquad\qquad (1)$$

Let the integral of equation (1) be $\phi(x, y, z) = c$

$$\therefore \quad \frac{\partial \phi}{\partial x} \, dx + \frac{\partial \phi}{\partial y} \, dy + \frac{\partial \phi}{\partial z} \, dz = 0$$

$$\therefore \quad \frac{\partial \phi}{\partial x} : \frac{\partial \phi}{\partial y} : \frac{\partial \phi}{\partial z} = P : Q : R$$

Since $\quad \dfrac{\partial \phi}{\partial x} , \dfrac{\partial \phi}{\partial y} , \dfrac{\partial \phi}{\partial z}$

are the direction ratios of the normal to the surface $\phi(x, y, z) = c$ at the point (x, y, z), then P, Q and R are also the direction ratios of *that* normal.

Next, the equations

$$\frac{dx}{P} = \frac{dy}{Q} = \frac{dz}{R} \qquad\qquad (2)$$

represent a system of curves such that the direction ratios of the tangent at any paint (x,y,z) on any curve of the family are P, Q, R. Hence P, Q and R may represent the direction ratios of the tangent to the curve (2). Hence equation (1) represents a singly infinite system of surfaces which cut the infinite system of curves (2) at right angles whenever they intersect

Example s of these are given in the case of lines of forces and equipotential surfaces in electrostatics.

Example 1   Find the family of curves orthogonal to the surfaces

$$x^2 + 2y^2 + 4z^2 = c$$

Solution:

since $x^2 + 2y^2 + 4z^2 = c$ is the primitive of the total differential equation x dx + 2y dy + 4z dz= 0, the differential equations of the family of orthogonal curves are:

$$\frac{dx}{x} = \frac{dy}{2y} = \frac{dz}{4z}$$

Solving $\dfrac{dx}{x} = \dfrac{dy}{2y}$ we have $y = c_1 x^2$

Solving $\dfrac{dy}{2y} = \dfrac{dz}{4z}$ we have $z = c_2 y^2$

The required family of curves has equations: $\quad y = c_1 x^2 , \; z = c_2 y^2$

12.1.4.5. Exercises on Total Differential Equations

386

Verify that the following differential equations are integrable and find their primitives

1) $yz \log z \, dx - zx \log z \, dy + xy \, dz = 0$

   Ans : $x \log z = cy$

2) $yz \, dx + (xz - yz^3) \, dy - 2xy \, dz = 0$

   Ans : $2xy = y^2 z^2 + cz^2$

3) $2(y + z) \, dx - (x + z) \, dy + (2y - x + z) \, dz = 0$

   Ans : $y + z = c(x + z)^2$

4) $2y(a - x) \, dx + [z - y^2 + (a - x)^2] \, dy - y \, dz = 0$

   Ans : $(a - x)^2 + z = y(c - y)$

5) $y(y + z) \, dx + z(x + z) \, dy + y(y - x) \, dz = 0$

   Ans : $y(x + z) = c(y + z)$

6) Find the form of the function $f(y)$ for which the equation
$y \, dx + z \, dy + f(y) \, dz = 0$ in integrable and solve the equation when the function has this form

   Ans : $f(y) = y \log y$; $\log y + \dfrac{x - c}{z} = 0$

Solve the following system of equations:

7) $\dfrac{dx}{x} = \dfrac{dy}{x + z} = \dfrac{dz}{-z}$

   Aus : $xz = c_1$, $y - x + z = c_2$

8) $\dfrac{dx}{1 + y} = \dfrac{dy}{1 + x} = \dfrac{dz}{z}$

   Ans : $z = c_1(2 + x + y) = \dfrac{c_2}{x - y}$

9) $dx = \dfrac{dy}{3x^2 \sin(2x + z)} = \dfrac{dz}{-2}$

   Ans : $z + 2x = c_1$, $y = x^3 \sin c_1 + c_2$

10) $\dfrac{dx}{x^2 + y^2} = \dfrac{dy}{2xy} = \dfrac{dz}{(x + y)^3 z}$

   Ans : $(x + y)^2 - 2 \log z = c_1$, $y = c_2(x^2 - y^2)$

11) $\dfrac{dx}{3y - 2z} = \dfrac{dy}{z - 3x} = \dfrac{dz}{2x - y}$

   Ans : $x + 2y + 3z = c_1$, $x^2 + y^2 + z^2 = c_2$

387

12) $\dfrac{dx}{x^2 - y^2 - z^2} = \dfrac{dy}{2xy} = \dfrac{dz}{2xz}$

Ans : $y = c_1 z$, $x^2 + y^2 + z^2 = c_2 z$

13) $\dfrac{dx}{x(2y^4 - x^4)} = \dfrac{dy}{y(z^4 - 2x^4)} = \dfrac{dz}{z(x^4 - y^4)}$

Ans : $xyz^2 = c_1$, $x^4 + y^4 + z^4 = c_2$

14) Find the family of curves orthogonal to the family of surfaces $x^2 + y^2 + 2z^2 = c$

Ans : $y = Ax$, $z = By^2$

15) Find a set of curves orthogonal to the surfaces $xyz = c^3$

Ans : $x^2 = z^2 + A$, $y^2 = z^2 + B$

16) Find a set of surfaces orthogonal to the set of curves given by

$\dfrac{dx}{x} = \dfrac{dy}{z} = \dfrac{dz}{y + 2z}$

Ans : $x^2 + 2yz + 2z^2 = c^2$

17) Find a set of surfaces orthogonal to the set of curves given by

$\dfrac{dx}{yz} = \dfrac{dy}{2zx} = \dfrac{dz}{-3xy}$ Ans : $xy^2 = cz^3$

18) Show that there exists no set of surfaces orthogonal to the curves whose differential equation is

$\dfrac{dx}{z} = \dfrac{dy}{x + y} = dz$

12.2. Partial Differential Equations of the First Order

An ordinary differential equation is formed by eliminating arbitrary constants, partial differential equations on the other hand are often formed by eliminating arbitrary functions.

<u>Example</u> (1) $z = f\left(\dfrac{y}{x}\right)$

Differentiating w.r.t. x and y we get

$\dfrac{\partial z}{\partial x} = f'\left(\dfrac{y}{x}\right) \times -\dfrac{y}{x^2}$

$\dfrac{\partial z}{\partial y} = f'\left(\dfrac{y}{x}\right) \times \dfrac{1}{x}$

$\therefore x\,\dfrac{\partial z}{\partial x} + y\,\dfrac{\partial z}{\partial y} = 0$

which is a partial differential equation of the first order.

388

Example (2)   Eliminate the arbitrary functions from
$$y = f(x - ct) + \varphi (x + ct)$$

Solution:

$$\frac{\partial y}{\partial x} = f'(x - ct) + \varphi'(x + ct)$$

$$\frac{\partial^2 y}{\partial x^2} = f'(x - ct) + \varphi''(x + ct)$$

$$\frac{\partial y}{\partial t} = - c\, f'(x - ct) + c\, \varphi'(x + ct)$$

$$\frac{\partial^2 y}{\partial t^2} = c^2\, f'(x - ct) + c^2\, \varphi'' (x + ct)$$

$$\therefore \frac{\partial^2 y}{\partial t^2} = c^2 \frac{\partial^2 y}{\partial x^2}$$

which is a partial differential equation of the second order. Partial differential equations can also be formed by eliminating arbitrary constants.

Example (3)   Eliminate the constants a and b from
$$z = ax^2 + by^2 + ab$$

Solution:

$$\frac{\partial z}{\partial x} = 2ax, \quad \frac{\partial z}{\partial y} = 2by$$

$$\therefore a = \frac{1}{2x} \frac{\partial z}{\partial x}, \quad b = \frac{1}{2y} \frac{\partial z}{\partial y}$$

Substituting in the given equation we get
$$z = \frac{x}{2} \frac{\partial z}{\partial x} + \frac{y}{2} \frac{\partial z}{\partial y} + \frac{1}{4xy} \frac{\partial z}{\partial x} \frac{\partial z}{\partial y}$$

$$\therefore 4xyz = 2px^2y + 2qy^2x + pq$$

Where

$$p = \frac{\partial z}{\partial x}, \quad q = \frac{\partial z}{\partial y}$$

12.2.1. Linear partial differential equations of the first order

They are of the form $Pp + Qq = R$

Where, P, Q and R are functions of x, y and z,

$$p = \frac{\partial z}{\partial x} \qquad q = \frac{\partial z}{\partial y}$$

The first systematic theory of equations of this type was given by Lagrange for that reason equation

(1) is frequently referred to as Lagrange's equation.

389

It should be observed that the term linear means that p and q appear to the first degree only, but P, Q, R may be any functions of x, y, z. This is in contrast to the situation in the theory of ordinary differential equations where z must also appear linearly, e. g., the equation

$$x \frac{\partial z}{\partial x} + y \frac{\partial z}{\partial y} = x^2 + z^2$$

is linear where as the equation

$$x \frac{dz}{dx} = x^2 + z^2$$

is not.

The general solution:

A linear partial differential equation of the first order involving a dependent variable z and two independent variables x and y is of the form

$$P\,p + Q\,q = R \quad (1)$$

where P, Q. R are functions of x, v, z. If P = 0 or Q 0 ; it may by solved easily. Thus, the equation

$$\frac{\partial z}{\partial x} = 2x + 3y$$

has a solution

$$z = x^2 + 3xy + \phi\,(y)$$

where ϕ is an arbitrary function of y. The geometrical meaning of equation (1) is that the normal to a certain surface is perpendicular to a straight line whose direction cosines are in the ratios P: Q: R

Note:
The direction cosines of the normal to a surface f (x, y, z) 0 at the point (x. y, z) are in the ratios

$$\frac{\partial f}{\partial x} : \frac{\partial f}{\partial y} : \frac{\partial f}{\partial z}$$

Since

$$p = \frac{\partial z}{\partial x} = - \frac{\frac{\partial f}{\partial x}}{\frac{\partial f}{\partial z}}, \qquad q = \frac{\partial z}{\partial y} = - \frac{\frac{\partial f}{\partial y}}{\frac{\partial f}{\partial z}}$$

The direction ratios can therefore be written p: q: -1.

The simultaneous equations

$$\frac{dx}{P} = \frac{dy}{Q} = \frac{dz}{R} \qquad (2)$$

represent a family of curves such that the tangent at any point (x, y, z) has direction ratios P ; Q: R. If u = constant and v = constant be two particular integrals of equations (2), then $\phi\,(u, v) = 0$ represents a surface through such curves.

Through every point of such a surface passes a curve of the family lying wholly on the surface. Hence the normal to the surface must be perpendicular to the tangent to this curve, i.e., perpendicular to a line whose direction ratios are P, Q, R. This is just what is required by the partial differential equation.

The equations (1) and (2) are equivalent for they define the same set of surfaces. When equation (1) is given, equations (2) are called the <u>subsidiary or auxiliary equations</u>.

Thus $\phi (u, v) = 0$ is an integral of (1) if u = constant and v = constant are any two independent solutions of the subsidiary equations (2) and P is an arbitrary function. This is called the <u>general integral of Lagrange's linear equation</u>.

Complete solutions:

If u = a and v = b are two independent solutions of (2) and if α , β are arbitrary constants,

$$u = \alpha v + \beta \qquad\qquad (3)$$

is called a complete solution of (1)
A complete solution (3) represents a two parameter family of surfaces which does not have an envelope (since partial differentiation with respect to β gives 0 = 1).

It is possible however to select one — parameter families of surfaces from among (3) which has envelopes. These envelopes (surfaces) are merely particular surfaces of the general solution.

<u>Example (4)</u>   Find the general solution of 2 p + 3 q = 1

The subsidiary equations are given by $\dfrac{dx}{2} = \dfrac{dy}{3} = \dfrac{dz}{1}$

Solution:

From $\quad \dfrac{dx}{2} = \dfrac{dz}{1} \quad$ we have x -2 z = a

From $\quad \dfrac{dx}{2} = \dfrac{dy}{3} \quad$ we have 3 x —2y = b

Thus the general solution is $\quad \phi (x - 2z, 3x - 2y) = 0$

The complete solution
$$x - 2z = \alpha (3x - 2y) + \beta$$

is a two parameter family of planes. The one parameter family determined by taking $\beta = \alpha^2$ has equation
$$x - 2z = \alpha (3x - 2y) + \alpha^2 \qquad\qquad (1)$$

Differentiating (1) with respect to a yields
$$0 = 3x - 2y + 2\alpha$$

or
$$\alpha = - \tfrac{1}{2} (3x - 2y)$$

Substituting from α in (1) we obtain the envelope and its equation is
$$x - 2z = - \tfrac{1}{4} (3x - 2y)^2$$
which is a parabolic cylinder. This envelope is clearly a part of the general solution.

<u>Example</u> (5)   Find the general solution of the differential equation

$$x^2 \frac{\partial z}{\partial x} + y^2 \frac{\partial z}{\partial y} = (x + y)\, z$$

The subsidiary equations are given by

$$\frac{dx}{x^2} = \frac{dy}{y^2} = \frac{dz}{(x+y)\, z}$$

Solution:
From

$$\frac{dx}{x^2} = \frac{dy}{y^2} \quad \text{we have} \quad \frac{1}{x} - \frac{1}{y} = a$$

Also

$$\frac{dx - dy}{x^2 - y^2} = \frac{dz}{(x + y)\, z}$$

Or

$$\frac{dx - dy}{x - y} = \frac{dz}{z}$$

$$\therefore \quad \log (x - y) = \log b\, z$$

$$\therefore \quad \frac{x - y}{z} = b$$

But $\dfrac{x - y}{xy} =$ constant , so that $\dfrac{xy}{z} =$ constant

The subsidiary equations are given by

$$\phi\left(\frac{xy}{z}\ ,\ \frac{x-y}{z}\right) = 0$$

Example (6)   If u is a function of x, y, z which satisfies the partial differential equation:

$$(y-z)\,\frac{\partial u}{\partial x} + (z-x)\,\frac{\partial u}{\partial y} + (x-y)\,\frac{\partial u}{\partial z} = 0$$

Show that u contains x, v and z in the combination

$$x + y + z \quad \text{and} \quad x^2 + y^2 + z^2$$

Solution:
The auxiliary equations are $\dfrac{dx}{y-z} = \dfrac{dy}{z-x} = \dfrac{dz}{x-y} = \dfrac{du}{0}$

Since du = 0 $\therefore$ u = constant

Also dx + dy + dz = 0 $\therefore$ x + y + z = constant

$$\frac{x\,dx}{xy - xz} = \frac{y\,dy}{yz - yx} = \frac{z\,dz}{xz - yz}$$

$\therefore$ $x\,dx + y\,dy + z\,dz = 0$

$\therefore$ $x^2 + y^2 + z^2 = $ constant

The general solution is given by

$$\phi\,(u,\ x + y + z,\ x^2 + y^2 + z^2) = 0$$

Or, $u = f(x + y + z,\ x^2 + y^2 + z^2)$

12.2.2. Integral surfaces passing through a given curve

Suppose that we have found two solutions u (x, y, z) = a, v (x, y, z) = b (1)
of the auxiliary equation:

$$\frac{dx}{P} = \frac{dy}{Q} = \frac{dz}{R}$$

Any solution of the corresponding linear equation P p + Q q = R is of the form

$\phi\,(u, v) = 0$ (2)

arising from a relation

$\phi\,(a, b) = 0$ (3)

between the constants a and b.

The problem we have to consider is that of determining the function ϕ in special circumstances. If we wish to find the integral surface which passes through the curve whose parametric equations are:

x = x(t), y = y(t), z = z (t) where t is a parameter, then the particular solution (1) must be such that:

u{ x(t), y(t), z(t)} = a,   v{ x(t), y(t); z(t)} = b

we therefore have two equations from which we may eliminate the single variable t to obtain a relation of type (3). The solution we are seeking is given by equation (2).

Example (7)   Find a solution of $xp + yq = z$ which represents a surface
containing the parabola $y^2 = 4x$, $z = 1$.

Solution:

The subsidiary equations are

$$\frac{dx}{x} = \frac{dy}{y} = \frac{dz}{z}$$

$$\therefore \quad x = az, \; y = bz \qquad\qquad\qquad (1)$$

The general solution is

$$\phi\left(\frac{x}{z} \; , \; \frac{y}{z}\right) = 0$$

$$y^2 = 4x \; , \; z = 1$$

Let $x = t^2$ then $y = 2t$ and $z = 1$. Substituting in (1)

$$\therefore \quad t^2 = a \; , \; 2t = b$$

Eliminating t $\therefore \quad b^2 = 4a$

and hence from (1)

$$\frac{y^2}{z^2} = 4\,\frac{x}{z} \qquad \therefore \quad y^2 = 4xz$$

Example (8)   Find the general solution of the equation

$$x\,(y^2 + z)\,p - y\,(x^2 + z)\,q = (x^2 - y^2)\,z$$

Also find the particular integral surface which contains the straight line $x + y = 0$, $z = 1$

Solution:

The auxiliary equations are:

$$\frac{dx}{x\,(y^2 + z)} = \frac{dy}{-\,y\,(x^2 + z)} = \frac{dz}{(x^2 - y^2)\,z}$$

$$\therefore \quad \frac{y\,dx}{xy^3 + xyz} = \frac{x\,dy}{-x^3y - xyz} = \frac{dz}{(x^2 - y^2)z}$$

$$\therefore \quad \frac{x\,dy + y\,dx}{-\,xy\,(x^2 - y^2)} = \frac{dz}{(x^2 - y^2)\,z}$$

$$\therefore \quad \frac{x\,dy + y\,dx}{xy} = -\,\frac{dz}{z}$$

$$\therefore \quad \log xy = -\,\log\frac{z}{a}$$

394

$$\therefore \quad x\,y\,z = a \qquad (1)$$

Also,

$$\frac{x\,dx + y\,dy}{z\,(x^2 - y^2)} = \frac{dz}{(x^2 - y^2)\,z}$$

$$\therefore \quad x^2 + y^2 - 2z = b \qquad (2)$$

The general solution is given by

$$\varphi\,(x\,y\,z,\ x^2 + y^2 - 2z) = 0$$

From the curve $x + y = 0$, $z\ 1$, we have the freedom equations

$$x = t,\ y = -t,\ z = 1$$

Substituting these values in equations (1) and (2)

$$\therefore \quad -t^2 = a\ ,\ 2t^2 - 2 = b$$

and eliminating t from the above two equations we get

$$2\,a + b + 2 = 0$$

showing that the desired integral surface is

$$2\,x\,y\,z + x^2 + y^2 - 2z + 2 = 0$$

12.2.3. Exercises on linear partial differential equations of the First Order

1) Find the general solution of the equation

$$y \frac{\partial z}{\partial x} + x \frac{\partial z}{\partial y} + x + y = 0 \qquad \text{Ans :} \quad \phi\,(x + y + z,\ x^2 - y^2) = 0$$

2) Find the general solution of the equation

$$(x^2 - y^2 - z^2)\,p + 2xy\,q = 2xz$$

Prove that a complete solution consists of spheres through the origin with centers on the plane yoz

$$\text{Ans :} \quad \phi\left(\frac{y}{z},\ \frac{x^2 + y^2 + z^2}{z}\right) = 0$$

a complete solution is $\quad x^2 + y^2 + z^2 = \alpha y + \beta z$

3) Find the equation of the surface satisfying the equation
$$4yz\,p + q + 2y = 0$$
and passing through $\quad y^2 + z^2 = 1, \quad z + z = 2$

$$\text{Ans :} \quad y^2 + z^2 + x + z = 3$$

4) Find the equation of the integral surface of the differential equation

$$2y\,(z-3)\,p + (2x - z)\,q = y\,(2x - 3)$$

which passes through the circle $z = 0, \quad x^2 + y^2 = 2x$

$$\text{Ans :} \quad x^2 + y^2 - 2x = z^2 - 4z$$

5) Find the equation of the integral surface of the equation

$$(x - y)\,y^2\,p + (y - x)\,x^2\,q = (x^2 + y^2)\,z$$

through the curve $\quad xz = c^3, \quad y = 0$

$$\text{Ans :} \quad z^3\,(x^3 + y^3)^2 = c^9\,(x - y)^3$$

12.2.4. Non linear partial differential equations of the first order

These are equations in which p and q appear other than in the first degree.

<u>Complete and singular solutions:</u>
Let the non linear partial differential equation of the first order.

$$f\,(x,\ y,\ z,\ p,\ q) = 0 \qquad\qquad (1)$$

be derived from

$$g\,(x,\ y,\ z,\ a,\ b) = 0 \qquad\qquad (2)$$

by eliminating the arbitrary constants a and b. Then (2) is called the complete solution of (1).
This complete solution represents a two—parameter family of surfaces which may or may not have an envelope. To find the envelope (if one exists) we eliminate a and b from

$$g = 0, \quad \frac{\partial g}{\partial a} = 0, \quad \frac{\partial g}{\partial b} = 0$$

If the eliminant $\lambda (x, y, z)$ (3)

satisfies (1) it is called a singular solution of (1) ;

If $\lambda (x, y, z) = u (x, y, z) . v (x, y, z)$

and if u = 0 satisfies (1) while v = 0 does not, u = 0 is the singular solution. As in the case of ordinary differential equations, the singular solution may be obtained from the partial differential equation by eliminating p and g from

$$f = 0, \quad \frac{\partial f}{\partial p} = 0, \quad \frac{\partial f}{\partial q} = 0$$

<u>Method of solutions:</u>

12.2.4.1. <u>Case I:</u> Equations containing p and q only

<u>Example</u> (1)   Find a complete solution for the equation
$$p^2 - q^2 = 1$$

Solution:

Let $q = a$ $\therefore \quad p = \sqrt{a^2 + 1}$

$$dz = \frac{\partial z}{\partial x} dx + \frac{\partial z}{\partial y} dy = p \, dx + q \, dy$$

$$\therefore \quad dz = \sqrt{a^2 + 1} \, dx + a \, dy$$

$$\therefore \quad z = \sqrt{a^2 + 1} \, x + ay + c$$

The equations for determining the singular solution are

$$z = \sqrt{a^2 + 1} \, x + ay + c , \quad \frac{a}{\sqrt{a^2 + 1}} + y = 0 ,$$

$1 = 0$
Thus there is no singular solution
If we put $a = \tan a$

$$\therefore \quad \sqrt{a^2 + 1} = \sec \alpha$$

so that

$z = x \sec a + y \tan a + c$

<u>Example</u> (2)   Solve the equation $q = \log p$

Solution:

Let $q = a$
$$\therefore \quad p = e^a$$
$$dz = p \, dx + q \, dy$$

$$= e^a \, dx + a \, dy$$
$$\therefore \quad z = xe^a + ay + c$$

There is no singular solution

12.2.4.2. <u>Case II</u>: Equations of the form $z = p\,x + q\,y + f(p, q)$

This is known as the extended Clairaut type of equation.
The complete solution can be found by putting $p = a$, $q = b$
so that $z = a\,x + b\,y + f(a, b)$

<u>Example</u> (3)   Find a complete solution of the equation
$$z = px + qy + p^2 + pq + q^2$$

   Find also the singular solution

Solution:

The complete solution is
$$z = ax + by + a^2 + ab + b^2$$
Differentiating the complete solution with respect to a and b we have
$0 = x + 2a + b, \ 0 = y + a + 2b$
Solving to obtain
$$a = \tfrac{1}{3}(y - 2x), \quad b = \tfrac{1}{3}(x - 2y)$$

and substituting in the complete solution, the singular solution is
$$3z = xy - x^2 - y^2$$

12.2.4.3. <u>Case III</u>: Only p, q, z present, the equation takes the form $f(z, p, q) = 0$

We assume as a trial solution $z = f(u)$ where $u = x + a\,y$

$$p = \frac{\partial z}{\partial x} = \frac{dz}{du}\frac{\partial u}{\partial x} = \frac{dz}{du}$$

$$q = \frac{\partial z}{\partial y} = \frac{dz}{du}\frac{\partial u}{\partial y} = a\,\frac{dz}{du}$$

when these are substituted in the given differential equation, we obtain an ordinary differential equation of the first order
$$f\!\left(z, \frac{dz}{du}, \ a\,\frac{dz}{du}\right) = 0$$
whose solution is the required complete solution.

<u>Example</u> (4)   Find a complete solution and the singular solution of the equation
$$z = p^2 + q^2$$

Solution:
Let $z = f(u)$ where $u = x + a\,y$ $\therefore \quad p = \frac{dz}{du}$, $q = a\,\frac{dz}{du}$

The equation is reduced to

$$z = \left(\frac{dz}{du}\right)^2 + a^2\left(\frac{dz}{du}\right)^2$$

$$= (1 + a^2) \left(\frac{dz}{du} \right)^2$$

$$\therefore \quad \frac{dz}{\sqrt{z}} = \frac{1}{\sqrt{1+a^2}} \, du$$

$$\therefore \quad 2\sqrt{z} = \frac{1}{\sqrt{1+a^2}} \, u + \frac{b}{\sqrt{1+a^2}} \quad \text{or} \quad 2\sqrt{z} = \frac{1}{\sqrt{1+a^2}} (u + b)$$

Thus a complete solution is
$$4(1 + a^2) \, z = (x + ay + b)^2$$

which is a family of parabolic cylinders. Taking the derivatives with respect to a and b we have
$$8\,a\,z - 2y \, (x + ay + b) = 0 \, , \, x + ay + b = 0$$
The singular solution is $z = 0$

12.2.4.4. <u>Case IV</u>: Equations of the form f(x, p) = F(y, q))

We put each side equal to an arbitrary constant a and solve to obtain
$$p = \phi(x, a)$$
and
$$q = \psi(y, a)$$
Since z is a function of x and y we have
$$dz = p \, dx + q \, dy$$
$$= \phi(x, a) \, dx + \psi(y, a) \, dy$$
$$\therefore \quad z = \int \phi(x, a) \, dx + \int \psi(y, a) \, dy + b$$

containing two arbitrary constants is the required complete solution.

<u>Example</u> (5) Find the complete solution of the equation
$$p - q = x^2 + y^2$$

Solution:

$$p - x^2 = q + y^2 = a \text{ say}$$
$$\therefore \quad p = a + x^2 \, , \, q = a - y^2$$
$$dz = p \, dx + q \, dy$$
$$= (a + x^2) \, dx + (a - y^2) \, dy$$
$$\therefore \quad z = ax + \tfrac{1}{3} x^3 + ay - \tfrac{1}{3} y^3 + b$$

There is no singular solution

12.2.5. Exercises on the non-linear partial differential equations of the first order

399

Find a complete solution for the following partial differential equations and the singular solution (S.S.) if any

1) $p^2 + q^2 = 9$

   Ans : $z = 3x \cos \alpha + 3y \sin \alpha + c$ no S.S.

2) $pq + p + q = 0$

Ans : $z = ax + by + c$ where $ab + a + b = 0$

or $z = ax - \dfrac{a}{a+1} y + c$, no S S.

3) $z = px + qy + 3 \sqrt[3]{pq}$

Ans : $z = ax + by + 3 \sqrt[3]{ab}$, S.S. $xyz = 1$

4) $z = px + qy + p^2 q^2$

Ans : $z = ax + by + a^2 b^2$, S.S. $z = -\frac{3}{4} \sqrt[3]{4} \, x^{\frac{2}{3}} y^{\frac{2}{3}}$

5) $4 (1 + z^3) = 9 z^4 pq$

   Ans : $a (1 + z^3) = (x + ay + b)^2$; S.S. $z^3 + 1 = 0$

6) $p (1 - q^2) = q (1 - z)$

Ans : $4 (1 - a + az) = (x + ay + b)^2$, no S.S.

7) $\sqrt{p} - \sqrt{q} + 3x = 0$

Ans : $z = -\dfrac{1}{9} (9 - 3x)^3 + a^2 y + b$; no S.S.

8) $q (p - \cos x) = \cos y$

Ans : $z = ax + \sin x + \dfrac{1}{a} \sin y + b$, no S.S.

12.3. Linear Partial Differential Equations with Constant Coefficients

12.3.1. Homogeneous partial linear equations with constant coefficients:

An equation which is linear in the dependent variable z and its partial derivatives is called a linear partial differential equation. The order is the order of the highest derivative which occurs. For Example ,

$$(x^2 + y^2) \frac{\partial^3 z}{\partial x^3} + 3y \frac{\partial^3 z}{\partial x \, \partial y^2} - \frac{\partial^2 z}{\partial y^2} + x \frac{\partial z}{\partial x} = e^{2x+y}$$

is a linear partial differential equation of the third order.

A linear partial differential equation in which the derivatives are all of the same order is called homogeneous. For Example ,

$$x^2 \frac{\partial^3 z}{\partial x^3} + 2xy \frac{\partial^3 z}{\partial x^2 \, \partial y} + 4 \frac{\partial^3 z}{\partial x \, \partial y^2} + \frac{\partial^3 z}{\partial x^3} = 2x + 3y^2$$

is a homogeneous partial differential equation of the third order.

The general form of the homogeneous linear partial differential equation of order n in two independent variables is

$$(D_x^n + a_1 D_x^{n-1} D_y + a_2 D_x^{n-2} D_y^2 + \cdots$$

$$\cdots + a_n D_y^n) \, z = \phi(x, y) \quad \cdots \cdots (1)$$

where $D_x \equiv \dfrac{\partial}{\partial x}$, $D_y \equiv \dfrac{\partial}{\partial y}$

$a_1, a_2, \ldots, a_n$ are real constants.

The simplest case is

$$(D_x - m \, D_y) \, z = 0$$

i.e. $p - m \, q = 0$

Subsidiary equations are

$$dx = \frac{dy}{-m} = \frac{dz}{0}$$

$$\therefore \quad z = a, \quad y + m \, x = b$$

so that the solution is

$$\phi(z, y + m \, x) = 0$$

i.e., $z = F(y + m \, x)$

This suggests what is easily verified that the general solution of (1) with $\phi(x, y) = 0$ is given by

$$z = F_1(y + m_1 x) + F_2(y + m_2 x) + \cdots + F_n(y + m_n x)$$

where $m_1, m_2, \ldots, m_n$ are the roots (supposed all different) of $m^n + a_1 m^{n-1} + a_2 m^{n-2} + \cdots + a_n = 0$

For Example , the equation

$$\frac{\partial^3 z}{\partial x^3} - \frac{\partial^3 z}{\partial x^2 \partial y} - 2 \frac{\partial^3 z}{\partial x \partial y^2} = 0$$

i - e. $(D_x^3 - D_x^2 D_y - 2 D_x D_y^2) \, z = 0$

Or,

$$D_x (D_x^2 - D_x D_y - 2 D^2 y) \, z = 0$$

i - e $\quad D_x (D_x - 2 D_y) (D_x + D_y) \, z = 0$

The roots are $0, 2, -1$.

Hence,

$$z = F_1(y) + F_2(y + 2x) + F_3(y - x)$$

Case when the auxiliary equation has equal roots:

Consider the equation
$$(D_x - m\, D_y)^2\, z = 0 \qquad (1)$$

Put $(D_x - m\, D_y)\, z = u$

Equation (1) becomes $(D_x - m\, D_y)\, u = 0$

giving $u = F_1\,(y + mx)$

Therefore $(D_x - m\, D_y)\, z = F_1\,(y + m\,x)$

or

$p - m\, q = F_1\,(y + m\,x)$

the subsidiary equations are $\quad dx = \dfrac{dy}{-m} = \dfrac{dz}{F_1(y + mx)}$

giving $y + m\, x = a$ and $\quad dz - F_1(a)\, dx = 0$

i-e $\quad z - x\, F_1\,(y + mx) = b$

so that the general integral is $\quad \phi\,[y + mx,\ z - x\, F_1\,(y + m\,x)] = 0$

Or $\quad z = F_o\,(y + mx) + x\, F_1\,(y + m\,x)$

Similarly we can prove that the integral of $\quad (D_x - m\, D_y)^k\, z = 0$

Is $\quad z = F_o\,(y + mx) + x\, F_1\,(y + m\,x) + x^2\, F_2\,(y + m\,x)$

$$+ \cdots + x^{k-1}\, F_{k-1}\,(y + m\,x)$$

where F_o, F_1, $\ldots$, F_{k-1} are arbitrary functions.

Case of imaginary roots:

If one of the numbers say m_1 is imaginary, then –another say m_2 is the conjugate of m_1.
Let $m_1 = a + i b$, then $m_2 = a - i b$, so that

$$f(D_x, D_y)\ z = [D_x - (a + i b)\ D_y][D_x - (a - i b)\ D_y] \times$$

$$(D_x - m_3\ D_y) \cdots (D_x - m_n\ D_y) = 0$$

The part of the general solution given by the first two factors is

$$\psi_1 \{ y + x (a + i b) \} + \psi \{ y + x (a - i b) \}$$

which can be written in the form

$$[\ \phi_1 (y + ax + ibx) + i\ \phi_2 (y + ax + ibx)] + [\phi_1 (y + ax - ibx) - i\ \phi_2 (y + ax - ibx)]$$

$$= \phi_1 (y + ax + ibx) + \phi_1'(y + ax - ibx) + i [\phi_2(y + ax + ibx) - \phi_2 (y + ax - ibx)]$$

where ϕ_1 and ϕ_2 are arbitrary real functions.

Example (1) Solve $(D_x{}^2 - 2 D_x D_y - 3 D_y{}^2)\ z = 0$

Solution:

$$(D_x - 3 D_y)\ (D_x + D_y)\ z = 0$$
$$m_1 = 3 \ '\ m_2 = - 1$$

The general solution is $z = \phi_1 (y + 3x) + \phi_2 (y - x)$

Example (2) Solve $(D_x{}^3 + 3 D_x{}^2 D_y - 4 D_y{}^3)\ z = 0$

Solution:
$$(D_x{}^3 - D_x{}^2 D_y + 4 D_x{}^2 D_y - 4 D_y{}^3)\ z = 0$$

$$\therefore\ [D_x{}^2 (D_x - D_y) + 4 D_y (D_x - D_y)\ (D_x + D_y)]\ z = 0$$

$$\therefore\ (D_x - D_y)\ (D_x{}^2 + 4 D_x D_y + 4 D_y{}^2)\ z = 0$$

or $(D_x - D_y)\ (D_x + 2 D_y)^2\ z = 0$
$$m_1 = 1,\ m_2 = m_3 = - 2$$

The general solution is

$$z = \phi_1 (y + x) + \phi_2 (y - 2x) + x\ \phi_3 (y - 2x)$$

<u>Example</u> (3)   Solve $(D_x^2 - 2 D_x D_y + 5 D_y^2) z = 0$

Solution:

$$D_x^2 - 2 D_x D_y + 5 D_y^2 = 0$$

$$D_x = \frac{2 D_y \pm \sqrt{4 D_y^2 - 20 D_y^2}}{2}$$

$$= D_y \pm 2i D_y = (1 \pm 2i) D_y$$

$$m_1 = 1 + 2i \ , \ m_2 = 1 - 2i$$

$$\therefore \ z = \phi_1 (y + x + 2ix) + \phi_1 (y + x - 2ix)$$

$$+ i \left[\phi_2 (y + x + 2ix) - \phi_2 (y + x - 2ix) \right]$$

where ϕ_1 and ϕ_2 are real functions.

The particular integral:

It can be easily seen that the general solution of the equation

$$f(D_x, D_y) z = \phi(x, y) \qquad\qquad (1)$$

where

$$f(D_x, D_y) \equiv D_x^n + a_1 D_x^{n-1} D_y + a_2 D_x^{n-2} D_y^2 + \cdots + a_n D_y^n$$

consists of the general solution. of the equation

$$f(D_x, D_y) z = 0$$

plus any particular integral of (I).

The following rules concerning the operators

$$D_x = \frac{\partial}{\partial x}, \qquad D_y = \frac{\partial}{\partial y}$$

are useful in determining the particular integral.

1) To prove that

$$F(D_x, D_y) e^{ax + by} = F(a, b) e^{ax + by}$$

The proof follows from the fact that $F(D_x, D_y)$ is made up of terms of the type $c_{rs} D_x^r D_y^s$ and

$$D_x^r D_y^s e^{ax+by} = D_x^r(D_y^s e^{ax+by}) = D_x^r(b^s e^{ax+by}) = b^s(D_x^r e^{ax+by})$$
$$= a^r b^s e^{ax+by}$$

so that $(c_{rs} D_x^r D_y^s) e^{ax+by} = c_{rs} a^r b^s e^{ax+by}$

and hence, $F(D_x, D_y) e^{ax+by} = F(a, b) e^{ax+by}$

Also

$$\frac{1}{F(D_x, D_y)}\left[F(D_x, D_y) e^{ax+by} \right] = e^{ax+by}$$

$$\therefore \quad \frac{1}{F(D_x, D_y)}\left[F(a, b) e^{ax+by} \right] = e^{ax+by}$$

$$\therefore \quad \frac{1}{F(D_x, D_y)} e^{ax+by} = \frac{1}{F(a, b)} e^{ax+by}$$

provided that $F(a, b) \neq 0$

If $F(a, b) = 0$ we use the following rule (2):

2) To prove that

$$F(D_x, D_y) \left\{ e^{ax+by} \ V \right\}$$

$$= e^{ax+by} \left\{ F(D_x + a, D_y + b) \ V \right\}$$

where V is a function of x and y.

The proof is direct, making use of <u>Leibnitz theorem</u> for the rth derivative of a product to show that

$$D_y^s (e^{by} V) = \sum_{p=0}^{s} {}^sC_p \ (D_y^p \ e^{by}) \ (D_y^{s-p} \ V)$$

$$= e^{by} \sum_{p=0}^{s} ({}^sC_p \ b^p \ D_y^{s-p})$$

$$= e^{by} (D_y + b)^s \ V$$

$$D_x^r \ D_y^s (e^{ax+by} \ V) = D_x^r \ e^{ax} \ e^{by} (D_y + b)^s \ V$$

$$= e^{by} \ D_x^r (e^{ax} \ U)$$

where $U = (D_y + b) \ V$

$$= e^{by} \ e^{ax} (D_x + a)^r \ U$$

$$= e^{ax+by} (D_x + a)^r (D_y + b)^s \ V$$

$$\therefore \quad F(D_x, D_y) \left\{ e^{ax+by} \ V \right\}$$

$$= e^{ax+by} \ F(D_x + a, D_y + b) \ V$$

Also:

$$\frac{1}{F(D_x, D_y)} \left[F(D_x, D_y) \ e^{ax+by} \ V_1 \right] = e^{ax+by} \ V_1$$

$$\therefore \quad \frac{1}{F(D_x, D_y)} \left[e^{ax+by} \ F(D_x + a, D_y + b) \ V_1 \right]$$

$$= e^{ax+by} \ V_1$$

Let $F(D_x + a, D_y + b) \ V_1 = V$

$$\therefore \ V_1 = \frac{1}{F(D_x + a, D_y + b)} \ V$$

$$\therefore \ \frac{1}{F(D_x, D_y)} \left[e^{ax+by} \ V \right] = e^{ax+by} \frac{1}{F(D_x + a, D_y + b)} \ V$$

Corollary:

$$\frac{1}{(\beta D_x - \alpha D_y)^r} \ e^{\alpha x + \beta y} = e^{\alpha x + \beta y} \ \frac{1}{\left\{ \beta(D_x + \alpha) - \alpha(D_y + \beta) \right\}^r} \tag{1}$$

$$= \frac{e^{\alpha x + \beta y}}{\beta^r \ D_x^r} \left(1 + \frac{\alpha D_y}{\beta D_x} + \dots \right) (1)$$

$$= \frac{1}{\beta^r} \ e^{\alpha x + \beta y} \ \frac{1}{D_x^r} \ (1)$$

$$= \frac{1}{\beta^r} \ e^{\alpha x + \beta y} \ \frac{x^r}{r!}$$

3) $F(D_x^2, D_x D_y, D_y^2) \ \begin{matrix} \sin \\ \cos \end{matrix} (ax + by) = F(-a^2, -ab, -b^2) \ \begin{matrix} \sin \\ \cos \end{matrix} (ax + by)$

$D_y^s \ \begin{matrix} \sin \\ \cos \end{matrix} (ax + by) = (-b^2)^s \ \begin{matrix} \sin \\ \cos \end{matrix} (ax + by)$

$D_x^{2r} \ D_y^{7s} \ \begin{matrix} \sin \\ \cos \end{matrix} (ax + by) = (-b^2)^s D_x^{2r} \ \begin{matrix} \sin \\ \cos \end{matrix} (ax + by)$

$$= (-a^2)^r (-b^2)^s \ \begin{matrix} \sin \\ \cos \end{matrix} (ax + by)$$

And $(D_x D_y)^p \ \begin{matrix} \sin \\ \cos \end{matrix} (ax + by) = (-ab)^p \ \begin{matrix} \sin \\ \cos \end{matrix} (ax + by)$

Hence

$$F(D_x^2, D_x D_y, D_y^2) \, \frac{\sin}{\cos} (ax + by) = F(-a^2, -ab, -b^2) \, \frac{\sin}{\cos} (ax + by)$$

Also

$$\frac{1}{F(D_x^2, D_x, D_y, D_y^2)} \left[F(D_x^2, D_x D_y, D_y^2) \, \frac{\sin}{\cos} (ax+by) \right] = \frac{\sin}{\cos} (ax + by)$$

$$\therefore \frac{1}{F(D_x^2, D_x D_y, D_y^2)} \left[F(-a^2, -ab, -b^2) \, \frac{\sin}{\cos} (ax+by) \right] = \frac{\sin}{\cos} (ax + by)$$

$$\therefore \frac{1}{F(D_x^2, D_x D_y, D_y^2)} \, \frac{\sin}{\cos} (ax+by) = \frac{1}{F(-a^2, -ab, -b^2)} \, \frac{\sin}{\cos} (ax + by)$$

provided that $F(-a^2, -ab, -b^2) \neq 0$

If $F(-a^2, -ab, -b^2) = 0$ we work as in example (6)

4) when $\phi(x,y)$ is a polynomial in x and y we expand
$\dfrac{1}{F(D_x, D_y)}$ in ascending powers of $\dfrac{D_y}{D_x}$ or $\dfrac{D_x}{D_y}$ as will be
seen in the solved examples.

5) General method for finding a particular integral:

Consider $(D_x - m D_y) z = \phi(x, y)$

The subsidiary equations are $\quad dx = \dfrac{dy}{-m} = \dfrac{dz}{\phi(x,y)}$

of which one integral is $y + m x = a$ (a is a constant)
Using this integral to find another,

$$z = \int \phi(x, a - mx) \, dx$$

where a is to be replaced by $y + m x$ after integration. Hence we may take

$$\frac{1}{D_x - m D_y} \phi(x,y) \qquad \text{As} \qquad \int \phi(x, a - mx) dx$$

where a is replaced by $y m x$ after integration. For <u>Example</u>, the particular integral of

$$(D_x + D_y) \ z = \cos(x - y)$$

is given by,

$$z = \frac{1}{D_x + D_y} \cos(x - y)$$

$$= \int \cos\left\{ x - (a + x) \right\} dx = \int \cos - a \ dx$$

$$= x \cos - a = x \cos(x - y)$$

<u>Example</u> (1)   Solve $(D_x - 2 D_y)^2 (D_x + 3 D_y) \ z = 10 \ e^{2x+y}$

Solution:

C. F. $z = \phi_1 (y + 2x) + x \ \phi_2 (y - 2x) + \phi_3 (y - 3x)$

P. I. $z = 10 \ \dfrac{1}{(D_x - 2D_y)^2 \ (D_x + 3D_y)} \ e^{2x+y}$

$$= 10 \ \frac{1}{(D_x - 2 D_y)^2} \left\{ \frac{1}{D_x + 3 D_y} \ e^{2x+y} \right\}$$

$$= 10 \ \frac{1}{(D_x - 2 D_y)^2} \left\{ \frac{1}{2 + 3} \ e^{2x+y} \right\}$$

$$= 2 \ \frac{1}{(D_x - 2 D_y)^2} \ (e^{2x+y} . \ 1)$$

$$= 2 \ e^{2x+y} \ \frac{1}{\left\{ (D_x + 2) - 2 (D_y + 1) \right\}^2} \quad (1)$$

$$= 2 \ e^{2x+y} \ \frac{1}{(D_x - 2 D_y)^2} \ (1)$$

$$= 2 \ e^{2x+y} \ \frac{1}{D_x^2} \ (1 - \frac{2 D_y}{D_x})^{-2} (1)$$

$$= 2 \ e^{2x+y} \ \frac{1}{D_x^2} \ (1)$$

$$= x^2 \ e^{2x+y}$$

$\therefore \ z = \phi_1 (y + 2x) + x \ \phi_2 (y + 2x) + \phi_3(y - 3x) + x^2 \ e^{2x+y}$

<u>Example</u> (2)   Solve $\dfrac{\partial^2 z}{\partial x^2} - 2 \dfrac{\partial^2 z}{\partial x \, \partial y} = e^{2x} + x^3 y$

Solution:

$$(D^2_x - 2\,D_x D_y)\,z = e^{2x} + x^3 y$$

$$D_x(D_x - 2D_y)\,z = e^{2x} + x^3 y$$

C. F. $\quad z = \phi_1(y) + \phi_2(y + 2x)$

P. I. $\quad z = \dfrac{1}{D_x(D_x - 2D_y)}\,e^{2x} + \dfrac{1}{D_x(D_x - 2D_y)}\,x^3 y$

$$= \dfrac{1}{2\times 2}\,e^{2x} + \dfrac{1}{D_x^2}\,\dfrac{1}{1 - \dfrac{2\,D_y}{D_x}}\,x^3 y$$

$$= \dfrac{1}{4}\,e^{2x} + \dfrac{1}{D_x^2}\left(x^3 y + \dfrac{2}{D_x}(x^3)\right)$$

$$= \dfrac{1}{4}\,e^{2x} + \dfrac{1}{D_x^2}\left(x^3 y + \tfrac{1}{2}x^4\right)$$

$$= \dfrac{1}{4}\,e^{2x} + \dfrac{1}{20}\,x^5 y + \dfrac{1}{60}\,x^6$$

$\quad z \quad = \quad$ C. F. + P. I.

Example (3) Solve

$$\frac{\partial^3 z}{\partial x^3} - 7\,\frac{\partial^3 z}{\partial x\,\partial y^2} - 6\,\frac{\partial^3 z}{\partial y^3} = \sin(x + 2y) + e^{3x+y}$$

Solution:

$$(D_x^3 - 7\,D_x D_y^2 - 6\,D_y^3)\,z = \sin(x + 2y) + e^{3x+y}$$

$\therefore \;(D_x + D_y)(D_x + 2\,D_y)(D_x - 3\,D_y)\,z = \sin(x+2y) + e^{3x+y}$

C. F. $\quad z = \phi_1(y - x) + \phi_2(y - 2x) + \phi_3(y + 3x)$

P. I $\quad z = \dfrac{1}{(D_x + D_y)(D_x^2 - D_x D_y - 6\,D_y^2)}\,\sin(x + 2y)$

$$+ \dfrac{1}{(D_x + D_y)(D_x + 2\,D_y)(D_x - 3\,D_y)}\,e^{3x+y}$$

$$z = \dfrac{1}{(D_x + D_y)(-1 + 2 + 24)}\,\sin(x + 2y)$$

$$+ \dfrac{1}{D_x - 3\,D_y}\left[\dfrac{1}{(3+1)(3+2)}\,e^{3x+y}\right]$$

$$= \dfrac{1}{25}\,\dfrac{D_x - D_y}{D_x^2 - D_y^2}\,\sin(x+2y) + \dfrac{1}{20}\,\dfrac{1}{D_x - 3\,D_y}\,e^{3x+y}$$

$$= \frac{1}{25} \frac{1}{-1+4} \left[\cos(x+2y) - 2\cos(x+2y) \right]$$

$$+ \frac{1}{20} e^{3x+y} \frac{1}{[(D_x+3)-3(D_y+1)]} \quad (1)$$

$$= -\frac{1}{75} \cos(x+2y) + \frac{1}{20} e^{3x+y} \frac{1}{D_x - 3D_y} \quad (1)$$

$$= -\frac{1}{75} \cos(x+2y) + \frac{1}{20} xe^{3x+y}$$

$$z = C.F. + P.I.$$

Example (4) Solve $(D_x + D_y)^2 (D_x - D_y) z = e^x \cos 2y$

Solution:

C. F. $z = \phi_1(y-x) + x\, \phi_2(y-x) + \phi_3(y+x)$

P. I. $z = \dfrac{1}{(D_x + D_y)^2 (D_x + D_y)} e^x \cos 2y$

$$= e^x \frac{1}{(D_x + 1 + D_y)^2 (D_x + 1 - D_y)} \cos 2y$$

$$= e^x \frac{1}{D_x + D_y + 1} \cdot \frac{1}{D_x^2 + 2D_x + 1 - D_y^2} \cos 2y$$

$$= e^x \frac{1}{D_x + D_y + 1} \cdot \frac{1}{2D_x + 1 + 4} \cos 2y$$

$$= e^x \frac{1}{(2D_x + 5)(D_x + D_y + 1)} \cos 2y$$

$$= e^x \frac{1}{2D_x^2 + 2D_x D_y + 7D_x + 5D_y + 5} \cos 2y$$

$$= e^x \frac{1}{7D_x + 5D_y + 5} \cos 2y$$

$$= e^x \frac{7D_x + 5D_y - 5}{49D_x^2 + 70D_x D_y + 25D_y^2 - 25} \cos 2y$$

$$= e^x (7D_x + 5D_y - 5) \frac{1}{-100 - 25} \cos 2y$$

$$= -\frac{1}{125} e^x (-10\sin 2y - 5\cos 2y) = \frac{1}{25} e^x (2\sin 2y + \cos 2y)$$

<u>Example</u> (5)

Find a real function V of x: and y, reducing to zero when y = 0 and satisfying

$$\frac{\partial^2 V}{\partial x^2} + \frac{\partial^2 V}{\partial y^2} = -4\pi (x^2 + y^2).$$

Solution:

Since a real function is required we consider only the particular integral given by

$$V = -4\pi \frac{1}{D_x^2 + D_y^2} (x^2 + y^2)$$

$$= -4\pi \frac{1}{D_x^2} \left[\frac{1}{1 + \frac{D_y^2}{D_x^2}} (x^2 + y^2) \right]$$

$$= -4\pi \frac{1}{D_x^2} \left[(1 - \frac{D_y^2}{D_x^2})(x^2 + y^2) \right]$$

$$= -4\pi \frac{1}{D_x^2} \left[(x^2 + y^2) - \frac{1}{D_x^2} \cdot 2 \right]$$

$$= -4\pi \frac{1}{D_x^2} (x^2 + y^2 - x^2)$$

$$= -4\pi \frac{1}{D_x^2} y^2 = -2\pi x^2 y^2$$

<u>Example</u> (6)   Solve $(D_x + 2 D_y)(D_x - 2 D_y)(D_x - 3 D_y) z = 20 \sin(2x+y)$

Solution:

C. F. $z = \phi_1 (y - 2x) + \phi_2 (y + 2x) + \phi_3 (y + 3x)$

P. I $z = 20 \frac{1}{D_x^2 - 4 D_y^2} \left[\frac{D_x + 3 D_y}{D_x^2 - 9 D_y^2} \sin(2x+y) \right]$

$$= 20 \frac{1}{D_x^2 - 4 D_y^2} \left[\frac{1}{-4 + 9} \right\} 2 \cos(2x+y) + 3 \cos(2x+y) \right]$$

$$= 20 \frac{1}{D_x^2 - 4 D_y^2} \cos(2x+y)$$

$$= 20 \ \text{Real} \left[\frac{1}{(D_x - 2 D_y)(D_x + 2 D_y)} e^{(2x+y)i} \right]$$

412

$$= 20 \text{ Real} \left[\frac{1}{D_x - 2 D_y} \frac{1}{4i} e^{(2x+y)i} \right]$$

$$= - 5 \text{ Real} \left[\frac{i}{D_x - 2 D_y} \right\} e^{(2x+y)i} \times 1 \right\} \right]$$

$$= - 5 \text{ Real} \left[i\, e^{(2x+y)i} \frac{1}{(D_x + 2i) - 2(D_y + i)} (1) \right]$$

$$= - 5 \text{ Real} \left[i\, e^{(2x+y)i} \frac{1}{D_x \left(1 - \dfrac{2 D_y}{D_x}\right)} (1) \right]$$

$$= - 5 \text{ Real} \left[i\, x\, e^{(2x+y)i} \right]$$

$$= - 5 \text{ Real} \left[i\, x \left\{ \cos(2x+y) + i \sin(2x+y) \right\} \right]$$

$$= 5x \sin(2x + y)$$

$$z = \text{C. F.} + \text{P.I.}$$

Note:

$$\frac{1}{D_x^2 - 4 D_y^2} \cos(2x+y)$$

can be easily evaluated by using the general method as follows:

$$\frac{1}{D_x^2 - 4 D_y} \cos(2x+y) = \frac{1}{D_x - 2 D_y} \left\{ \frac{1}{D_x + 2 D_y} \cos(2x+y) \right\}$$

$$= \frac{1}{D_x - 2 D_y} \int \cos\{ 2x + a + 2x \} dx$$

$$= \frac{1}{4} \frac{1}{D_x - 2 D_y} \sin(4x + a)$$

$$= \frac{1}{4} \frac{1}{D_x - 2 D_y} \sin(4x + y - 2x)$$

$$= \frac{1}{4} \frac{1}{D_x - 2 D_y} \sin(2x+y)$$

$$= \frac{1}{4} \int \sin(2x + a - 2x) dx$$

$$= \frac{1}{4} x \sin a = \frac{1}{4} x \sin(2x + y)$$

Example (7) Solve $\dfrac{\partial^2 z}{\partial x^2} + 2 \dfrac{\partial^2 z}{\partial x\, \partial y} - 8 \dfrac{\partial^2 z}{\partial y^2} = \sqrt{2x + 3y}$

Solution:

C. F. $D_x^2 + 2 D_x D_y - 8 D_y^2 = 0$

$(D_x - 2 D_y) (D_x + 4 D_y) = 0$

$\therefore \ z = \phi_1 (y + 2x) + \phi_2 (y - 4x)$

P. I. $z = \dfrac{1}{(D_x - 2 D_y)(D_x + 4 D_y)} (2x + 3y)^{\frac{1}{2}}$

Using the general method we have

$$\frac{1}{D_x + 4 D_y} (2x + 3y)^{\frac{1}{2}} = \int \{ 2x + 3(a + 4x) \}^{\frac{1}{2}} dx$$

$$= \int (14x + 3a)^{\frac{1}{2}} dx = \frac{1}{21} (14x + 3a)^{\frac{3}{2}}$$

$$= \frac{1}{21} \{ 14x + 3 (y - 4x) \}^{\frac{3}{2}}$$

$$= \frac{1}{21} (2x + 3y)^{\frac{3}{2}}$$

$$\therefore \ z = \frac{1}{21} \frac{1}{D_x - 2 D_y} (2x + 3y)^{\frac{3}{2}}$$

$$= \frac{1}{21} \int \{ 2x + 3(a - 2x) \}^{\frac{3}{2}} dx$$

$$= \frac{1}{21} \int (3a - 4x)^{\frac{3}{2}} dx = \frac{1}{21} \times \frac{2}{5} \times - \frac{1}{4} (3a - 4x)^{\frac{5}{2}}$$

$$= - \frac{1}{210} \{ 3 (y + 2x) - 4x \}^{\frac{5}{2}}$$

$$= - \frac{1}{210} (2x + 3y)^{\frac{5}{2}}$$

12.3.2. Exercises on homogeneous partial linear equations with constant coefficients

Solve each of the following equations

1) $(D_x^2 + 5 D_x D_y + 6 D_y^2) z = e^{x - y}$

Ans : $z = \phi_1 (y - 2x) + \phi_2 (y - 3x) + \frac{1}{2} e^{x - y}$

2) $(D_x^2 + D_y^2) z = 360 \, x^2 y^2$

414

Ans : $z = \phi_1(y+i\,x) + \phi_1(y - i\,x)$

$$+ i\{\phi_2(y+i\,x) - \phi_2(y-i\,x)\} + 2(15x^4y - x^6)$$

3) $(D_x^3 - 3\,D_x^2\,D_y + 4\,D_y^3)\,z = 6\,e^{y+2x}$

Ans : $z = \phi_1(y-x) + \phi_2(y+2x) + x\,\phi_3(y+2x)$

$$+ x^2\,e^{y+2x}$$

4) $(D_x^3 + 2\,D_x^2\,D_y - D_x\,D_y^2 - 2\,D_y^3)\,z = (y+2)\,e^x$

Ans : $z = \phi_1(y+x) + \phi_2(y-x) + \phi_3(y-2x) + y\,e^x$

5) $(D_x^3 - 3\,D_x\,D_y^2 - 2\,D_y^3)\,z = 27\cos(x+2y) - e^y(3+2x)$

Ans : $z = \phi_1(y-x) + x\,\phi_2(y-x) + \phi_3(y+2x)$

$$+ \sin(x+2y) + x\,e^y$$

6) $(D_x^3 - 3\,D_x^2\,D_y - 4\,D_x\,D_y^2 + 12\,D_y^3)\,z = 8\sin(y+2x)$

Ans : $z = \phi_1(y-2x) + \phi_2(y+2x) + \phi_3(y+3x)$

$$+ 2x\sin(y+2x)$$

7) $D_x^3 - 7\,D_x\,D_y^2 - 6\,D_y^3)\,z = 4\cos(x-y) + x^2 + xy^2 + y^3$

Ans : $z = \phi_1(y-x) + \phi_2(y-2x) + \phi_3(y+3x) + x\cos(x-y)$

$$+ \frac{5}{72}x^6 + \frac{1}{60}x^5(1+21\,y) + \frac{1}{24}x^4\,y^2 + \frac{1}{6}x^3\,y^3$$

8) $(D_x^3 - 3\,D_x\,D_y^2 + 2\,D_y^3)\,z = \sqrt{x+2y}$

Ans : $z = \phi_1(y+x) + x\,\phi_2(y+x) + \phi_3(y-2x)$

$$+ \frac{8}{525}(x+2y)^{\frac{7}{2}}$$

12.3.3. Non homogeneous linear equations with constant coefficients

A non homogeneous linear partial differential equation with constant coefficients such as

$$(2\,D_x^2 - 3\,D_x\,D_y + D_y^2 + 3\,D_x - 2\,D_y + 1)\,z$$

$$\equiv (D_x - D_y + 1)(2\,D_x - D_y + 1)\,z = x^3 y$$

is called reducible, since the let member can be resolved into factors each of which is of the first degree in Dx, Dy,

While $(D_x\,D_y - 3\,D_y^3)\,z \equiv D_y(D_x - 3\,D_y^2)\,z = \sin(2x+5y)$

which cannot be resolved is called irreducible.

12.3.3.1. Reducible non - homogeneous equations

Consider reducible non homogeneous equation

$$f(D_x, D_y)\, z \equiv (a_1\, D_x + b_1\, D_y + c_1)\, (a_2\, D_x + b_2\, D_y + c_2) \cdots$$

$$(a_n\, D_x + b_n\, D_y + c_n)\, z = 0 \quad .. \quad (1)$$

where a_i, b_i, c_i are constants. Any solution of

$$(a_i\, D_x + b_i\, D_y + c_i)\, z = 0 \qquad \cdots \cdots \cdots (2)$$

is a solution of (1). The subsidiary equations of (2) are

$$\frac{dx}{a_i} = \frac{dy}{b_i} = \frac{dz}{-c_i z}$$

The general solution of (2) is given by

$$\phi \left\{ z\, e^{\frac{c_i}{a_i}\, x}, \; a_i\, y - b_i\, x \right\} = 0$$

i.e.,

$$z = e^{-\frac{c_i}{a_i}\, x} \, \psi\, (a_i\, y - b_i\, x) \qquad a_i \neq 0 \quad (3)$$

Or

$$z = e^{-\frac{c_i}{b_i}\, y} \, F\, (a_i\, y - b_i\, x) \qquad b_i \neq 0 \quad (4)$$

Thus if no factor of (1) is a mere multiple of another, the general solution of (t) consists of the sum of n arbitrary functions of the type (3) and (4). Therefore the general solution is given by

$$z = e^{-\frac{c_1}{a_1}\, x} \, \phi_1\, (a_1\, y - b_1 x) + e^{-\frac{c_2}{a_2}\, x} \, \phi_2\, (a_2\, y - b_2 x)$$

$$+ \cdots + e^{-\frac{c_n}{a_n}\, x} \, \phi_n\, (a_n\, y - b_n x)$$

Or

$$z = e^{-\frac{1}{b_1}\, y} \, F_1\, (a_1 y - b_1 x) + e^{-\frac{c_2}{b_2}\, x} \, F_2(a_2 y - b_2 x)$$

$$+ \cdots + e^{-\frac{c_n}{b_n}\, y} \, F_n\, (a_n\, y - b_n\, x)$$

If $\quad f(D_x, D_y)\, z = (a_1\, D_x + b_1\, D_y + c_1)^k (a_{k+1}\, D_x$

$$+ b_{k+1}\, D_y + c_{k+1})\, (a_n\, D_x + b_n\, D_y + c)\, z = 0$$

The part of the general solution corresponding to the k repeated factors is

$$e^{-\frac{c_1}{a_1} x} \left[\phi_1 (a_1 y - b_1 x) + x \, \phi_2 (a_1 y - b_1 x) \right.$$

$$\left. + \cdots + x^{k-1} \, \phi_k (a_1 y - b_1 x) \right]$$

The general solution of

$$f(D_x, D_y) \, z = (a_1 D_x + b_1 D_y + c_1)(a_2 D_x + b_2 D_y + c_2) \cdots$$

$$\cdots (a_n D_x + b_n D_y + c_n) \, z = \phi (x, y)$$

is the sum of the general solution of (1) (the complementary function) and a particular integral given by

$$z = \frac{1}{f(D_x, D_y)} \, \phi (x, y)$$

The methods of evaluating the particular integral are those given before.

Example (1)   Solve

$$(D_x + D_y + 5)(D_x - 2 D_y + 1) \, z = e^{2x+y} + e^{x+y}$$

Solution:
The complementary function is

$$z = e^{-5x} \, \phi_1 (y - x) + e^{-x} \, \phi_2 (y + 2x)$$

The particular integral corresponding to the first term is

$$z = \frac{1}{(D_x + D_y + 5)(D_x - 2D_y + 1)} \, e^{2x+y}$$

$$= \frac{1}{(2+1+5)(2-2+1)} \, e^{2x+y} = \frac{1}{8} \, e^{2x+y}$$

The particular integral corresponding to the second term is

$$z = \frac{1}{D_x - 2 D_y + 1} \left(\frac{1}{D_x + D_y + 5} \, e^{x+y} \right)$$

$$= \frac{1}{D_x - 2 D_y + 1} \left(\frac{1}{1+1+5} \, e^{x+y} \right)$$

$$= \frac{1}{7} \frac{1}{D_x - 2 D_y + 1} \, (e^{x+y} \times 1)$$

$$= \frac{1}{7} \, e^{x+y} \frac{1}{(D_x + 1) - 2(D_y + 1) + 1} \, (1)$$

$$= \frac{1}{7} \, e^{x+y} \frac{1}{D_x (1 - 2\frac{D_y}{D_x})} \, (1)$$

$$= \frac{1}{7} \, e^{x+y} \frac{1}{D_x} \, (1) = \frac{1}{7} \, x \, e^{x+y}$$

$$\therefore \quad z = e^{-5x}\,\phi_1\,(y-x) + e^{-x}\,\phi_2\,(y+2x)$$

$$+ \frac{1}{8}\,e^{2x+y} + \frac{1}{7}\,x\,e^{x+y}$$

Example (2) Solve

$$(D_x{}^2 - D_x\,D_y - 2D_y{}^2 + 2D_x - 4D_y)\,z = y\,e^x + 3x\,e^{-y}$$

Solution:

$$[(D_x - 2D_y)(D_x + D_y) + 2(D_x - 2D_y)]\,z = y e^x + 3x\,e^{-y}$$

$$\therefore \quad (D_x - 2D_y)(D_x + D_y + 2)\,z = y\,e^x + 3x\,e^{-y}$$

The complementary function is

$$z = \phi_1\,(y + 2x) + e^{-2x}\,\phi_2\,(y - x)$$

The particular integral _corresponding to the first term is

$$z = \frac{1}{(D_x - 2D_y)(D_x + D_y + 2)}\,y\,e^x$$

$$= e^x\,\frac{1}{(D_x + 1 - 2D_y)(D_x + 1 + D_y + 2)}\,y$$

$$= \tfrac{1}{3}\,e^x\,\frac{1}{(1 + D_x - 2D_y)(1 + \frac{D_x + D_y}{3})}\,y$$

$$= \tfrac{1}{3}\,e^x\,(1 - D_x + 2D_y)(1 - \tfrac{1}{3}D_x - \tfrac{1}{3}D_y)\,y$$

$$= \tfrac{1}{3}\,e^x\,(1 - \frac{4}{3}\,D_x + \frac{5}{3}\,D_y \ldots)\,y$$

$$= \tfrac{1}{3}\,e^x\,(y + \frac{5}{3})$$

The particular integral corresponding to the second term is given by

$$z = 3\,\frac{1}{(D_x - 2D_y)(D_x + D_y + 2)}\,(e^{-y}x)$$

$$= 3e^{-y}\,\frac{1}{(D_x - 2D_y + 2)(D_x + D_y + 1)}\,x$$

$$= \frac{3}{2}\,e^{-y}\,(1 - \tfrac{1}{2}D_x \ldots)(1 - D_x \ldots)\,x$$

$$= \frac{3}{2}\,e^{-y}\,(1 - \frac{3}{2}\,D_x)\,x$$

$$= \frac{3}{2}\,e^{-y}\,(x - \frac{3}{2})$$

$$\therefore \; z = \phi_1(y + 2x) + e^{-2x}\,\phi_2(y - x) + \tfrac{1}{3}\,e^x\,(y + \tfrac{5}{3})$$

$$+ \tfrac{3}{2}\,e^{-y}\,(x - \tfrac{3}{2})$$

Example (3)   Solve

$$(2\,D_x\,D_y + D_y{}^2 - 3\,D_y)\; z = 50\,\cos(3x - 2y)$$

Solution:

$$D_y\,(2\,D_x + D_y - 3)\; z = 50\,\cos(3x - 2y)$$

The complementary function is

$$z = \phi_1(x) + e^{3y}\,\phi_2(2y - x)$$

The particular integral is given by

$$z = 50\;\frac{1}{2\,D_x\,D_y + D_x{}^2 - 3\,D_y}\;\cos(3x - 2y)$$

$$= 50\;\frac{1}{2\times 6 - 4 - 3\,D_y}\;\cos(3x - 2y)$$

$$= 50\;\frac{1}{8 - 3\,D_y}\;\cos(3x - 2y)$$

$$= 50\;\frac{8 + 3\,D_y}{64 - 9\,D_y{}^2}\;\cos(3x - 2y)$$

$$= 50\;\frac{8 + 3\,D_y}{64 + 36}\;\cos(3x - 2y)$$

$$= \frac{1}{2}\left[\,8\,\cos(3x - 2y) + 6\,\sin(3x - 2y)\,\right]$$

$$= 4\,\cos(3x - 2y) + 3\,\sin(3x - 2y)$$

$$z = \text{C. F.} + \text{P. I.}$$

12.3.3.2. Irreducible equations with constant coefficients

Consider the linear equation with constant coefficients

$$f(D_x, D_y)\, z = 0 \tag{1}$$

Since $D_x^r\,D_y^s\,(c\,e^{ax + by}) = c\,a^r\,b^s\,e^{ax+by}$, where a, b, c are constants, the result of substituting

$$z = c\,e^{ax+by} \tag{2}$$

in (1) is $c\,f(a, b)\,e^{ax+by} = 0$

Thus (2) is a solution of (1) provided that $f(a, b) = 0$ (3)

where c is arbitrary. Now let any chosen value of a (or b) one or more values of b (or a) are obtained by means of (3). Thus there exist infinitely many pairs of numbers (a_i, b_i) satisfying (3). Moreover

$$z = \sum_{i=1}^{\infty} c_i\, e^{a_i x + b_i y} \tag{4}$$

where $f(a_i, b_i) = 0$

is a solution of (1)

If $f(D_x, D_y)\, z = (D_x + h D_y + k)\, g(D_x, D_y)\, z$

then any pair (a, b) for which $a + h b + k = 0$ satisfies

(3). Consider all such pairs $(a_i, b_i) = (-h b_i - k, b_i)$.

By (4):

$$z = \sum_{i=1}^{\infty} c_i\, e^{-(h b_i + k) x + b_i y}$$

$$= e^{-kx} \sum_{i=1}^{\infty} c_i\, e^{b_i (y - hx)}$$

is a solution of (1) corresponding to the linear factor $(D_x + h D_y + k)$ of $f(D_x, D_y)$ This is of course

$$e^{-kx}\, \phi\,(y - hx).$$

Thus if f (Dx, Dy) has no linear factor, (4) will be called the solution of (1) ; however if f (Dx, Dy) has m < n linear factors, we shall write part of the solution involving arbitrary functions (corresponding to the linear factors) and the remainder involving arbitrary constants.

Example (1) Solve $(D^2_x + D_x + D_y)\, z = 0$

Solution:

The equation is irreducible
Here

$$f(a, b) = a^2 + a + b = 0$$

so that for any $a = a_i$

$$b_i = -a_i (a_i + 1)$$

Thus the solution is

$$z = \sum_{i=1}^{\infty} c_i\, e^{a_i x + b_i y} = \sum_{i=1}^{\infty} c_i\, e^{a_i x - a_i (a_i + 1) y}$$

with c_i and a_i arbitrary constants.

<u>Example</u> (2)   Solve $(2 D_x{}^2 - D_y{}^2 + D_x) z = x^2 - y$

Solution:

The complementary function is

$$z = \sum_{i=1}^{\infty} c_i \, e^{a_i x + b_i y}$$

where

$$2a^2{}_i - b^2{}_i + a_i = 0$$

The particular integral is given by

$$z = \frac{1}{2 D_x{}^2 - D_y{}^2 + D_x} \, (x^2 - y)$$

$$= -\frac{1}{D_y{}^2} \cdot \frac{1}{1 - \dfrac{D_x + 2D_x{}^2}{D_y{}^2}} \, (x^2 - y)$$

$$= -\frac{1}{D_y{}^2} \left[1 + \frac{1}{D_y{}^2} (D_x + 2D_x{}^2) + \frac{1}{D_y{}^4} (D_x{}^2 + 4D_x{}^3 \right.$$
$$\left. + 4 D_x{}^4) \right] (x^2 - y)$$

$$= -\frac{1}{D_y{}^2} \left[1 + \frac{1}{D_y{}^2}(D_x + 2D^2{}_x) + \frac{1}{D_y{}^4} D_x{}^2 \cdots \right] (x^2 - y)$$

$$= -\frac{1}{D_y{}^2} \left(x^2 - y + xy^2 + 2y^2 + \frac{1}{12} y^4 \right)$$

$$= -\frac{1}{2} x^2 y^2 + \frac{1}{6} y^3 - \frac{1}{12} xy^4 - \frac{1}{6} y^4 - \frac{1}{360} y^6$$

The required solution is given by

$$z = \sum_{i=1}^{\infty} c_i \, e^{a_i x \pm \sqrt{2a_i{}^2 + a_i}\, y} \quad - \frac{1}{2} x^2 y^2 + \frac{1}{6} y^3$$
$$- \frac{1}{12} xy^4 - \frac{1}{6} y^4 - \frac{1}{360} y^6$$

<u>Example</u> (3)
Solve
$$(D_x + 2 D_y)(D_x - 2 D_y + 1)(D_x - D_y{}^2) z = 0$$

Solution:

Corresponding to the linear factors we have

$\phi_1 (y - 2x)$ and $e^{-x} \phi_2 (y + 2x)$

For the irreducible factor $D_x - D_y{}^2$ we have $a - b^2 = 0$

or $a = b^2$

The required solution is

$$z = \phi_1 (y - 2x) + e^{-x} \phi_2 (y + 2x) + \sum_{i=1}^{\infty} c_i \, e^{b_i{}^2 x + b_i y}$$

(c_i and b_i and constants)

421

12.4. Practical Applications of differential Equations

12.4.1. Damped mechanical oscillations

2\12\71.   Applications.

Damped oscillations

I. Mechanical illustrations :—

a spring of stiffness ᴍᴍᴍᴍᴍᴍ $\overset{k}{}$ $\overset{M}{}$
k is placed on a rough horizontal
table. The stiffness is the force required
to produce unit extension. one end of the
spring is fixed at a pt o. on the table. & the other

end is attached a mass M, the resisting force to
motion is proportional to the velocity. Let x be
the displacement of mass m at time t the eqn of
motion of M is.

$$M\ddot{x} = -r\dot{x} - kx \cdot \text{ where } r \text{ the resistance of}$$
$$\text{friction} \propto \text{velocity}$$
$$\cdot K \text{ the stiffness} \propto \text{displace}$$

$$\therefore \boxed{M\ddot{x} + r\dot{x} + kx = 0}$$

12.4.2. Electrical oscillations- RCL circuit

II. Electrical illustration

A condenser of capacity c is
discharged through a coil of
Resistance R and inductance L
drops are

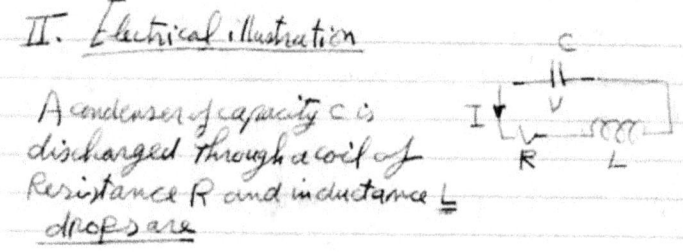

422

$$IR + L\frac{di}{dt} = V$$

$$\therefore \quad I = -\frac{dQ}{dt} \quad \& \quad Q = CV$$

$$\therefore \quad I = -C\frac{dV}{dt}$$

$$\therefore \quad RC\frac{dV}{dt} - CL\frac{d^2V}{dt^2} - V = 0$$

$$\boxed{CL\frac{d^2V}{dt^2} + RC\frac{dV}{dt} + V = 0}$$

Since $Q = CV$

$$\therefore \quad L\frac{d^2Q}{dt^2} + R\frac{dQ}{dt} + \frac{Q}{C} = 0$$

$$\boxed{LC\frac{d^2Q}{dt^2} + RC\frac{dQ}{dt} + Q = 0}$$

Differentiating the eqn in V

$$CL\dot{V} + RC\ddot{V} + \dot{V} = 0$$

as $I \propto \dot{V}$

$$\boxed{CL\ddot{I} + RC\dot{I} + I = 0}$$

Thus we see that V, Q, I all satisfy the same diff^l eq^n. Hence we have :—

$$M\ddot{x} + \Gamma\dot{x} + kx = 0 \quad —①$$

$$\& \quad L\ddot{Q} + R\dot{Q} + \frac{Q}{C} = 0 \quad —②$$

Comparing ① & ② we see that.

The inductance L takes place of the mass M

" Resistance R " " , resistance r

" capacitance C " , , extensibility $\frac{1}{k}$

i.e. the extension produced by unit force. - eqs ①
& ② can now be written in the form.

$$x'' + 2bx' + n^2 x = 0 \quad —③$$

where $2b = \frac{r}{M}$ $n^2 = \frac{k}{M}$ for mechanical p.
 & $2b = \frac{R}{L}$ $n^2 = \frac{1}{CL}$ " electrical p

The Auxiliary eqn: $m^2 + 2bm + n^2 = 0$

$$m = -b \pm \sqrt{b^2 - n^2}$$

Case 1   Roots are real and distinct. $b^2 > n^2$
$2b = \frac{r}{M} > \frac{R}{L}$ $\frac{R^2}{4L^2} > \frac{1}{CL}$

(High resistance) $\frac{r^2}{4M^2} > \frac{k}{M}$

Here the two roots are real and negative. hence

$$x = Ae^{-m_1 t} + Be^{-m_2 t}. \quad —①$$

as $x \to 0$, $\tau \to \infty$. also $0 = Ae^{-m_1 t} + Be^{-m_2 t}$
 ∴ $Ae^{-m_1 t} = -Be^{-m_2 t}$
$e^{(m_2 - m_1)t} = -\frac{B}{A}$

The exponential is positive hence we have a real
soln on A & B are of opposite signs.

i) $x = 0$ ∴ $Ae^{-x} + Be^{-x} = 0$
 $t = \infty$ The pendulum takes infinite
        time to return.

ii) $x = o$

$\therefore e^{(m_2-m_1)t} = -\frac{B}{A}$

The time taken to complete
a cycle is positive there is
an oscillation.

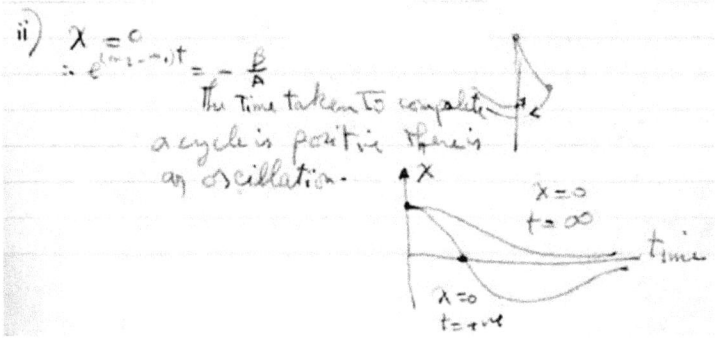

This means that however the motion is started the mass
cannot pass through its equilibrium position
more than once to its finally creeps
asymptotically; j x This is a type of motion is
    ni isn't ↑
realised in the case of the simple pendulum immersed
in a very viscous medium. This type of motion
is called   Aperiodic s...ic

	Aperiodic
$b^2 > n^2$	
$\dfrac{r^2}{4M^2} > \dfrac{k}{M}$	$\dfrac{r^2}{4M} > k$
$\dfrac{R^2}{4L^2} > \dfrac{1}{cL}$	$\dfrac{R^2}{4L} > \dfrac{1}{c}$

2 Roots of Auxiliary eq^n are equal.

The two root are $-b \lessgtr b$ $b = n$

Soln $x = e^{-bt}(A+Bt)$

$x \to o$ as $t = \infty$

& $x = o$ if $t = -\dfrac{A}{B}$

The motion here is similar to the previous case. The damping is said to be __critical__ & the resistance is the __least resistance to prevent oscillation.__

The least resistance to prevent oscillation

$$b = n = \frac{r}{2M} = \sqrt{\frac{k}{M}}$$

$$\frac{R}{2L} = \sqrt{\frac{1}{cL}}$$

3. $\underline{b^2 < n^2}$

put $b^2 - n^2 = -q^2$

i.e. $q^2 = n^2 - b^2$ (law resistance)

The roots of the aux. eqⁿ are $-b \pm q$

and the solⁿ is

$$x = R e^{-bt} \sin(qt + \epsilon)$$

This motion may be described as a simple harmonic oscillation of amplitude $R e^{-bt}$ which tends $\to$ as $t \to \infty$

the trigonometric fⁿ $\sin(qt + \epsilon)$ oscillates between ± 1 and hence the space time curve oscillates between $x = \pm R e^{-bt}$

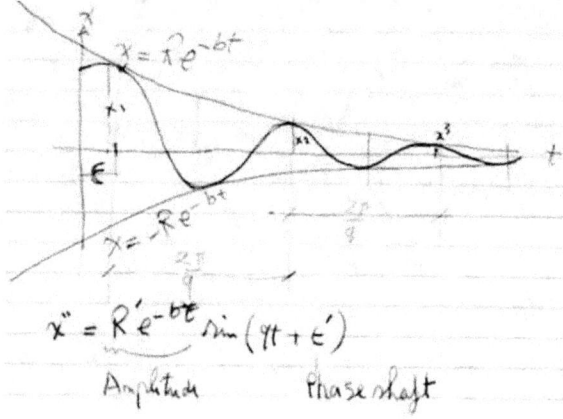

$$\underbrace{x'' = R' e^{-bt}}_{\text{Amplitude}} \underbrace{\sin(qt + \epsilon')}_{\text{Phase shift}}$$

From this it follows that successive maximum on the same side occur at regular intervals its

interval $= \dfrac{2\pi}{q} = \dfrac{2\pi}{\sqrt{n^2-b^2}}$ as this is taken as the period of oscillation.

interval of el. osc. $= \dfrac{2\pi}{\sqrt{\frac{1}{cL} - \frac{R^2}{4L^2}}} = \dfrac{2\pi L}{\sqrt{\frac{L}{c} - \frac{R^2}{4}}}$

Resonance $w_0 = \dfrac{1}{\sqrt{cL}}$ $\qquad t_0 = 2\pi\sqrt{cL}$

Damped osc. $t_D = \dfrac{2\pi}{\sqrt{\frac{1}{cL} - \frac{R^2}{4L^2}}} = \dfrac{2\pi}{\sqrt{\frac{4L - cR^2}{4L^2c}}}$

$$t_D = 2\pi\sqrt{\frac{4L^2c}{4L - cR^2}} = 2\pi\sqrt{\frac{Lc}{1 - \frac{cR^2}{4L}}}$$

$\therefore t_D > t_0$. $\therefore$ the interval of oscillation in Damping is longer than that of Resonance

$$\frac{x_1}{x_2} = \frac{R\,e^{-bt}\sin(qt+\epsilon)}{R\,e^{-b(t+\frac{2\pi}{q})}\sin\left[q(t+\frac{2\pi}{q}) + \epsilon\right]}$$

$\therefore \dfrac{x_1}{x_2} = e^{\frac{2\pi b}{q}} = $ const. This means that x_1, x_2, x_3 form a geometric progression.

$\boxed{\log_e \dfrac{x_1}{x_2} = \dfrac{2\pi b}{q}}$ This is called the Logarithmic decrement.
$\underset{k}{}$

If the resistance is small the period of oscillation is approximately $\dfrac{2\pi}{n}$ hence the resistance slowen the motion.

A small amount of friction hardly affect the period of osc. but its main effect is on the ampl. since it is due to the continual diminution of ampl. that the motion finally dies away

427

ex.

A condenser of capacity 1.5 $\mu.F.$ is charged to $50V.$ & is discharged through a serious circuit The inductance $L = 36$ $m.H.$ & The current oscillates with frequency 900 $c/sec.$ calculate the Resistance.

$$CLI'' + CRI' + I = 0$$

$$(CLD^2 + CRD + 1)I = 0$$

$$D = \frac{-CR \pm \sqrt{C^2R^2 - 4CL}}{2CL}$$

$$D = -\alpha \pm i\beta$$

$$\text{soln} \quad I = Ae^{-\alpha t}\sin(\beta t + \epsilon)$$

$$\beta = \omega \qquad \therefore \text{Frequency} = \frac{\beta}{2\pi} = \frac{1}{2\pi}\sqrt{\frac{-C^2R^2 + 4CL}{4C^2L^2}}$$

$$400 = \frac{1}{2\pi}\sqrt{\frac{1}{CL} - \frac{R^2}{4L^2}}$$

$$= \frac{1}{2\pi}\sqrt{\frac{1}{1.5\times10^{-6}\times 36\times10^{-3}} - \frac{R^2}{4(36\times10^{-3})^2}}$$

$$R \cong 250 \ \Omega$$

12.4.3. Oscillation of a spring

$9 \setminus 12$ $^8/71$

I. Damped oscilations of a verticle spring

$$M\ddot{x} = -r\dot{x} - kx + Mg$$

الرسم البياني / (x°0y)

$$M\ddot{x} + r\dot{x} + kx = Mg.$$

natural length

$$\text{C.F.} \quad x = Ae^{-\alpha x}\sin(\beta t + \epsilon)$$

428

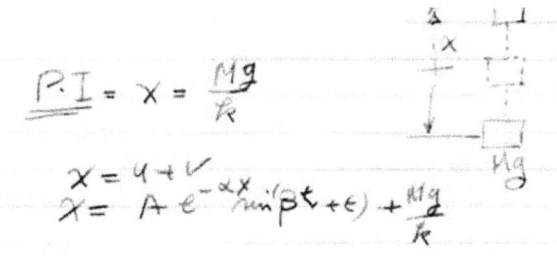

$$\underline{P.I} = x = \frac{Mg}{k}$$

$$x = y + v$$
$$x = A e^{-\alpha x} \sin(\beta t + \epsilon) + \frac{Mg}{k}$$

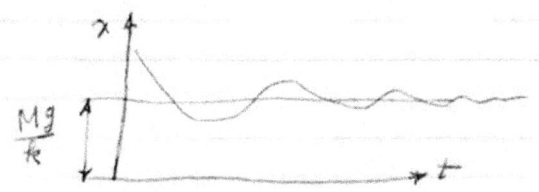

II The electrical analogue

The change of a condeser capacity c through a coil of resistance R, the inductance L by means of a battery of e.m. $V = E$

$$IR + L\frac{dI}{dt} + V = E \qquad I \quad E$$

$$Q = cV \qquad I = \frac{dQ}{dt} = c\frac{dV}{dt}$$

$$cL\frac{d^2V}{dt^2} + cR\frac{dV}{dt} + V = E$$

i.e. $\quad c\cdot E - V = A e^{-\alpha t} \sin(\beta t + \epsilon)$

$\quad\quad x \to 0 \text{ as } t \to \infty \quad$ transient

$$\underline{P.I} \qquad V = E \quad \text{steady state}$$

$$V = E$$

III   An alternating e.m.f $= E_0 \sin \omega t$ is applied to a series circuit of zero capacity, resistance R and inductance L. Find the current I at any time.

$\underline{\text{Drops}}:-$

$$I R + L \frac{dI}{dt} = E_0 \sin \omega t$$

$$(R + LD) I = E_0 \sin \omega t$$

$\underline{C.F.} \quad I = A e^{-\frac{R}{L}t} \quad \text{tends to } 0 \text{ as } t \to \infty$

     transient

$\underline{P.I} \quad I = \frac{1}{(R + LD)} E_0 \sin(\omega t - \epsilon)$

$$= \frac{R - LD}{R^2 - L^2 D^2} E_0 \sin(\omega t - \epsilon)$$

$$= \frac{R - LD}{R^2 + \omega^2 L^2} E_0 \sin(\omega t - \epsilon)$$

$$= \frac{R E_0 \sin(\omega t - \epsilon) - L\omega E_0 \cos(\omega t - \epsilon)}{R^2 + \omega^2 L^2}$$

$$I = \frac{E_0}{\sqrt{R^2 + L^2 \omega^2}} \left(\sin \omega t \cos \epsilon - \sin \frac{\omega \omega t}{R^2 + L^2} \right)$$

$$I = \frac{E_0}{\sqrt{R^2 + L^2 \omega^2}} \sin(\omega t - \epsilon)$$

where $\epsilon = \tan^{-1} \frac{\omega L}{R}$

$$\therefore I = A e^{-\frac{R}{L}t} + \frac{E_0}{\sqrt{R^2 + L^2 \omega^2}} \sin(\omega t - \epsilon)$$

12.4.4. Forced oscillations without damping

$$M \ddot{x} = - k (x - y)$$

$$M \ddot{x} + k x = k y = k a \sin \omega t$$

$$\ddot{x} + \frac{k}{M} x = \frac{k}{M} a \sin \omega t \quad \text{suppose } \frac{k}{M} = n^2$$

430

$$\therefore \ddot{x} + n^2 x = n^2 a \sin \omega t$$

C.F $\quad x = A \cos nt + B \sin nt \qquad \underline{free\ osc^n}$

P.I $\quad x = \dfrac{1}{D^2 + n^2} \cdot n^2 a \sin \omega t \qquad \underline{forced\ osc^n}$

$$= \dfrac{-n^2 a}{-\omega^2 + n^2} \sin \omega t \qquad \omega \neq n$$

$$\therefore x = A \cos nt + B \sin nt + \dfrac{n^2 a}{-\omega^2 + n^2} \sin \omega t$$

Interpretation

2 - S.H.M² composed on each other the motion is a superposition of 2 s.th osc²s one free osc² which are oscilations with which the system osc²s freely without an external agent.

The other are forced osc² due to the external osc² imposed on the system.

The special case of Resonance

$\omega = n$ that is $\therefore \dfrac{\omega}{2\pi} = \dfrac{n}{2\pi}$

i.e. The emposed osc²s have the same frequency as the natural frequency now we have already show that the solution of the eq².

The term $\left(-\dfrac{n a}{2} t \cos nt\right)$ may be interpreted as a simple harmonic oscilation is which which

the amplitude $\dfrac{n a t}{2}$ increases indifinively w.r.t time and the space time curve has osc² oscilate bet. $x = \pm \dfrac{n a t}{2}$

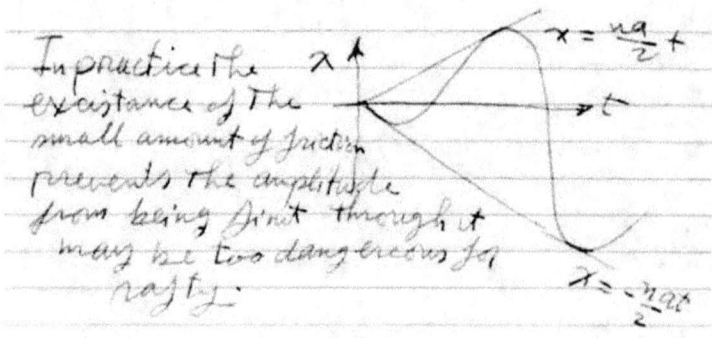

In practice the existance of the small amount of friction prevents the amplitude from being infinit though it may be too dangerous for safty.

$x = \frac{na}{2}t$

$x = -\frac{na}{2}t$

12.4.5. Forced oscillations with damping

Forced Osc.^{n} with Damping.

ex. An alternative e.m.f. $E_0 \sin \omega t$ is applied to a series circuit resistance R, inductance L and capacitance C. calculate the steady voltage and steady current.

$$IR + L\frac{dI}{dt} + V = E_0 \sin \omega t$$

$$I = C\frac{dV}{dt} = \frac{dx}{dt}$$

$$\therefore LC\frac{d^2V}{dt^2} + CR\frac{dV}{dt} + V = E_0 \sin \omega t$$

$$(CLD^2 + CRD + 1)V = E_0 \sin \omega t$$

P.I. $$V = \frac{1}{(CLD^2 + CRD + 1)} E_0 \sin \omega t$$

$$= \frac{1}{1 - CL\omega^2 + CRD} E_0 \sin \omega t.$$

$$\boxed{V = \frac{1}{\sqrt{(1 - CL\omega^2)^2 + C^2R^2\omega^2}} E_0 \sin(\omega t - \epsilon)}$$

where $\epsilon = \tan^{-1} \frac{\omega CR}{1 - CL\omega^2}$

This gives the steady voltage.

432

• Steady current $= c\dfrac{dV}{dt} = cV'$

$$\boxed{I = cV' = \dfrac{c\,E_0\,\omega\,\sin\left(\omega t - \epsilon + \dfrac{\pi}{2}\right)}{\sqrt{(1 - Lc\omega^2)^2 + c^2R^2\omega^2}}}$$

Now consider the particular that important case
$1 - Lc\omega^2 = 0$ i.e. $\omega = \dfrac{1}{\sqrt{cL}}$

Here $\epsilon = \tan^{-1}\infty = \dfrac{\pi}{2}$
hence the steady current is given by

$$I = \dfrac{\omega c\,E_0\,\sin\omega t}{cR\omega} \quad\text{i.e.}\quad \dfrac{E_0\,\sin\omega t}{R}$$

This shows that the inductance L and capacitance c neutralize one another & now the steady current with phase with applied e.m.

Mechanical ex :-

A 50 gm mass hangs on a vertical spring, stiffness 20 gm/cm. resistance to motion is 1.4 gm/cm/sec. The support has a point oil see & amplitude 1 cm calculate the amplitude of the forced oscie.

$y = a\sin\omega t \quad a = 1\,cm$
period $\dfrac{2\pi}{\omega} = 0.1$
$\omega = 20\pi$
$y = \sin 20\pi t$.

and $50\,x'' = -1.4 \times 981\,x'$
$\quad - 20 \times 981(x - y)$
$\quad + 50 \times 981$

$x'' + 27.5\,x' + 392.4\,x = 981 + 392.4\,\dfrac{y}{gm}$
$\qquad\qquad\qquad\qquad \sin 20\pi$

$$\text{Amplitude} = \frac{392.9}{\sqrt{(3558)^2 + (27.5 \times 20\pi)^2}}$$

$$= 0.1 \text{ approx} = 1 \text{ mm}$$

$$\text{...}$$

$$\underline{\text{P.I.}} \quad x = \frac{981}{392.9} + \frac{1}{(D^2 + 27.5D + 392.9)} \, 392.9 \sin \omega t$$

$$x = 2.5 + \frac{392.9 \sin(20\pi t)}{-(20\pi)^2 + 392.9 + 27.5D}$$

$$x = 2.5 + \frac{1}{-3558 + 27.5D} \, 392.9 \sin 20\pi t$$

$$= 2.5 + \frac{392.9 \sin(20\pi t - \epsilon)}{\sqrt{(3558)^2 + (275 \times 2.\pi)^2}}$$

$$\text{where} \quad \epsilon = \tan^{-1} \frac{27.5 \times 20\pi}{-3558}$$

$\boxed{1}$ Solve if $y = e^x$ is a sol^n of
$$x y'' - (2x+1) y' + (x+1) y = 0 \qquad ①$$
Then find the complete sol^n of
$$x y'' - (2x+1) y' + (x+1) y = (x^2 + x - 1) e^x$$

Sol^n.

form
$$y'' + P y' + Q y = R \qquad ②$$
where
$$P = -\left(\frac{2x+1}{x}\right), \quad Q = \frac{x+1}{x} \quad \Big\}$$
$$R = \frac{x^2 + x - 1}{x} e^x \qquad \Big\} ③$$

Assume
$$y = UV$$
$$y' = u' V + v' u$$
$$y'' = u'' V + v'' u + 2 u' v' \quad \Big\} ④$$

so eq^n ② become
$$u'' V + v'' u + 2 u' v' + P(u' V + v' u) + Q \, uv$$
$$= R$$

434

or $V(u'' + Pu' + Qu) + v''u + 2u'v'$
$\quad\quad + \enspace P \enspace v'u = R$

as $u'' + Pu' + Qu = 0$, $u = e^x$

so $v''e^x + 2e^x v' + P e^x v' = R$

$V'' + V'(2 + P) = \dfrac{R}{e^x}$ $\qquad$ —⑤

Put $W = V'$, $W' = V''$

so $\quad w' + W(2 + P) = \dfrac{R}{e^x}$

or $\dfrac{dW}{-W(2+P) + \dfrac{R}{e^x}} = dx$

or by multiplying by

integrating factor $\left[e^{\int (2+P)\,dx} \right]$

$d\underline{W} \cdot e^{\int (2+Px)\,dx} + (2+P)We^{\int (2+P)\,dx}\,dx = \dfrac{R}{e^x} e^{\int (2\cdot \text{--}x)\,dx}\,dx$

$d\left[We^{\int (2+P)\,dx} \right] = \dfrac{R}{e^x} e^{\int (2+P)\,dx}\,dx$

so $We^{\int (2+P)\,dx} = \displaystyle\int \dfrac{R}{e^x} e^{\int (2+P)\,dx}\,dx + C$

$\qquad\qquad\qquad\longrightarrow$ ⑥

as $2 + P = 2 - \dfrac{2x+1}{x} = -\dfrac{1}{x}$

$e^{-\int \frac{1}{x}\,dx} = e^{-\ell_n x} = \dfrac{1}{x}$

$\& \dfrac{R}{e^x} = \dfrac{x' + x - 1}{x e^x} e^x$ $\quad\left. \right]$ —⑦

so

$$\frac{W}{x} = \int \frac{x^2 + x - 1}{x} \cdot \frac{1}{x} \, dx + C$$

$$= x + \ln x - \frac{1}{x} + C$$

$$V = \int (x^2 + x(\ln x - 1 + cx)) \, dx$$

$$= \frac{x^3}{3} + \frac{1}{2}\int \ln x \, dx^2 - x + c'x^2$$

$$= \frac{x^3}{3} + c'x^2 - x + \frac{1}{2}x^2\ln x - \frac{1}{2}\int x^2 \, d\ln x$$

$$= \frac{x^3}{3} + c'x^2 - x + \frac{1}{2}x^2\ln x - \frac{1}{2}\int \frac{x^2}{x} \, dx$$

$$= \frac{x^3}{3} + c'x^2 - x + \frac{1}{2}x^2\ln x - \frac{1}{2}\cdot\frac{x^2}{2}$$

$$V = \frac{x^3}{3} + Ax^2 - x + \frac{1}{2}x^2\ln x$$

so

$$Y = VU = \left(\frac{x^3}{3} + Ax^2 - x + \frac{1}{2}x^2\ln x\right) e^x$$

$\int \frac{1}{x} \, dx = \ln x = \frac{dy}{y} = d\ln y$

3) Transform to Normal form

$$\frac{d^2y}{dx^2} + 2x\frac{dy}{dx} + (x^2 - 8)y = 81x^2 e^{\frac{x^2}{2}}$$

0

$$\begin{cases} y = Ve^{-\frac{x^2}{2}} \\ y' = V'e^{-\frac{x^2}{2}} + V(-x)e^{-\frac{x^2}{2}} \\ y'' = V''e^{-\frac{x^2}{2}} + (-x)e^{-\frac{x^2}{2}}V' + V'(-xe^{-\frac{x^2}{2}}) \\ \qquad + (-Ve^{-\frac{x^2}{2}}) + (-Vx)(-xe^{-\frac{x^2}{2}}) \\ = e^{-\frac{x^2}{2}}\left[V'' - 2xV' - V + x^2V\right] \end{cases}$$

$$\begin{cases} y = Ve^{-\frac{x^2}{2}} \\ y' = V'e^{-\frac{x^2}{2}} + V(-x)e^{-\frac{x^2}{2}} \\ y'' = V''e^{-\frac{x^2}{2}} + (-x)e^{-\frac{x^2}{2}}V' + V'(-xe^{-\frac{x^2}{2}}) \\ \qquad + (-Ve^{-\frac{x^2}{2}}) + (-Vx)(-xe^{-\frac{x^2}{2}}) \\ = e^{-\frac{x^2}{2}}\left[V'' - 2xV' - V + x^2V\right] \end{cases}$$

so eqn become

$$e^{-\frac{x^2}{2}}\left[V'' - 2xV' + V(x^2-1)\right] + 2x e^{-\frac{x^2}{2}}(V' - xV)$$

$$+ (x^2-8)Ve^{-\frac{x^2}{2}} = 81x^2 e^{-\frac{x^2}{2}}$$

$$\boxed{V'' - 9V = 81x^2}$$ C.F + P.I

<u>C.F</u> $V'' - 9V = 0$
   $(D^2 - 9)V = 0$
   $(D-3)(D+3)V = 0$
   $V = Ae^{3x} + Be^{-3x}$

P. I $V = \dfrac{1}{D^2 - 9}\, 81x^2$

   $= -\dfrac{1}{9\left(1 - \frac{D^2}{9}\right)}\, 81x^2$

   $= -\dfrac{1}{9}\left[1 + \dfrac{D^2}{9} + \dfrac{\left(\frac{D^2}{9}\right)^2}{2!} + \cdots\right] 81x^2$

   $= -\dfrac{1}{9}\left[81x + \dfrac{81 \times 2}{9}\right]$

   $= -\dfrac{1}{9}\left[81x + 18\right] = -9x - 2$

Sol: $V = -9x - 2 + Ae^{3x} + Be^{-3x}$
   $y = uV = e^{-\frac{x^2}{2}}\left[Ae^{3x} + Be^{-3x} - 9x - 2\right]$